BREWERY SAFETY

PRINCIPLES, PROCESSES, AND PEOPLE

"No matter where you are on your safety journey, this book will help you improve. Matt has been my mentor for many years, and I'm thrilled that his knowledge, wisdom, and wit are now available to everyone. *Brewery Safety* should become a well-worn reference guide in every brewery to help educate and empower both new and seasoned employees."

—RACHEL BELL
Health and Safety Manager, CANarchy Craft Brewing Collective
and Brewers Association Safety Subcommittee Chair

"*Brewery Safety* is both an excellent reference manual and an indispensable resource for forming, building, improving, and maintaining an effective brewery safety program. This book is a one-stop shop for information and resources that should be on every brewery manager's bookshelf. I guarantee it will not gather dust, as you will refer to it often. Matt Stinchfield's years of safety experience, depth of knowledge, and writing style make for an easy-to-understand-and-apply resource and reference manual. I highly recommend this invaluable publication!"

—JEFF MASON
Plant Manager, Ska Brewing Co.

"There are so many things that go into the production and sale of great beer, but there is nothing more important than protecting the physical, emotional, and mental health of all staff members. Safety is complicated work, and the efforts must be continuous. Matt has done a phenomenal job providing an in-depth, demystifying overview of the work necessary to send your staff home every night happy, healthy, and safe."

—JASON PERKINS
Brewmaster, Allagash Brewing Company

"Whether a brewery has already established safety policies, is starting from scratch, or is somewhere in between, this book provides a comprehensive framework for ensuring the safety and health of workers in the brewery. While regulatory requirements are covered, they are not the focal point—the holistic health and safety of people is center stage. This book will look fantastic on your shelf and will be one of the most reached-for resources as your approach to safety grows and matures."

—ABBY FERRI
CSP, ARM, Chief Safety Officer, Insurate

"In my experience, safety is a core element that is essential for successful brewery operations. The absence of safety priorities reduces the ability of team members to focus on quality and efficiency, and thus it is a foundational element for breweries of all sizes. Compiling theory, resources, and best practices, this book provides a relevant source to assure that safety protocols and systems meet the expectations of staff, customers, regulators, and, most importantly, yourself."

—JOHN MALLETT

former VP of Operations, Bell's Brewery Inc.

"'Safety is no accident' is an old saying that is 100 percent true. We owe it to our employees and our businesses to anticipate work hazards, formulate procedures and training that mitigate those hazards, and communicate early and often about these issues. This book will help you and your team build and manage a safety program designed to ensure that you and your coworkers get home safely from the brewery every night."

—LARRY HORWITZ

Brewmaster and Director of Brewing Operations, Crooked Hammock Brewery

BREWERY SAFETY

PRINCIPLES, PROCESSES, AND PEOPLE

BY MATT STINCHFIELD

BREWERS PUBLICATIONS®

Brewers Publications®
A Division of the Brewers Association℠
PO Box 1679, Boulder, Colorado 80306-1679
BrewersAssociation.org
BrewersPublications.com

Proudly Printed in the United States of America.
10 9 8 7 6 5 4 3 2 1
ISBN-13: 978-1-938469-74-9
ISBN-10: 1-938469-74-7
EISBN: 978-1-938469-79-4

Library of Congress Control Number: 2023938613

Publisher: Kristi Switzer
Technical Editors: Rachel Bell, Jeff Mason
Copyediting: Iain Cox
Indexing: Doug Easton
Art Director, cover and interior design, and production: Jason Smith
Production: Justin Petersen
Cover Photo: Luke Trautwein at Holidaily Brewing Co., Golden, Colorado

To Carleton Paul Stinchfield,
my first example that safety is love.

TABLE OF CONTENTS

FOREWORD

I first met Matt Stinchfield in Atlanta at a Georgia Craft Brewers Guild meeting. I had just started my brewery safety journey, and it was pure luck that our brewery happened to be hosting the guild meeting that month. Back then, Matt was the Brewers Association's Safety Ambassador and would travel around to state guild meetings trying to educate and, it was hoped, save us all. I was drowning in a confusing fog of OSHA standards and teaching myself from the ground up through frantic googling. It was wonderful to have a real brewery safety professional to translate those standards into practice.

I do not think I talked too much, still being extremely shy at the time, but I must have said a few of the right things because Matt asked if I would be interested in joining the Brewers Association Safety Subcommittee. I told him I would think about it, but I never pursued it. Life went on, and several months later I found myself moving to Salt Lake City, Utah. Nearly two years after our first meeting, I found myself at a Utah Brewers Guild meeting where Matt was giving his safety lecture. Being a little surer of my place in the industry, I made it a point to talk to Matt afterward and see if he remembered me from Atlanta. He said the invitation to join the Safety Subcommittee was still open—this time I followed through with it.

Five years later, I find myself writing a foreword for my friend and mentor's book, a book I wish had existed at the start of my career. Matt's invitation to join the Safety Subcommittee was not only a gateway to knowledge, friendship, and professional growth, but also a gateway to developing a key part of myself.

I recently remarked to a colleague that I have become so much more extroverted and driven, a transformation that came about from speaking up on behalf of brewery employees to try and protect them from getting hurt on the job. My colleague responded by saying that the truest form of love is when our love for someone or something helps us discover who we really are. I cannot think of a better way to describe my journey, or a better legacy for Matt. One of Matt's famous quotes is, "Safety is love." There are many aspects of craft beer that I love, but the pursuit of safety is the endeavor that protects all the rest. It is easy to get caught up in regulation and compliance if you are new to the subject, but this book reminds us why regulation exists: to protect who and what we care about.

Matt's knowledge and passion have already benefited so many of us, and now this legacy will extend even further with *Brewery Safety*. He and I have discussed that we will never actually know how many employees have benefited from something they heard us say or something their coworkers learned from us and passed on. It is impossible to know how many minor injuries or catastrophes we have helped prevent; sometimes it feels like we aren't doing anything at all. But, every so often, someone will tell me that they hear my voice in their head when they are cleaning a tank or getting a keg for the taproom, and they just take a second to ask themselves if they are doing whatever they're doing as safely as possible. Little moments like that make me feel like we are making a difference.

Brewery Safety has something to teach everyone, regardless of their experience level. As you, dear reader, start or continue your own journey with safety, remember that you ARE making a difference. It is always worthwhile to admit what you do not know and look for ways to improve. Even though you will never know the full extent of your success, your voice in someone's head could be the reason they do not need stiches that day, or get scarred by third degree burns, or their kids don't have to grow up without a parent.

This book contains not only Matt's extensive knowledge and lived experience but also his passion and love for what he does. On paper, safety is a daunting array of regulatory guidelines, statistics, and good management practices. It is up to us, the people, to interject our passion and our motivation to put them into practice. Safety doesn't have to be your primary job for it to be one of the values you include in everything you do. The section on culture is arguably the most important part of this book, because so much of safety is about changing thoughts to change actions. Culture is the framework of beliefs and practices that influence the choices that we make, and any attempt to make changes and improvements to safety at your brewery with this book will start with understanding why people think and act the way they do.

Translate the technical knowledge you will learn from this book and put it into practice at your brewery. Some days this will feel difficult or impossible, so remember to be kind, be patient, and remember your motivation is wanting to keep people safe. Be transparent not only about why changes are needed, but why you care. Ask others to find their personal "why" to motivate them. Everyone is a stakeholder in safety, so always share knowledge and listen to ideas. Sharing stories of both wins and losses is a crucial part of learning how to "do" safety, as you will see from Matt's numerous personal examples.

At the end of the day, we work in the craft beer industry because we love it. If you have picked up this book, it is because you feel compelled to protect the thing that you love, protect yourself, and protect your colleagues. You do not have to make safety your life's work to appreciate *Brewery Safety*, but you can and should use it to honor your passion within craft beer. Passionate people are the biggest asset to our industry, and protecting our people is the best way I can think of to protect that passion and continue to elevate craft beer.

Rachel Bell
Health and Safety Manager,
CANarchy Craft Brewing Collective
Brewers Association Safety Subcommittee Chair

ACKNOWLEDGMENTS

This book would not have been written without the moral and intellectual support of many people. First and foremost is Peter Whalen, who I met at the Craft Brewers Conference® in Phoenix in 1997. Affable and conversational, Peter is one of the first insurance agents in the country to build programs specifically for craft beverage producers. It was he who suggested I apply my safety and training background to the brewing industry. In that way, this volume was twenty-five years in the making. Peter and his lovely family have been my friends ever since. Some chapters of this book were written in a beautiful studio above his garage.

The list of associates at the Brewers Association is long, but it must include Chuck Skypeck, Chris Swersey, Bob Pease, and Paul Gatza, all of whom drove the creation of the Safety Subcommittee in 2013. Ann Obenchain, Luke Trautwein, and Julia Herz all helped spread the word of brewery safety with their polished promotional skills.

My colleagues on the Brewers Association Safety Subcommittee and the Master Brewers Brewery Safety Committee have taught me much and acted as a supportive presence during the development of this book. I hold them all in high regard for their safety aptitude and years of tireless volunteerism. While there have been too many to mention over the last decade, I do wish to call out Russell "Tony" McCrimmon, Ken Anderson, Bill Lenczuk, Chris Bogdanoff, Chris LaPierre, Rachel Bell, Andy Clearwaters, David Currier, Kevin Walter, Reva Golden, and Andrew Dagnan.

Over the years, I have met and been influenced by many of the brightest brewers and owners in the land. These are people who embraced safety and were never shy about having those potentially awkward conversations. Thank you to John Mallett, James "Otto" Ottolini, Larry Horwitz, Eric Wallace, Jason Perkins, Mark Kamarauscus, Tod Mott, Matt Meadows, Michael "Mufasa" Ferguson, and Teri Fahrendorf. Additionally my collaborations with John Olaechea from Occupational Safety and Health Administration (OSHA) Region VIII, Nick Donofrio from OSHA Region II, and Beth Embry, PhD, from the Leeds School of Business at the University of Colorado have certainly benefited this volume.

Finally, I wish to thank some amazing women in my life who have given counsel, support, and encouragement. As far as bringing out the best in my brewery safety messaging, I remain permanently improved by Acacia Coast, Dr. "J" Nikol Jackson-Beckham, MacKenzie Staples, Kaylyn Kirkpatrick, Katie Fromoth, and my publisher, Kristi Switzer. And most of all, to my astonishing daughter, Lydia.

SECTION I
PRINCIPLES

The early 1970s was a time when many Americans were experiencing feelings of helplessness. The Vietnam War had divided the people. Pollution and workplace hazards were rampant. President Richard Nixon's administration and Nixon himself were to become embroiled in controversy. In the middle of this period of turmoil, the Occupational Health and Safety Administration (OSHA) was created. Whatever else history says about Richard Nixon, the creation of both OSHA and the Environmental Protection Agency came under his watch. People are complicated.

A colleague in the 1990s told me, "There is no such thing as 'environmental.' There is only social, political, and economic." For America's woes, with evidence of DDT accumulation in eagles' eggs, rivers on fire from combustible discharges, and occupational injuries peaking, there was clearly a deeper socioeconomic underpinning. As I write this some 50 years on from the creation of OSHA, so much seems the same. Workers are injured, the environment is polluted, and business selfishness has somehow resurged as the norm.

Please don't fight safety. If you are a militant brewer or one with an edgy brand identity that somehow portrays the illusion that safety isn't sexy, get over yourself. When my conspiracy-theorist neighbor Barnacle Bob said it was "him against the man," I replied, "Bob, you are the man. It's you against yourself. Stop fighting it." He never did like my advice. A classic literature theme for a modern time.

Brewers, please don't be "the man." Please don't fight against something that is so fundamental to your business. Safety is a value, like quality, sustainability, diversity, profitability, and community involvement. Safety builds equity in your company and helps ensure its survival. It lowers business expenses. You can say, "Screw OSHA!" all you want, but that doesn't give you permission to operate an unsafe brewery. Brewers, in particular craft brewers, need not feel that OSHA is out to get them. The rules laid down by OSHA are the minimum suggested/mandated protocols. The rules are sensible and legislation requires they be periodically vetted for their effectiveness; that is, are the rules saving lives? And they are and they do.

I visited with a brewery manager in the UK, which, to its credit, has one of the greatest safety and health agencies in the world. What I learned speaking to him was how, in all of Western Europe, the Health and Safety Executive (HSE) is lauded as an exemplary workplace safety program, better than those in Germany or Austria. When back Stateside, I looked into the agency further and read this: the HSE is the body "responsible for the encouragement, regulation and enforcement of workplace health, safety and welfare, and for research into occupational risks in Great Britain" (Howard 2019, 34). The word that stands out here is *encouragement*. That is the sentiment I wish to bring to this book. Let's be honest, workplace safety is a work in progress. You're probably never going to have all the bells and whistles, all the written programs, nor a flawless workplace safety track record. But you can work toward that. It's like being a good person: you have your faults, but you're working on understanding them and reducing them.

Somewhere along the line I was quoted as saying, "Safety is love." I don't disagree with it. My brewers guild handler at the time, Acacia Coast, was the person who made something of it. And I embraced it, even though it took some getting used to. I created the subtitle that goes beneath the expression, that safety is love for yourself and love for those around you. Some of you related to it—wore the buttons, repeated the sentiments, tweeted, and even sought me out to say thank you for the time I

got a nurse to attend to your torn hand. I'm talking to you, Dallas in Omaha.

The culture of safety is not about mandating compliance, legislating behavior, or even knowing all there is to know about safety. It is about caring for yourself and for others. Encouragement—to inspire courage—is more powerful than negativity or condescension. As the movie line goes, "If you haven't got your health, you haven't got anything." I wholeheartedly agree. The work we're doing as brewers, cider makers, vintners, distillers, and kombucha brewers isn't worth it if we're broken. When we act to instill a safety culture, we recognize the courage it takes to embrace safety in the workplace and show we care for ourselves and our fellow workers.

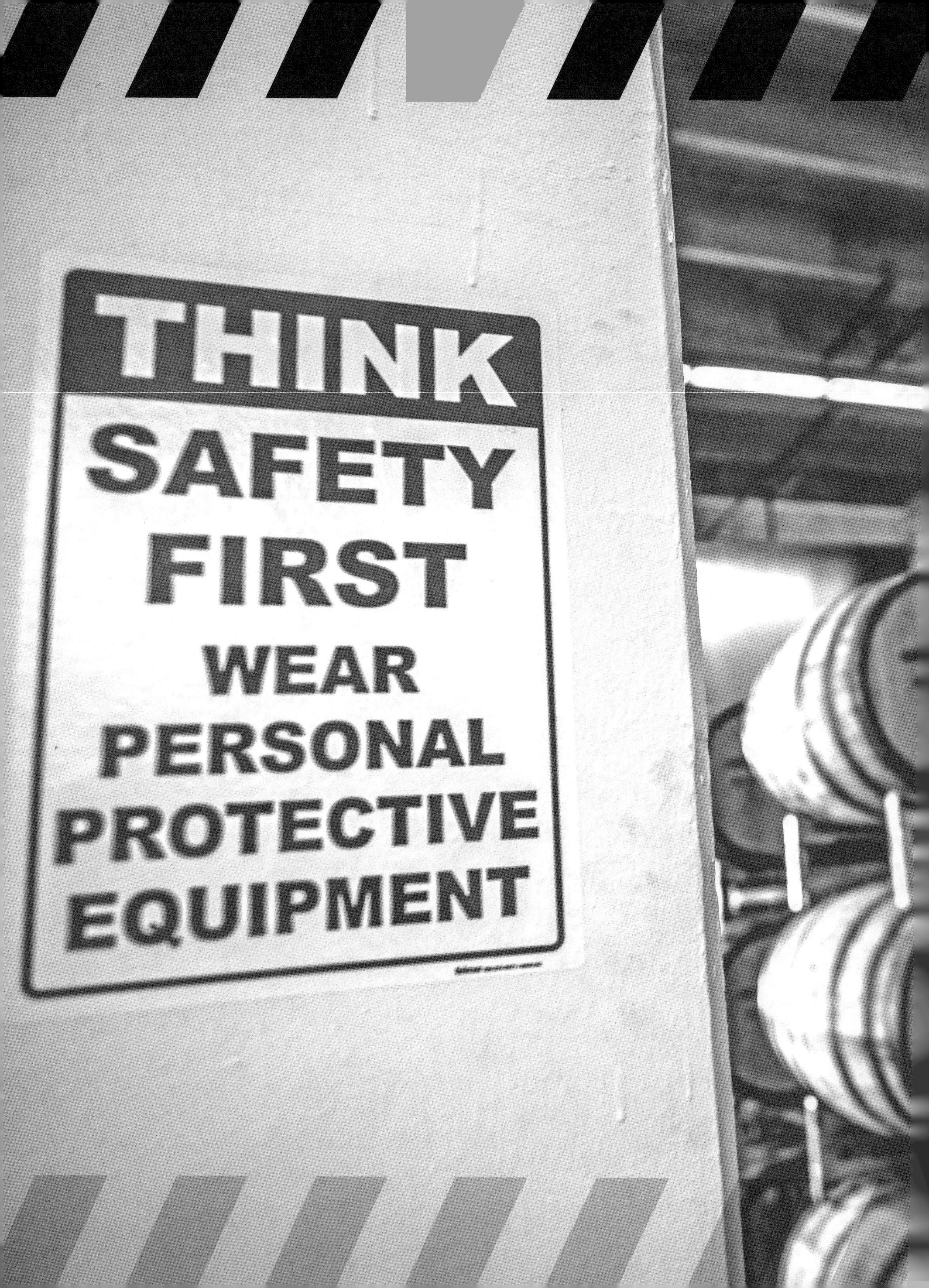
THINK
SAFETY
FIRST
WEAR
PERSONAL
PROTECTIVE
EQUIPMENT

1

SAFETY IS A VALUE

Safety is freedom from harm, plain and simple. That is, freedom from occurrences that lead to short-term injury or long-term illness or lack of well-being; it is avoiding products, equipment, and environments that lead to such occurrences.

If you look at the way *safety* is said in other languages, you will see it as some version of *security*, whether *securité* (French), *seguridad* (Spanish), or *sekuriteit* (Afrikaans). The most direct definition comes in German, where security is *sicherheit*, translating literally to "for sureness."

Most Americans would not use the word security when talking about safety. In the US, we tend to think of security as "measures taken to guard against espionage or sabotage, crime, attack, or escape." Actually, that definition is the eighth out of nine listed by Merriam-Webster.com. The first definition given is "freedom from danger: safety."

SAFETY EVOLVES

Safety is a value, not an encumbrance. Its worth has grown over time. Take, for instance, the thousands who died building the Great Wall of China or the hundred or so who died of heat stroke and asphyxiation building the Hoover Dam. Nowadays, even a single life lost at work is cause for great concern, public scrutiny, and costly reparations.

And why is it that safety is now more important than ever? As a society, we have placed a higher consideration on the individual. After all, we're not talking about the slaves that built ancient edifices or the thin, starving, desperate Depression-era men who built the Hoover Dam. Society compensates people better now—maybe not as well as it should, but better.

SAFETY IS A TEAM EFFORT

The Occupational Safety and Health Administration (OSHA) was created in 1971. One of the agency's keystone principles is called the General Duty Clause (GDC). The GDC predates the writing of OSHA regulations: it is language in the original Occupational Safety and Health Act of 1970 (Pub. L. No. 91–596, § 5, 84 Stat. 1593).

The GDC is the origin of the idea that the employer's responsibility is to maintain a "safe and healthful workplace," words that code for injury-free and disease-free, respectively (Ferguson 2019). But the other part of the precept is that employees must act appropriately, use protective equipment, adhere to training, and comply with safety regulations. Thus, in our time, the concept of safety is a combination of greater individual worth and greater individual responsibility.

There are many regulations promulgated by OSHA, some of which may seem complicated, even onerous. The agency has made attempts to simplify the wording in recent years, but it is still easy to become confused with the number of specialized terms used. Some regulations tell the employer specifically how to achieve compliance, while others list compliance endpoints and require employers to chart their own course.

These regulations have proven their worth in saving lives and reducing injuries across OSHA's first 50 years. It is improbable OSHA is going to go away in the foreseeable future. Its rules are the minimum legal requirement. If you are running a business, then understanding these rules and doing your best to comply with them is your legal responsibility.

However, through most of this volume, the emphasis is on how to understand hazards in the brewery as they relate to common brewery processes, and how to reduce or eliminate the risks posed by these processes and related chemicals and equipment. While there are going to be discussions of various OSHA requirements as part of this,

the key skill every owner and employee needs to master is how to recognize and eliminate a hazardous situation.

We still have a ways to go if we're going to involve everyone in safety. Brewers are notoriously independent. But don't confuse an edgy brand identity with a disrespect for employee safety and well-being. In the brewery, we have hazards of falling, burning, electrocution, loud noise, and injury by equipment—any of these hazards can be abetted by ignorance, insolence, or intoxication. We have to get our heads in the right place.

Remember that safety is a value. Safety is security. Safety is freedom from harm. Safety is love for yourself and others.

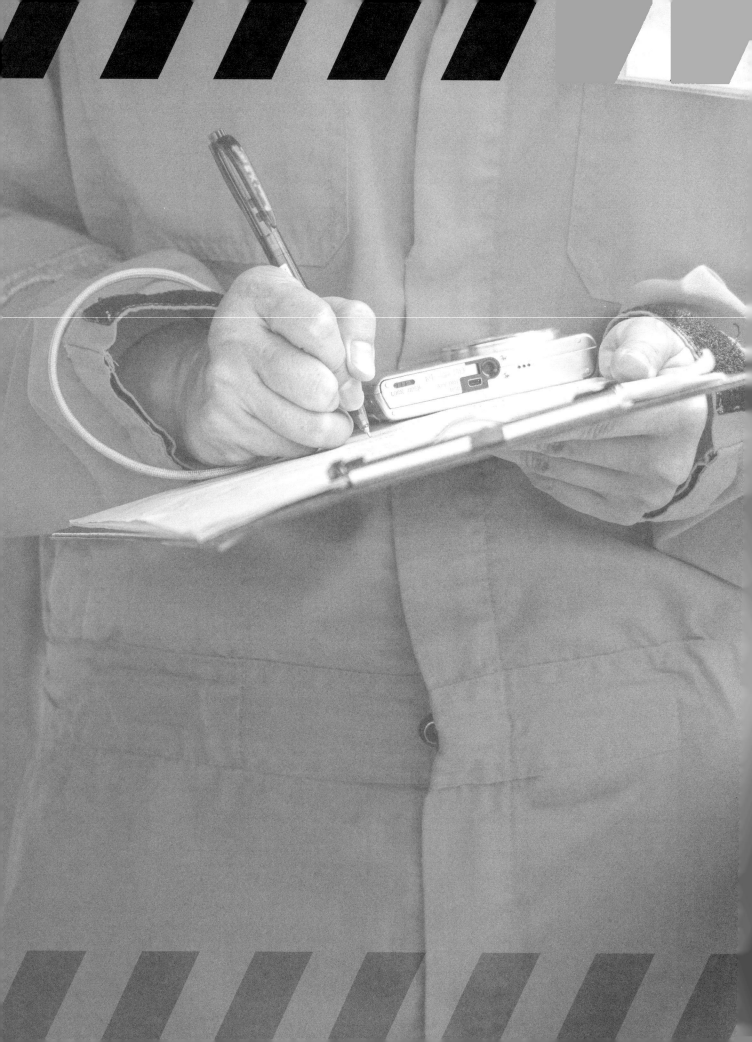

2

UNDERSTANDING OSHA

The acronym *OSHA* stands for the Occupational Safety and Health Administration, also called the *agency*. The agency was established by the 1970 Occupational Safety and Health Act, typically abbreviated as the OSH Act.

The first thing to know about OSHA in your jurisdiction is whether it operates as a state or federal agency. Instead of federal OSHA, some states implement OSHA-approved job safety and health programs run by state agency staff rather than federal employees. When a state implements its OSHA program itself, we refer to it as a *state plan*. Any such state-run OSHA agency can enact rules that are different from the federal regulations, as long as the state rules are not less stringent than the federal ones. Consequently, even if your workplace operates under a state plan, the rules in your state may be the same as the federal rules; however, the rules can vary some or, as with California, vary greatly.

The map in figure 2.1 indicates which US jurisdictions run state plans, which run federal, and which have some sort of hybrid plan where some classes of employees are overseen by federal enforcement, while others are under state enforcement. An interactive version can be found at https://www.OSHA.gov/stateplans, which also provides links to further details for any state or territory.

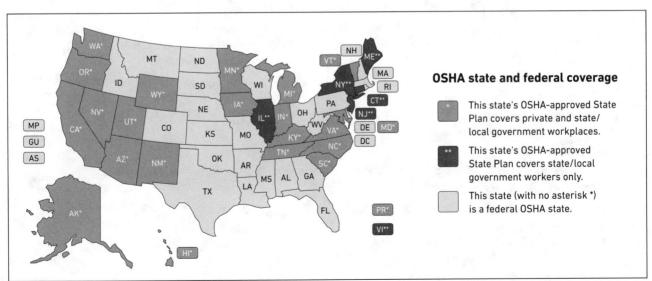

Figure 2.1 Map of OSHA state and federal coverage.

THE BIG PICTURE

Despite everything that will be mentioned in this chapter about OSHA and required recordkeeping, compliance is only one part of a health and safety program. Know that OSHA is actually on your side, meaning that it is on the side of employees' welfare. As part of this, OSHA provides employers with essential guidance on the minimum legal requirements of workplace safety. OSHA also provides compliance assistance through on-site consultation programs in every state, provides rich resources, including guidance documents, bulletins, training aids, online simulators and calculators, and a number of other digital resources.

The Brewers Association and the Master Brewers Association of the Americas, along with state brewers guilds, have forged alliances with OSHA compliance and consultation branches in several states. These alliances open channels of communication, provide shared resources, and have already shown positive effects on safety and health programs in the craft beer industry.

OSHA was created out of concern for worker welfare. Some employers were skimping on workplace hazard control, personal protective equipment, and training. In the half century since OSHA was created, American businesses have continued to succeed at the same time as employee protections have increased, proving that safety isn't the problem. It's bad operators that are the problem—don't be one.

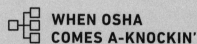

WHEN OSHA COMES A-KNOCKIN'

Why Might OSHA Drop By?

- Imminent danger situations
- Severe injuries or illness report
- Complaints from a variety of sources
- Referral from another agency

- Employer requests inspection
- Targeted inspection*
- Emphasis program†
- Follow-up inspections

In response to a complaint, but not an inspection, OSHA might first send a letter or email asking about a known or suspected hazard and give the employer a fixed, short timeframe to respond in writing.

So, OSHA Wants a Look Around

- Advance notice is rare and only happens if one or more of the following circumstances apply:
  - In situations of apparent imminent danger
  - Inspection requires an after-hours visit or special preparations
  - Managers or employee representatives will not be on site
  - The OSHA Area Director believes it will result in a more complete inspection, e.g., in a fatality investigation
- An OSHA employee who otherwise provides advance notice of an inspection will be subject to a criminal enforcement action
- The attending OSHA representative will show a numbered, photo ID

The Inspection Itself

- There are three parts to an OSHA inspection. Once opened, OSHA has six months to close an inspection and issue any citations.

Opening Conference
  - Held with employer and employee reps‡
  - Kept brief
  - Addresses the scope of the inspection
  - Inspector may review OSHA 300 logs and other required recordkeeping

Walkaround
- Focuses on known hazards or complaints
- Inspector can note hazards outside of inspection focus
- Employer fixes apparent hazards on the spot, when feasible
- Inspector may take industrial hygiene measurements or photographs
- Inspector may speak with employees privately away from employer
- Employer and employee rep should take notes and their own photographs
- Be respectful of inspector's role and regulatory expertise

Closing Conference
- Inspector identifies "apparent violations" observed
- Inspector gives advice of remedies for hazards
- Inspector may discuss possible fines
- Whistleblowers are protected by law
- Employees can seek "party status" if employer contests citation(s)[**]
- Employer must notify employees when the company contests citation or asks for different abatement date
- Employees may contest an abatement date; such contest must be submitted in writing within 15 days of receipt of the citation

Now for the Bad News
- OSHA sends written citations/penalties within six months of opening an inspection
- Citations will reference the specific OSHA standard or the General Duty Clause
- Citations are not levied against employees; OSHA holds the employer responsible for unsafe conditions
- Penalties can be reduced by OSHA for small businesses and good-faith efforts (except in the case of a willful violation)
- Citations are classified according to severity and frequency; maximum penalties shown are per violation:[††]
  - Serious: $15,625
  - Other than serious: $15,625
  - Willful or repeated: $156,259 [‡‡]
  - Posting requirements: $15,625
  - Failure to abate: $15,625 per day beyond the abatement date

Notes
[*] Workplace experiencing high rate of injuries.

[†] Inspections aimed at specific high-hazard industries or activities.

[‡] Opening and closing conferences are normally held with management and employee representatives together but can be held apart. Typically, employees go first.

[**] Party status allows employees and their representatives to attend meetings with the Occupational Safety and Health Review Commission independent of the employer.

[††] 2023 penalties are shown; check https://www.OSHA.gov/penalties for latest figures.

[‡‡] Willful or repeated 'less-than-serious' penalties are less.

OSHA REGULATIONS IN THE LEGAL CODE

OSHA regulations are found in Title 29 of the *Code of Federal Regulations* (abbreviated C.F.R. in notes). The *Code of Federal Regulations* has a specific indexing system. Each title uses the following hierarchy: title, subtitle, chapters, subchapters, parts, subparts, sections, paragraphs, and subparagraphs. Depending on complexity, some entries will not make use of items like subtitles, subchapters, or subparts.

There are two main parts in Title 29 of interest to breweries. Part 1910 is devoted to workplace safety in general industry, which is any industry that is not construction related. Part 1926 addresses safety and health regulations specific to construction. Where the same sorts of hazards exist in both occupational spheres, duplicate regulations are usually found in both in parts. Unless you are involved in building or modifying your brewery, you will look to Part 1910 for your requirements.

One of the best online search tools for the *Code of Federal Regulations* is the Legal Information Institute run by the Cornell University (Cornell Law School n.d.). This is a most useful searchable regulatory database.

Table 2.1 lists important subparts and sections of Title 29 that will be relevant to most breweries. Subparts that do not pertain to the manufacturing setting of a brewery have been omitted. It is up to you to review each section and determine whether it applies in your workplace. For example, 29 C.F.R. § 1910.1096 "Ionizing radiation" would apply to a brewery with an X-ray can-fill inspection device, but not to a brewery without such a hazard.

CORE CONCEPTS IN OSHA STANDARDS

Part 1910 of Title 29 promulgates the occupational safety and health standards pertaining to OSHA regulations. Regulations or regulatory regimes can be categorized into two types, prescriptive and performance oriented. OSHA regulations use both.

Prescriptive versus Performance-Oriented Regulations

Prescriptive regulations provide detailed instructions with little latitude. The other type is called performance-oriented regulations. As pertains to 29 C.F.R. Part 1910, performance-oriented regulations require the employer to identify hazards for themselves and implement appropriate planning, training, and hazard controls relative to their specific workplace. While performance-oriented regulations allow for more flexibility for the employer,

---

C.F.R. INDEXING EXAMPLE

Let's say you want to know if there is a requirement to lock the legs open on a stepladder in the brewery. Your practice has been to simply lean it up against the fermentor. The hierarchy below shows how you can drill down all the way through Title 29. In actual practice, you can start at Part 1910 and just read down through the sections from there.

Title 29—Labor

  Subtitle B—Regulations Relating to Labor

    Chapter XVII—Occupational Safety and Health Administration, Department of Labor

    Part 1910—Occupational Safety and Health Standards

      Subpart D—Walking-Working Surfaces

      Section 1910.23—Ladders

        Paragraph (c)—Portable Ladders

          Subparagraph (2)—[The employer must ensure:] Each stepladder or combination ladder used in a stepladder mode is equipped with a metal spreader or locking device that securely holds the front and back sections in an open position while the ladder is in use.

The citation for this rule is written:

29 C.F.R. § 1910.23(c)(2) (2023)

---

they can be more frustrating because they don't always tell you exactly what to do or how to do it.

OSHA regulations are a hybrid of both performance-oriented and prescriptive language. For example, when an OSHA regulation states you have to implement some means of fall protection above four feet, it is performance oriented; but when it specifies the height of two guardrails if you choose to use railings as your fall protection system, it is prescriptive.

Looking at the six right-hand columns in table 2.1, in general, the written programs, required plans, procedures, other documentation, and personal protective equipment

(PPE) requirements will be mostly performance-oriented regulations. By contrast, technical specifications for equipment and facilities, worker monitoring, and site testing tasks usually reflect prescriptive rules.

Program versus Plan

OSHA uses the terms *program* and *plan* in specific ways. A program, often called a written program, is an overarching document that describes how the employer will administer a safe workplace with regard to a specific class of hazards. For instance, for respiratory protection, the employer must have a written program that details medical screening, training, and protective equipment.

A plan, on the other hand, is usually a more specific written document focused on a single activity. Examples include emergency action plans, fire prevention plans, confined space permits, and energy control procedures.

Written programs and plans are essential compliance documents, but also assist the employer in creating employee training materials and consistent written procedures in the workplace.

The Rebuttable Presumption

The General Duty Clause (GDC) and specific OSHA standards are written following a philosophy called the *rebuttable presumption*. The workplace is presumed to have hazards and it is up to the employer to document they have created a safe workplace. In other words, the workplace is deemed hazardous until proven to be not hazardous. Guilty until you prove yourself innocent.

A canning line is presumed to have electrical, mechanical, pneumatic, noise, and chemical hazards. It is up to the employer to reduce or eliminate those hazards and that usually includes a form of documentation known as the hazard assessment (HA), also known as job hazard analysis (JHA). Chapters 3 and 4 lay the groundwork for assessing hazards, and chapter 5 explains the process of conducting and documenting HAs. Hazard assessments form an essential written record showing that the employer has evaluated the hazards associated with a task. Defining how to control those hazards leads directly into creating robust standard operating procedures (SOPs).

The GDC is far-reaching. There are certain well-known hazards in breweries that are not specifically mentioned in Title 29. Some examples include kettle boilovers, dry hop volcanos, overpressurized vessels, heat stress, and repetitive motion disorders. When there is no specific

---

GENERAL DUTY CLAUSE

On its enactment, the original legislation that authorized the creation of OSHA stated the government's intent that workplace safety was the responsibility of both the employer and the employee:

Duties

Sec. 5 (a) Each employer—

(1) shall furnish to each of his employees employment and a place of employment which are free from recognized hazards that are causing or are likely to cause death or serious physical harm to his employees;

(2) shall comply with occupational safety and health standards promulgated under this Act.

(b) Each employee shall comply with occupational safety and health standards and all rules, regulations, and orders issued pursuant to this Act which are applicable to his own actions and conduct.

Source: Occupational Safety and Health Act of 1970, Pub. L. No. 91–596, § 5, 84 Stat. 1593 (codified at 29 U.S.C. § 654)

OSHA standard, then the employer and employee are held to the GDC as a catch-all. Knowing that hazards exist but taking no measures to prevent them or protect against them can be grounds for citation under 29 C.F.R. § 1910.9 (training and PPE) or be considered a violation of the law according to Section 5(a) of the original Occupational Safety and Health Act (codified at 29 U.S.C. § 654(a)).

To be fair, an OSHA inspector has to clear a pretty high bar to make a citation under the GDC. The burden of proof must meet four criteria:

- The employer failed to keep the workplace free of a hazard to which its employees were exposed.
- The hazard was recognized.
- The hazard was causing or was likely to cause death or serious physical harm.
- A feasible and useful method to correct the hazard was available.

Table 2.1 **Title 29 subparts and sections important for breweries**

Subpart	Subpart Title	Section(s)	Section Title	Written Program(s)	Plan, Procedure, or other Documents	Training Req'd.	PPE Req'd.	Technical or Equip. Specs.	Employee Monitoring or Site Testing
A	General	§ 1910.9	Compliance duties owed to each employee			X	X		
D	Walking-Working Surfaces	§§ 1910.21–.30	(all indicated sections)			X	X	X	
E	Exit Routes and Emergency Planning	§§ 1910.33–.39	(all indicated sections)		X	X		X	
F	Powered Platforms, Manlifts, and Vehicle-Mounted Work Platforms[a]	§§ 1910.66–.68	(all indicated sections)		X	X		X	
G	Occupational Health and Environmental Control	§§ 1910.95	Occupational noise exposure	X		X	X	X	X
H	Hazardous Materials[b]	§ 1910.101	Compressed gases (general requirements)					X	
I	Personal Protective Equipment	§§ 1910.132–.140	(all indicated sections)	X	X	X	X	X	X
J	General Environmental Controls	§ 1910.141	Sanitation					X	
		§ 1910.144	Safety color code for marking physical hazards					X	
		§ 1910.145	Specifications for accident prevention signs and tags					X	
		§ 1910.146	Permit-required confined spaces	X	X	X	X	X	X
		§ 1910.147	The control of hazardous energy (lockout/tagout)	X	X	X		X	
K	Medical and First Aid	§ 1910.151	Medical services and first aid			X	X		
L	Fire Protection	§§ 1910.157–.158	Portable fire suppression equipment			X	X	X	X
		§ 1910.159	Automatic sprinkler systems					X	X
		§§ 1910.164–.165	Other fire protection systems						X
M	Compressed Gas and Compressed Air Equipment	§ 1910.169	Air receivers					X	

Table 2.1 **Title 29 subparts and sections important for breweries** (cont.)

Subpart	Subpart Title	Section(s)	Section Title	Written Program(s)	Plan, Procedure, or other Documents	Training Req'd.	PPE Req'd.	Technical or Equip. Specs.	Employee Monitoring or Site Testing
N	Materials Handling and Storage	§ 1910.176	Handling materials - general						
		§ 1910.178	Powered industrial trucks	X	X	X		X	
		§ 1910.184	Slings					X	
O	Machinery and Machine Guarding	§ 1910.212	General requirements for all machines					X	
		§ 1910.219	Mechanical power-transmission apparatus					X	
P	Hand and Portable Powered Tools and Other Hand-Held Equipment	§§ 1910.241-.244	(all indicated sections)					X	
Q	Welding, Cutting and Brazing	§§ 1910.251-.255	(all indicated sections)			X	X	X	
R	Special Industries	§ 1910.272	Grain handling facilities		X	X		X	X
S	Electrical	§§ 1910.302-.308	Design safety standards for electrical systems					X	
		§§ 1910.331-.335	Safety-related work practices			X	X	X	
Z	Toxic and Hazardous Substances	§ 1910.1000	Air contaminants					X	X
		§ 1910.1020	Access to employee exposure and medical records		X				X
		§ 1910.1030	Bloodborne pathogens		X	X	X		X
		§ 1910.1053	Respirable crystalline silica		X	X	X	X	X
		§ 1910.1096	Ionizing radiation		X	X		X	X
		§ 1910.1200	Hazard communication	X	X	X	X	X	X
		§ 1910.1201	Retention of DOT markings, placards and labels						
		§ 1910.1450	Occupational exposure to hazardous chemicals in laboratories		X	X	X	X	X

ª This subpart is only relevant if operating any of these types of lifting equipment.

ᵇ If operating a distillery or using ammonia for refrigeration, other sections in Subpart H may apply. See also Subpart Z for other hazardous material requirements.

Even if OSHA can adequately gather enough testimonials, documents, or other proof to meet these criteria, GDC citations are often contested in court. Consequently, OSHA may issue the employer a written notice that improvement is warranted or suggest the employer seek assistance from an OSHA or private safety consultant (Ferguson 2019).

CLASSIFYING YOUR BUSINESS

OSHA and many federal statistical agencies, like the Bureau of Labor Statistics, group businesses according to the North American Industry Classification System (NAICS), using a five- or six-digit code. The "nakes" or "nacks" code, as it is variably pronounced, allows agencies to collate and analyze a wide range of data related to the economy. Data for occupational injuries, illnesses, amputations, and fatalities are published according to the NAICS.

It is up to each business to select the NAICS code that best represents its primary business activity. For most breweries, if beer manufacturing is the main source of income, the employer will declare a NAICS code of 312120, "Brewery." Brewpubs, where sit-down meals are served and where the sales from food exceed the sales of beer, may choose a code of 722511, "Full-Service Restaurants." A list of NAICS codes relevant to beer producers can be seen in the sidebar.

Choosing a NAICS code has far-reaching ramifications. Businesses that fall under the "Drinking Places" NAICS code have historically been considered by OSHA to be low hazard compared to those designated under "Beverage Manufacturing" and "Restaurants and Other Eating Places," while restaurants have more leeway than breweries when it comes to safety requirements. These differences become apparent when contemplating injury and illness reporting, which is discussed below.

A few words of caution here. The NAICS codes are occasionally modified; new subdivisions may be introduced or categories may be collapsed. It is best to check the current NAICS classifications when first defining a new business. Do not try to game the system by choosing a restaurant code if your business is primarily a brewery. Understand that NAICS codes evolve over time, so review the most current listings when you choose your most appropriate code.

INJURY AND ILLNESS RECORDKEEPING AND REPORTING

There are many OSHA standards that require some form of recordkeeping, whether for injuries, medical evaluations, documentation of training, forklift inspection, or otherwise. Most of these will be dealt with as the individual standards shown in table 2.1 and are referenced in chapters 6–18.

There are two types of injury and illness reports that are required by OSHA regardless of whether you are governed by a state plan or under federal jurisdiction. The first is the urgent reporting of critical workplace injuries involving fatalities, amputations, or stays in hospitals. The second is annual tracking and reporting of recordable workplace injuries and illnesses, OSHA's so-called 300 forms.

Reporting a Severe Injury or Fatality

All employers, even if having just a single employee, must report directly to OSHA in the case of a workplace fatality, in-patient hospitalization, amputation, or loss of an eye. This requirement is found in 29 C.F.R. § 1904.39. There are no exemptions from this requirement based on size or industry type.

Employers report any such injury or fatality occurrences either by filing on online report at https://www .osha.gov/report, or by calling either the national hotline at 800-321-6742 (OSHA) or their nearest area state or federal OSHA area office during regular business hours. It is a good idea to put a copy of the chart shown in figure 2.2 and means of notifying OSHA into either the company's overall safety and health program, the emergency action plan required in 29 C.F.R. § 1910.38, or both.

NAICS CODES FOR BEER PRODUCERS

3121 Beverage Manufacturing
312120 Brewery
312130 Winery
312140 Distillery

7224 Drinking Places
722410 Drinking Places (Alcoholic Beverages)

7225 Restaurants and Other Eating Places
722511 Full-Service Restaurant (formerly 7221)
722513 Limited-Service Restaurant
        (formerly 7222)

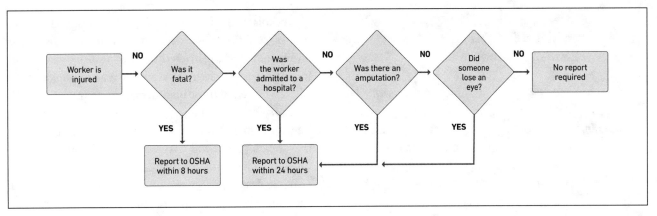

Figure 2.2. Required OSHA reporting for a severe injury or fatality.

OSHA'S INSPECTION PRIORITIES

- Imminent danger situations
- Severe injuries or illness report
- Worker complaints
- Referrals from other agencies
- Targeted inspections*
- Emphasis programs†
- Follow-up inspections

Notes

* Workplace experiencing high rate of injuries.

† Inspections aimed at specific high-hazard industries.

If you are wondering whether reporting such a case to OSHA will result in an inspection, the answer is yes, quite possibly. OSHA's priority list of triggers for an inspection are shown in the sidebar (US Department of Labor 2016a).

The 300 Forms

The 300 forms are used by employers to record workplace injuries and illnesses. There are three forms in the series: the 300, 300A, and 301. All the 300 forms are available at https://www.osha.gov/recordkeeping/RKforms.html and include detailed instructions. Although the 300 forms are fairly easy to complete, there are a lot of embedded terms that you should know before you begin recording workplace injuries or illnesses.

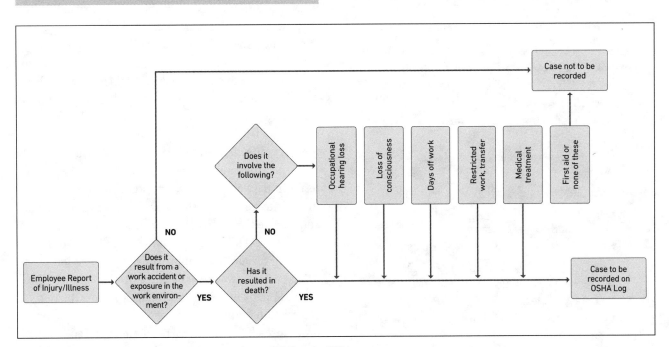

Figure 2.3. Injury or illness events requiring reporting on OSHA series 300 logs.

WE'RE HERE TO HELP. NO, REALLY!

It doesn't matter if you are an employee, an owner, a customer, an insurer, or a regulatory agent, a safe business is in your best interest. There is lots of help available and much of it is free.

- OSHA On-Site Consultation: Get general or targeted practice inspections or industrial hygiene monitoring with no fear of receiving citations, just be sure to make the recommended improvements. Contact your state's On-Site Program.
- Business Insurer or Workers' Comp Provider: Safer businesses have fewer insurance claims, so many providers offer safety consultation for free or at a reduced price.
- Trade Association Resources: The Brewers Association, Master Brewers Association of the Americas, and the American Society of Brewing Chemists all have written, audio, and video safety resources—bookmark your browser with their addresses.
- Professional Consultants: Sure, these will cost the business something, but professional consultants sure come in handy for developing written programs, conducting training, and helping the business after an OSHA inspection.

Going the Extra Mile
The following special OSHA-run programs go beyond the normal compliance inspection focus.

- OSHA Challenge: businesses are paired with a mentoring agency ("Challenge Administrator") and work to build a better health and safety management system.
- OSHA Strategic Partnerships and Alliances: OSHA teams up with employers or trade associations and tackles serious hazards or widespread safety concerns.
- Safety and Health Achievement Recognition Program (SHARP): bragging rights for a few businesses with exceptional safety programs.
- Voluntary Protection Program (VPP): this OSHA program is kind of like TSA Pre-Check, but for businesses.

Workplace injuries or illnesses that result in death, loss of consciousness, occupational hearing loss, days away from work, restricted work or transfer to another job, or medical treatment are known as *recordable cases* (fig. 2.3). Each of these qualifiers has a definition specific to injury recordkeeping. On these forms, the employer only documents recordable cases, with the exception that the employer will also record qualifying injuries or illnesses of a contractor injured on the property.

While death and loss of consciousness may be obvious, the other four qualifiers are subject to OSHA's specific definitions. Occupational hearing loss is typically a result of gradual deterioration of auditory range due to long-term exposure to loud workplace sound. Under some conditions, short-term noise exposure can also result in deterioration of hearing. OSHA's definition for hearing loss relies on first having a baseline audiogram of the worker's hearing, which is then compared to a recent audiogram. If there is a quantitative loss of overall hearing, or loss of hearing at specific frequency ranges, then the illness is recordable. More on auditory injury is provided in chapter 10.

When counting days away from work, begin with the day after the injury was sustained or the illness discovered. In the event that a worker is told by a licensed healthcare provider to stay home and the employee returns to work anyway, the employer still records this as days away from work. Beyond 180 days away from work, the employer is given the option to record "180 days away" or to continue keeping track of the days.

Although somewhat less severe than days away from work, OSHA also considers restricted work or reassignment of duties to be recordable. Whenever an employer or healthcare provider designate that one or more routine functions shall not be performed, the employee is said to be work restricted or transferred. Routine functions are defined as work the employee regularly performs at least once weekly. Injuries and illnesses that result in days away from work, job duty restrictions, or transfer to another position are often lumped together and referred to as DART (for days away, restrictions, transfers). DART cases are rounded up to full days, not counting the day of the injury or illness detection.

Medical treatment is understood to be anything beyond first aid, visits to an healthcare provider for observation and/or evaluation only, and for diagnostic procedures. The sidebar lists first aid treatments that are not considered recordable medical treatments by OSHA.

OSHA'S DEFINITION OF FIRST AID

- Using a nonprescription medication at nonprescription strength
- Tetanus immunization*
- Cleaning, flushing, or soaking surface wounds
- Wound coverings, e.g., bandages, gauze pads, butterfly bandages†
- Hot or cold therapy
- Non-rigid means of support, e.g., elastic bandages, wraps, non-rigid back belts
- Temporary immobilization devices for victim transport
- Draining or relieving pressure from blister or nail
- Eye patch
- Removing foreign bodies from eye using irrigation or cotton swab
- Removing splinters or foreign material (other than in eye) by irrigation, tweezers, cotton swabs, or other simple means
- Finger guards
- Massage therapy
- Drinking fluids for relief of heat stress

Notes

* Other immunizations, e.g., hepatitis or rabies, are medical treatments

† Sutures, staples, etc. are medical treatments

Source

29 C.F.R. § 1904.7(b)(5)(ii)

Contrary to the numbering system, the first form the employer usually completes is Form 301, "Injuries and Illnesses Incident Report." Form 301 contains information about the specific employee, what treatments were rendered and details about the incident. It must be completed within seven calendar days of the occurrence of, or knowledge of, the recordable incident.

Since personal medical information is on Form 301, the employer should maintain records at the worksite for a minimum of five years and in a secure manner, whether with a locked file cabinet, a secure server, or through a qualified third-party medical records management provider. In small breweries, the owner might identify a designated individual to maintain records in a secure filing system. This could be the owner themselves, the human resources (HR) manager, or the safety manager.

The employer keeps a summary log of recordable injuries and illnesses on Form 300, "Log of Work-Related Injuries and Illnesses." As with Form 301, Form 300 contains confidential employee data and should be maintained in a confidential recordkeeping system. Some employers may make an entry on Form 300 first, then complete the more detailed Form 301 afterward.

Finally, using data from Form 300, the employer completes Form 300A, "Summary of Work-Related Injuries and Illnesses." Form 300A does not include employee names or confidential medical information. It summarizes the number of cases in the previous year, the days of work lost, and the number and types of injuries and illnesses noted. This form is posted in the workplace for all employees to see from February 1 through April 30 of the year following the previous calendar reporting year. Even if there were no recordable cases during the previous year, the employer still completes and posts Form 300A.

For some business types and sizes, the information on Form 300A is then submitted electronically to OSHA by March 2 each year. Table 2.2 shows the annual reporting requirements for different sized employers in the beverage manufacturing and restaurant sectors. Exactly which form(s) to submit and by whom has varied over the past decade. The requirements shown here were those required at the time of publishing this book, but the reader should stay abreast of current requirements because these rules do occasionally change.

As mentioned previously, some businesses do not have to collect or report injury and illness data. These businesses are said to be "partially exempt for size" or "partially exempt establishments in certain industries." The word *partially* is used because this exemption applies only to the collecting and reporting of data for the 300 forms. It is not permissible to go without creating a safe workplace or abiding by other health and safety regulations under OSHA.

"Partially exempt for size" applies to all businesses whose number of employees in the prior calendar year was ten or fewer. Part-time employees are counted as an employee just like a full-time worker. Exempt-for-size employers do not have to collect injury and illness data, and consequently have no reporting to fulfill.

If you are operating a brewpub or taproom business

Table 2.2 **Injury and illness recordkeeping and reporting requirements by NAICS code**

NAICS code	No. of Employees, n	Required Data to Collect	Post in Workplace	Submit to OSHA	Title 29 Citation
312120	n ≥20	301, 300, 300A	300A	300A	§ 1904.1(a)
	>10 n ≤19	301, 300, 300A	300A	not reqd.	
	n ≤10	not reqd.	not reqd.	not reqd.	
722410	n is any number	not reqd.	not reqd.	not reqd.	§ 1904, App. A
722511	n is any number	not reqd.	not reqd.	not reqd.	§ 1904, App. A [b]
722513	n is any number	not reqd.	not reqd.	not reqd.	§ 1904, App. A [b]

[a] (US Department of Labor 2019)

[b] Exempt from recordkeeping even though not listed in Appendix A (US Department of Labor n.d.[a])

model, be sure to count all employees, not just those involved in beer production activities. If you operate multiple locations under one business identity, you count the total workforce of all locations.

The exemption for certain industries applies to those business types that OSHA has set aside as not very hazardous. Among the possible classifications for a beer business shown in the NAICS sidebar, only those identifying as 722410, "Drinking Places (Alcoholic Beverages)," can claim the recordkeeping exemption.

Businesses identifying as a beverage manufacturer under NAICS code 3121- have the most impactful injury and illness reporting requirements. When the number of employees in the prior calendar year exceeds 10 but is less than or equal to 20, the employer collects data for all the 300 forms and posts the 300A summary in the workplace, but these employers do not have to submit any data to federal OSHA.

If your business is a small brewery start-up with fewer than 10 employees, recording injuries still makes sense. These records allow ownership to track trends in injury rates throughout the life of the business. Injury and illness rates are an important key performance indicator (KPI) and are discussed again in section 3 of this book. Starting early also enables transition to required reporting as your workforce comes to exceed 10 employees.

3

HAZARD CLASSIFICATION AND IDENTIFICATION

This chapter and the two that follow will take us on a stepwise path to performing a hazard assessment (HA). This starts with being able to identify and classify hazards (ch. 3), then learning how to control or eliminate them (ch. 4), and, finally, documenting the safe ways to do things in a way that clearly and correctly informs others of the way to perform the job in accordance with our understanding of all of the above (ch. 5). This knowing, controlling, and documenting is the basis of safety.

We cannot reach this point if we don't have a firm grasp on the hazards associated with work activities. Only once we identify a hazard associated with a task can we avoid it by substitution or control it with equipment and protective gear. Knowing the hazard is also necessary for us to manage the risk with standard operating procedures and reliable training.

HAZARDS AND OUTCOMES

If safety is freedom from harm, then hazards are those circumstances that can cause harm. There is always risk when hazards are present. Thus, hazards exist as a potential, a chance, a liability. As we will explore in chapter 7, hazards can summarily be described as situations that can transfer an undesirable form or amount of energy from any source to our bodies, our equipment, our product, or the environment.

The result of negatively experiencing a hazard can be called the outcome. That is, the outcome occurs after this exchange of energy. An overfilled kettle of boiling wort is a hazard; the resulting boilover and third-degree burns is the outcome.

As the scale of a brewery increases, say, from a home brewery to a microbrewery to a regional brewery, the quantity of energy required to run brewery systems increases accordingly. Pressure, heat capacity, the height of elevated workplaces, mechanized equipment—these can all increase in their energy as breweries grow. This is not to say that a homebrewer can't get burned by an overflowing kettle or electrocuted by bad wiring, but hazardous energy of all sorts generally increases with brewery scale.

Automation and the use of remote systems can remove people from this equation. So, while we might decry replacement of the worker with machines, the use of an automated weighing system or a palletizer, for example, does tend to reduce workplace injury.

Most small breweries cannot afford the types of automation that regional and multinational breweries utilize. The burden of creating a safe workplace in a small brewery is more intense than in a large workplace. While this is partly due to the larger brewery's more extensive automation, this is partly due to the fact that small breweries probably do not have a safety expert on staff, whereas large breweries may have an entire safety department. Instilling a small workforce with the correct training, attitude, and behavior to keep up with industry norms is not always easy. We will return to this topic in section 3.

Classifying Hazards

Safety experts have traditionally classified hazards as either physical, chemical, or biological. The idea behind these classifications is based on both the type of energy transferred to an individual and the type of impact it has on the body.

Physical hazards involve forces like gravity, mechanical motion, cutting, pinching, crushing, electric shock, and fire. In general, such hazards cause physical trauma to the body. Examples of physical outcomes include lacerations, strains, broken bones, and burned tissues.

Chemical hazards are different from physical hazards in that they relate to a substance reacting with bodily organs or tissues. This could mean a corrosive chemical irritating the skin or eyes; a gas absorbed into the bloodstream and displacing oxygen, as with carbon monoxide; or ingestion of a disinfectant causing digestive disorders.

Biological hazards are usually defined as infectious agents that can negatively affect the individual. In a brewery, outcomes might include transmissible illnesses brought by a customer, mold growing on tanks, droppings left by vermin and birds, or a bloodborne disease transmitted by contact with a coworker's blood during an injury response.

There are those additional classes of hazards that don't fit so neatly into the three categories mentioned. These include radiation, ergonomic, generalized "safety hazards," and psychosocial hazards. Radiation exposure is a toss-up between physical and chemical: in this book, radiation sources are classified as physical hazards.

Ergonomic hazards are well known in the brewing industry. These can include repetitive motion, heavy lifting, vibration, and awkward movements. They can result in long-term injury to the nervous and musculature systems. Incidentally, to date, OSHA has not published a standard on ergonomic injury prevention. In the absence of specific guidance, it is up to the employer and workforce to create a safe and healthful workplace, as directed by the General Duty Clause.

There are also hazards generally referred to as "safety hazards," but these are mostly physical injury risks. They may have a predominant behavioral root cause, but the outcomes are typically physical. Some examples include tripping over a brewery hose, falling off a ladder, contacting machinery because of a lack of guarding, or using a forklift improperly. In this book, both ergonomic and general safety hazards will mainly be classified as physical hazards and are discussed in chapters 10–15.

Psychosocial hazards are primarily emotional or behaviorally related. These can include workplace stress, workload, stressful customer interactions, discrimination, workplace violence, and resistance to wearing PPE. Psychosocial hazards often combine with other workplace hazards to make accidents more likely and, potentially, more severe (see sidebar).

INFLUENCE OF PSYCHOSOCIAL HAZARDS

Psychosocial hazards related to emotional or behavioral issues are often predominant in the root causes of an accident. At a previous employer, I was involved in investigating a serious workplace injury. An employee suffered a compound fracture of his forearm and hand because his glove got caught on a rotating soil auger. There were three colleagues on the job using a piece of rental equipment they hadn't used before. When the employee in question's glove got caught on the bolt attached to a shaft extension, the operators made a mistake with the throttle, speeding the motor instead of shutting it down. This led to the disastrous outcome.

When I individually interviewed the three of them to get to the root cause of the accident, one employee told me he hadn't seen his parents in two years and they were coming to town for Christmas, the second told me his mind was on interviewing for a job at another firm, and the third told me he was having marital problems. Each employee took responsibility that his distraction was the reason for the accident. The emotional and behavioral issues due to their personal circumstances meant each employee was not paying attention to the job at hand.

Table 3.1 **Examples of classes of hazard**

Classification	Hazard type	Example outcome
Physical	Thermal	Contact with hot liquid burns tissue
	Potential energy	Fall from a ladder results in broken bones
	Potential energy	Trip and fall on a brewery hose causes broken tailbone
	Kinetic energy	Getting caught in moving machinery results in amputation
	Kinetic energy	Exposure to loud noise causes hearing loss
	Electrical	Contact with an energized circuit causes nerve damage
	Non-ionizing radiation	Eye exposure to laser light results in sight loss
	Ionizing radiation	Exposure to gamma radiation from X-ray inspection device causes cellular mutation
	Ergonomic	Back injury results from repetitive lifting
Chemical	Corrosive	Caustic cleaning chemical splashes on skin or into eyes
	Simple asphyxiant	Oxygen deficiency in cellar causes unconsciousness
	Chemical asphyxiant	Carbon monoxide from forklift exhaust causes death
Biological	Infectious agent	Norovirus contracted from a sick coworker in kitchen
	Bloodborne pathogen	Hepatitis from exposure to injured coworker's blood
	Allergenic	Anaphylactic shock from consuming allergen in beer
Psychosocial	Toxic workplace	Stress from angry coworker leads to physical conflict
	Fatigue	Distraction caused by overwork leads to physical injury
	Substance use disorder	Drinking on the job causes injury of a coworker
	Workplace violence	Angry ex-worker returns to work with a firearm

In this book, chapters 9 through 18 will address physical, chemical, biological, and psychosocial hazards as the main hazard categories. These chapters are bookended by chapters 8 and 19, which address broader hazard types: systemic hazards and combined hazards. Systemic hazards are those hazard themes that pervade the entire operation, regardless of task (for example, PPE, training, and housekeeping). Combined hazards can be thought of as more complex hazards, where two or more distinct hazards occur within one routine task. Think about a confined space entry activity that could involve mechanical, atmospheric, thermal, and gravitational hazards. Ultimately, when defining hazards, it is best to just think about the means and potential that a hazard has to cause harm to us, our equipment, or the world around us.

Outcomes

Safety people use the words incident and accident in a specific way. An *incident* is the opportunity for a hazard to manifest, while an *accident* is the manifestation of the opportunity. In other words, the incident is a near miss or a chance, while the accident is the result: injury, illness, or damage to something.

The unfortunate stigma that comes with the word *accident* is that most people think of an accident as being an event without culpability, that accidents are caused by a random event. This is a fallacy. All accidents are caused by a series of events that stem from either human or mechanical failure or both. Ignorance is not an excuse. To blame an incident or accident on your lack of awareness—your ignorance—is to admit you have ignored your responsibility for your well-being, the well-being of others, or the care of equipment and facilities.

Parallel to the terms incidents and accidents, are *hazards* and *outcomes*. The hazard is the potential source of harm. The outcome is the result of the hazard being realized. A utility knife is a hazard. A cut from slipping with the knife is the outcome. Outcomes may be an injury to a worker, or damage to product, equipment, or the environment. In this book, you will see *hazards* and *outcomes* used more often than *incident* and *accident* since there is less confusion about the former pair's meaning.

Injuries that result in immediate outcomes are *acute* injuries, like the knife example just mentioned. Outcomes that take a while to show up, and usually after repeated, low-level exposure to the hazard, result in *chronic* injuries. Take a chronically sore back—you can pick up a 40-pound box once in a while without injury, but if you do this more often, over a long period of time, you may end up with a chronic back injury.

We describe outcomes, or injuries, as either acute or chronic for several reasons. Acute outcomes may benefit from first aid or other immediate action, such as flushing a chemical off of the skin, deploying a fire extinguisher, or escaping a building. Acute outcomes are more "real," more visible, more believable. There is a high likelihood that swiping a sharp knife blade against your skin will result in a laceration: we immediately understand that.

Hazards that can result in chronic workplace injuries or illnesses may be harder to measure. Hearing loss is one such example. The hazards that cause hearing loss may be at work or at home. How do we determine which location was at fault? Furthermore, we can measure the noise levels in a particular environment, but how that noise affects different individuals is a matter of workplace behaviors, use of protective equipment, and, to some degree, genetic makeup.

Hazards that result in chronic injuries may have more of a probabilistic feel to them, as opposed to appreciating the more immediate outcome of an acute injury. For many of us, chronic injuries can seem far off and less concerning. With chronic injuries, we tend to negotiate the odds. You may have heard someone say something like, "Smoking isn't that bad, my grandma smoked two packs a day her whole life and lived to be 95 years old."

We can consider outcomes in terms of the chance that they will occur and, to the extent that they do occur, how bad the results could be. The first consideration is likelihood. We can look at the way hoses are left about on the cellar floor and surmise that a tripping accident is a pretty likely outcome. By contrast, we might consider if a worker could be engulfed in the bulk grain silo. If the hatch and ladder were locked, the key was held only by the manager, there was proper confined space training, and the silo functioned as designed, then there is a low likelihood of an employee entering the hazardous space and suffering such a consequence.

The second consideration is the severity of the possible outcome. With the utility knife example, a cut across the thumb will not be life threatening. After administering first aid, or perhaps a few stitches, the employee can resume work. Contrast this with the hazard of driving a forklift off the edge of the loading dock. Even if the employee is buckled in, the outcome could involve a serious back or head injury. Without a seat belt, there is a good chance the accident results in a fatality.

Lastly, think about how well we can detect the hazard. A hazardous iodine-based sanitizing agent has a distinctive color and odor. We detect the hazard simply by observing the situation. Carbon monoxide is far more insidious: it is a colorless, odorless, poisonous gas with a vapor density nearly the same as air (meaning it doesn't sink to the floor). Exposure to carbon monoxide slows down thinking and motor control, before causing loss of consciousness and death.

Once we begin applying the hazard assessment process to our workplace, we can prioritize which tasks seem the most hazardous based on the likelihood and severity of the outcomes and the detectability of the related hazards.

IDENTIFYING HAZARDS

As the advice from any psychologist goes, you can't improve until you first recognize the problem. It's the same for safety—you can't work in a safe way until you first identify and defeat the inherent hazards. Some hazards are obvious, like the risk of falling from a ladder. Others hazards are more nuanced, such as long-term hearing damage from a noisy work environment. There are two main ways we learn about workplace hazards: *process knowledge* and *measurements and monitoring*.

Process Knowledge

Process knowledge is expertise we gain from various learning experiences, or it can be a prediction that a certain hazard will exist if certain processes take place. Process knowledge is something you hold within you that (one hopes) won't be forgotten. It is experience-based and can be influenced by training, workplace experience, and even "gut feeling."

Process knowledge is also prone to thought and behavioral interferences, such as subjectivity, workplace mythology or superstition, and resistance to authority. The good news is that process knowledge can be wired into your psyche. The bad news is, you could have been programmed with the wrong information. Since anybody who wants to start a brewery and has assembled some financial backing can start producing beer, there are plenty in the brewing industry who have not been mentored.

Where do we get these learning experiences? It used to be there was a tradition in the trades to apprentice and work your way up in the workforce based on experience. Some brewers still acquire their knowledge this way, though usually by less formal mentoring than the apprentice-journeyman-master approach. Brewers new to the industry may learn through on-the-job training (OJT), being mentored, or working an apprenticeship or internship. Existing employees can increase their experience through further training, attending safety meetings ("toolbox talks"), participating in a safety committee, or being tasked with accident investigation. While some people pursue a formal education in safety, the majority most likely learn about safety informally.

There is also exposure to workplace documentation. A prepared workplace has signs, labels, placards, standard operating procedures (SOPs), safety data sheets, and other sources of information that provide convenient, reliable information to employees. These resources are usually quite candid about the hazards at hand and are vetted by safety professionals. However, they can be overlooked and underutilized by staff.

It is probably safe to say that most US brewers working today have learned how to do their job from an undocumented absorption of preferred practices from senior brewers. We can call these preferred practices *habits*. Chapter 20 describes habits in more detail and discusses normalization and habituation.

Habits can be good or bad. A good habit is wearing your seatbelt every time you are in a vehicle. A bad habit might be not wearing your seatbelt on a forklift because you're typically only in the seat for a minute to move some kegs. Even with some version of mentorship, one can pick up bad habits, unsafe procedures, shortcuts, and sketchy procedures. An award-winning garage brewer can scale up their operation, but do they understand the necessity of commercial wiring, plumbing, and emergency preparedness?

Measuring and Monitoring

The other way we learn about hazards is by conducting measurements of the current conditions, then comparing those results to some boundary condition. Examples include pH measurements to detect corrosivity, an air monitor to document carbon dioxide levels, or taking a person's temperature to help diagnose heat stress. Some methods provide a numerical value (quantitative) that can be qualified by equipment calibration, zeroing, and measurements of concentration standards. Other methods might give a go/no-go result or change color (qualitative), as with chemical test strips.

Measuring and monitoring can be classified by three purposes: 1) process control, 2) workplace or ambient conditions, and 3) monitoring of the worker's physiological condition.

Process Control Measurements

Process control measurements are most often thought of as quality and cost parameters associated with certain tasks, rather than as safety information. Examples include using a sanitizer test strip to ensure sanitizer effectiveness, measuring pH of the mash to optimize enzymatic conversion of starches, or calculating dissolved CO_2 concentrations in beer while knowing the temperature and tank pressure of the storage vessel. Often, the acceptable range of results will be documented in an SOP or manufacturer's specification sheet. Results from this type of testing can be used to modify process conditions when results indicate an out-of-control condition.

Process control measurements can also improve the safety of the workplace. Let's say it was your team's job to perform a hazard assessment of manual cleaning of the exterior of returned kegs. As a result of measuring the pH of the washing solution, you determine that a pH result of 14 is unacceptable, so you end up replacing the cleaning solution with a less alkaline, but still acceptable, cleaning solution. A process control measurement helped you understand how dangerous this chemical was, so you opted for substituting a less corrosive product.

Workplace Monitoring

Workplace monitoring most often involves measuring some kind of ambient environmental condition that could be hazardous to workers if the exposure exceeds a certain threshold. We might measure concentrations of a gas, dust, or mist, or other conditions like temperature, voltage, ambient light, or sound pressure levels. For workplace monitoring to be valuable to us, first we have to know the physical state or condition we are measuring. We also require a numerical standard to compare against the results to help us determine if a hazard exists.

If we make preemptive measurements and use the results to help us avoid a hazardous condition, then workplace monitoring is a valuable preventive tool. In the context of a hazard assessment, monitoring can inform us whether a perceived hazard is at hazardous levels or not worth worrying about. We often specify

Table 3.2 **Hazardous states to monitor**

State/form of hazard		Definition	Brewery hazard example(s)
Gas/vapor agents	Gas	Substance normally in the gas phase at ambient conditions; a formless fluid	CO_2 in a fermentor; propane from a forklift
	Vapor	Gas-phase material emanating from a liquid or solid while still below its boiling point	Acetic acid above pail of peracetic acid; solvent from cleaning of labeler
Particulate agents	Aerosol	A general term that can apply to dust, mists, fumes, i.e., a fine suspension of solid or liquid particles, usually in air	Sprayed paint, adhesive, or cleaning product; malt dust during milling; hot water spray during vessel washdown
	Dust[a]	Fine suspension of solid particles in air caused by abrasive or destructive processes such as cutting, grinding, filing, frictional wear, rapid impact, or detonation	Grain dust; dusts from fabrication and maintenance, powdered cleaning agents, or diatomaceous earth filter aid
	Fume	Particulate resulting from vapor condensation, usually from the heating of metal; may be accompanied by smoke or vapor	Welding or soldering fume from equipment repair; heated rubber
	Mist	Fine suspension of microscopic liquid droplets generated by mechanical force such as spraying or splashing. Size varies, but visible mist is typically 30–400 µm	Disinfectant from a cleaning sprayer
	Fog	Suspended microscopic liquid droplets generated by vapor condensation. Fogs are often visible, whereas most vapors are not[b]	Kettle steam condensing in a cold brewery work space
	Smoke	Fine suspension of liquids, gases/vapors, and solid particulates resulting from incomplete combustion; includes soot, hydrocarbons, ash, carbon monoxide, nitrogen oxides, and water vapor	Propane forklift exhaust; smoke from a building fire; gases from cooking devices or burners
	Fiber	Solid material, often crystalline, with a length at least three times its width/diameter[c]	Asbestos on old steam pipes
Energy	Acoustic	Sound waves transmitted through air	Noise near a centrifuge
	Thermal	Temperature of air, surface, or process material	Radiant heat near brew kettle
	Electrical	Movement of electrons through a conductive medium, e.g., wire, surfaces, fluids, and air	Exposed live wiring
	Gravitational	Attractive force between Earth's mass and other objects	Falling objects; person falling from height
	Kinetic	Energy of an object in motion	Repetitive motion injury from data entry
	Potential	Energy of an object due to its position	Keg stored in a raised position
	Luminescent	Energy of visible light; hazardous when too low or too high	Inadequate lighting
	Radiation, ionizing	Energy that causes displacement of electrons it comes in contact with, hence ionizing materials	X-ray inspection system
	Radiation, non-ionizing	Electromagnetic radiation insufficient to ionize atoms, but which can transmit hazardous energy to eyes and other tissues	UV light disinfection system
	Ergonomic	Condition that can injure musculoskeletal system	Lifting sacks of grain

Table 3.2 **Hazardous states to monitor** (cont.)

State/form of hazard		Definition	Brewery hazard example(s)
Biological hazards	Infectious agent	Diseases caused by a viruses, bacteria, or fungi that can be transmitted person to person (or animal to person) either directly or through contamination of food or water	Worker-to-worker transmission (e.g., common cold, influenza, COVID-19, meningitis); indirect spread due to poor sanitation of brewery surfaces, water-damaged wood and drywall structures, unclean kitchen equipment, time/temperature violations of food safety standards
	Bloodborne pathogen	Communicable disease transmitted by contact with blood or other bodily fluids	Hepatitis from coworker's open wound
	Biotoxin	Toxic substance produced by virus, bacteria, mold, fungi, etc.	Botulism toxin, tetanus toxin
	Allergen	Agent of biological or chemical origin causing allergic response or sensitivity, usually in respiratory system or the skin	Lactose, nuts, seashells used in beer; allergies caused by yeast, mold, mushrooms, mildew, and fungal rust (e.g., fungal sinusitis and mold-induced asthma)
	Vermin	Animal pests that carry disease or produce toxic substances	Tick-borne bacterial disease; histoplasmosis (a fungal disease associated with bat and bird droppings)
Psychosocial hazards	Bullying	Bullying, harassment, prejudice	Sexual harassment by coworker
	Violence	Actual or threatened violence	Angry customer becomes active shooter
	Substance use disorder	Habitual drug or alcohol use	Daily consumption during work
	Fatigue	Fatigue, exhaustion, burnout, distraction	Tired operator takes hazardous shortcut
	Anxiety	Fear, anxiety, lack of understanding	Job security threatened by management turnover

[a] Heavier particles will deposit on surfaces: suspended atmospheric dust, 0.003–1.0 µm; settling dust, 1–100 µm; heavy dust, >100 µm.

[b] Fog particles are generally about 1–30 µm.

[c] World Health Organization (2002, 379)

monitoring after a worker becomes concerned about the conditions they may have been exposed to. This could be due to some symptomatic evidence (like shortness of breath or difficulty hearing), the recommendations of a safety audit, or simply a general worry.

A professional who specializes in collecting and interpreting these results is called an industrial hygienist (IH). A brewery employee who is not an IH can conduct workplace monitoring if they are trained in specific calibration and testing procedures.

Monitoring devices are designed to monitor one or more hazardous conditions. We are not only concerned with the substance or hazardous condition, but also what form it is in. A nitric acid hazard can be present as a bulk liquid or, if it is being sprayed in the workplace, as a mist (i.e., an aerosol). Both require different monitoring equipment. Terms relating to what we are monitoring are defined in table 3.2. Definitions in the table are limited to how these hazards might exist in a brewery situation; broader definitions may exist in a wider range of settings.

Reliable and Accurate Monitoring

Being able to name a specific hazard allows us to plan for the physical form or state it is in (table 3.2). That form could be a mixture in air, a physical or chemical state, or a living agent. Knowing the form informs us about the appropriate measurement technology. Measurements may involve evaluation of a sample at a laboratory or be instantly available to us with a direct-reading or real-time device.

For example, dust, fibers, and fumes can be measured by first collecting a sample on a very fine filter, then analyzing the material retained on the filter. The levels could also be measured in real-time by measuring how much light the airborne particles scatter using a device called a laser nephelometer.

These methods won't work for gases, which cannot be captured on the filter and will not scatter laser light. For gases, an electronic sensor that is specific to a certain gas is often used. Caution must be exercised. Are we measuring the right thing with the right device? Using a carbon *dioxide* monitor to see if carbon *monoxide* has accumulated due to forklift exhaust will not yield helpful results and may give a false sense of security that could lead workers into a hazardous environment.

Direct-reading devices are reliable because they are calibrated in accordance with the manufacturer's recommended frequency and with agents of known concentration. Calibration is a quantitative quality check showing that the reading is accurate. It is achieved by sampling a known quantity of hazard and dialing in the instrument to reflect that known result. Another important part of calibration is zeroing the instrument. Let's say you intend to measure the airborne dust concentration during grain mill operation. You would zero the device by sampling air with no discernable dust concentration, say, in an office with air filtration. Finally, there is the bump test. This is where you informally measure a high concentration sample to make sure the device's sensor and alarm is recognizing the intended hazard. For example, a carbon dioxide (CO_2) monitor can be bumped by sampling air exhaled into a paper bag or by releasing CO_2 gas directly from a fitting into the device's sensor. With a bump test, the only concern is that there is enough of the hazard present to be detected and trigger the alarm; the concentration isn't really important.

Monitoring results can be collected one time, periodically, or continuously. Noise exposures around a beer centrifuge might be measured over the course of a typical day and the results used to inform the employer about whether hearing protection or a full-blown hearing conservation program would be required.

More frequent monitoring is called for in the case of monitoring CO_2 removal from a cellar vessel. This periodic monitoring might be scheduled for every time an employee will put their head through the tank manway to conduct a cleanliness inspection. If routine monitoring shows consistently safe CO_2 levels, it is often reasonable to

reduce the frequency of testing to a periodic confirmation that the tank inspection procedure remains safe.

Continuous monitoring can be used in cases where hazard levels are dynamic or where there is a potential for exceeding a designated threshold concentration. A common brewery example is a continuously operating CO_2 monitor in a cellar, packaging area, or walk-in cooler. Levels exceeding a preset concentration automatically turns on an exhaust fan and/or sounds an alarm.

Defining Hazards by Exposure

One of the most argued aspects of workplace monitoring is the question of whether a certain level of substance or condition actually constitutes a hazard. Substance lists can be incomplete or updated infrequently. Fundamental assumptions about toxic pathways and disease may be outdated. Sadly, most of OSHA's published exposure limits go back to the agency's formation in 1970. It literally takes an act of Congress to change these limits. Surely epidemiology, risk assessment, and monitoring technologies have changed over half a century. Couple this with the fact that individual workers all have different genetic predispositions, health and fitness levels, and lifetime exposures to hazards. Although industrial hygiene is a legitimate science, the legal exposure limits that IHs use to define an unacceptable exposure are packed with assumptions and colored by politics and lobbying.

Under OSHA, a "safe" level is called a permissible exposure limit (PEL), which is a matter of regulation. If a worker is exposed above the PEL without adequate protective equipment and other controls, then the employer can be cited on the presumption that the employee has been put at risk from the exposure.

There are two types of PEL. The most commonly used is PEL-TWA: the time-weighted average exposure over an eight-hour shift without mitigation by any sort of PPE. With PEL-TWA, the levels of exposure may have gone above the PEL in any one moment, but exposure levels averaged over the whole eight-hour workday will still be below the PEL-TWA and are permissible.

Some substances will also have an established PEL-C. Here "C" stands for ceiling. The PEL-C is a concentration that cannot be exceeded at any point during the day. Above the PEL-C, the worker is expected to be at grave risk or compromised so much that they are unable to leave the hazardous area. If real-time monitoring is not available for a substance subject to a PEL-C, then the results from a 15-minute interval can be used.

APPLYING PERMISSIBLE EXPOSURE LIMITS IN THE WORKPLACE

- If any PEL is exceeded, measures to reduce worker exposure must be put into place.
- Typically, if the PELs are not exceeded then the employer is not required to provide PPE or workplace controls.
- Most PELs, except for workplace noise, are constrained to airborne hazards.
- Hazardous exposures due to heat, scalding liquids, electrical shock, repetitive motion, lifting, contact with allergens or infectious agents, and many other hazards, have no established exposure limits.

Where PELs are defined, monitoring the relevant hazardous substances and agents is vital to comply with OSHA requirements. Due to the limitations inherent in defining hazards by exposure, however, it is not best to rely entirely on the PELs. Workplaces should be looking for ways to avoid occupational exposures to the greatest extent practicable, up to and including outright elimination of hazards through careful choice of supplies, processes, and control technologies. Thus, there are several important reasons for involving an IH in the development of a monitoring plan:

1. Determining the correct monitoring device based on the condition being measured
2. Establishing which regulatory criteria or other health guidance to be used in evaluating the results
3. Recommending effective hazard controls based on findings

Physiological Monitoring

The third type of monitoring that can help to describe workplace hazards is physiological monitoring of the worker themselves. Physiological monitoring usually involves a medical professional, such as a medical doctor, pulmonary technologist, radiologist, or audiologist. From a safety perspective, there are two aspects to physiological monitoring: recording and categorizing the exposure level in the workplace, and determining how that exposure might affect workers' health to inform the decisions on how to control the hazard with equipment and PPE.

There are chemical exposures that can be detected by measuring the substance or its metabolites in blood, urine, or even hair and fingernails. Drug testing for blood alcohol, THC, prescription narcotics, and illicit drugs also falls under physiological monitoring.

Cardiopulmonary assessment is performed on workers who will be wearing any type of tight-fitting respirator at work. This is because respirators cause stress to the heart and lungs. Tight-fitting respirators include half-face and full-face cartridge-type respirators and supplied-air respirators such as those used by sandblasters or fire and hazmat personnel. The half-face disposable dust mask, usually just a fiber or fabric covering over the nose and mouth, does not require medical assessment of the worker, provided the employer has determined that the worker is not being exposed above the PEL and has not instituted a respiratory protection program as a result.

For hearing conservation, we can measure noise in the workplace but we must determine if that noise is causing long-term degeneration of hearing in the exposed worker. The workplace monitoring is called noise monitoring, while evaluation of the worker is called audiometric testing.

Finally, there are cases where a worker may have been exposed to an occupational hazard that could have affected them in some other way that does not have an established OSHA requirement. For example, if an employee in the brewery warehouse experiences back pain, they may be evaluated by an orthopedist to find out if the issue is a simple back strain or something more serious like structural damage to the spine. All medical evaluation and monitoring must remain strictly confidential in compliance with the Health Insurance Portability and Accountability Act of 1996 (HIPAA).

Keep in mind that measuring the effects of a hazardous exposure on the body is a trailing key performance indicator (KPI). That is, it tells you something important after the fact. Chapter 21 is devoted to KPIs. With occupational exposure to hazardous conditions, your best approach as the employer is to know how to identify and describe the hazard so you are well situated to begin thinking about how to eliminate, prevent, or reduce exposure.

CAUTION

WATCH FOR LIFT TRUCKS

4

HAZARD CONTROLS

In the grand scheme of the hazard assessment (HA) process, probably the most challenging step is determining which hazard controls to use. What preventive and protective measures will you employ to reduce the totality of hazards that come with any brewery task? In fact, this task isn't as hard as you might think because earlier in the HA process you will have already outlined each task and identified possible hazards for each step of the task (ch. 3).

WHAT ARE HAZARD CONTROLS?

Hazard controls are steps taken to reduce or eliminate any known or anticipated hazard that could cause harm to persons, equipment, processes, or the environment. To select the best controls, you need to think about each of the possible hazards related to the task and come up with one or more solutions to reduce or eliminate those hazards.

The hazards might be common, even obvious, or they might be rare and more theoretical. Hazards have known attributes. If you understand those attributes, you can select the best countermeasures.

It is important to understand what hazardous energy transfer is involved, which we delve into in greater detail in chapter 7. For instance, if an ultraviolet light source can injure the eye, blocking the ultraviolet light from hitting the eye is an obvious solution. But to properly implement a hazard control requires more than that. Think creatively about how to block the light.

 HAZARDS ARE LIKE ANIMALS

One way to think about hazards is to compare them to animals. You might think of an asphyxiating gas creeping along the floor like a python, ready to sneak up on you and suffocate you. Getting strong caustic on your skin is like being bitten by a venomous spider: it hurts and then the tissue deteriorates and dies. Forklifts might be like rhinos: heavy, powerful, and unforgiving.

What animal, real or fictitious, do you associate with a kettle boiling over and spewing hot wort? Did you think of a dragon? What about the rotating knives in a lauter tun? Perhaps a shark or a tiger ready to tear you apart?

Relating to hazards by comparing them to things we understand can help us appreciate their potential impact, and helps explain hazards to those unfamiliar with them.

Does it involve shielding the light source or wearing special safety glasses, or is there a way to avoid using the ultraviolet light altogether? When you have decided which approach(es) you plan to use, then you have selected your hazard controls.

The ultraviolet light example illustrates how the best hazard control is to avoid the hazard altogether, but this is not always possible. In cases where the hazard remains, we need protection. In a brewery, you are necessarily going to have ladders and heavy bags of ingredients, and boiling liquids and asphyxiating gases. Legally, it is up to the employer to learn what these hazards are and to minimize them.

 HAZARD CONTROL CORNERSTONES

- Eliminate the hazard(s) by revamping your process.
- Reduce the likelihood or frequency of a hazard with task planning, job rotation, and safe work practices.
- Inform workers about the hazard with training, signs, labels, and standard operating procedures.
- Reduce the severity or magnitude of hazardous energy transfer with engineering controls and personal protective equipment.
- Increase detectability by using sensors and monitoring.

SELECTION OF CONTROLS

Why we select some controls over others is based on a value equation. This equation, or rationale, considers the effectiveness and reliability of the control strategy, as well as the cost to implement it. We also have to recognize the severity of the hazard. Common hazards that cause loss of life warrant significant controls, even if they might never be called into use.

Prevention and Protection

As the saying goes, the best defense is a good offense. With hazard controls, the best way to reduce risk is to understand the hazard and preempt it before it can do harm. If you can manipulate the situation so that the hazard ceases to exist or the hazard cannot affect you, you have used *prevention*. If you accept that the hazard exists, or if you can't do your job without the hazard being present, but you use controls to lower the chance of a bad outcome, you have used *protection*.

Hierarchies of Control

There are numerous factors that play into the selection of a hazard control. How expensive is the control? Is it effective and reliable? Does it require workers to act in a precise way, to read a sign or a label, or have specific training? Will people employ the chosen control strategy? Does the control impact the time or cost of a certain process? Who is legally responsible for safety?

One control may be cheap and easy, but does it have the reliability we want? Another could be expensive, being effective but requiring regular maintenance, for example. It could be that using these controls together provides the best chance of avoiding harm.

These sorts of questions have led safety professionals to create formulaic approaches to hazard controls, to try to make the selection of controls adhere to a system. Such a system, termed a hierarchy of controls (HoC), is a starting point but not a panacea, for the very reason that we often choose two or more hazard controls to ensure safe operations.

Employer's Responsibility

The simplest view is that the employer protects the employee from whatever hazards exist in the workplace. This falls short on two fronts. If the employer is not keenly aware of workplace hazards and not sufficiently motivated to keep them in check, the employee is bound to suffer. This view also runs contrary to OSHA's original intent in the OSH Act, which clearly requires both employers and employees to work together to create a safe business (29 U.S.C. § 654).

OSHA's Hierarchy of Controls

OSHA's hierarchy of controls is widely cited and repeated throughout the entire global safety universe. Its principles pre-date the creation of the agency in 1970, going as far back as the 1930s (Bauto 2022; US Office of Technology Assessment 1985, 176). The OSHA hierarchy of controls is an effective way of prioritizing workplace controls, although it has been criticized by proponents of behavioral safety and effective safety culture, including this author.

Historically, standards agencies (including OSHA) have taken the view that the employer is primarily responsible for workplace safety, following the "control at the source" mantra. This approach views workplace behaviors and personal protective equipment (PPE) as supplemental controls in the event that an engineering

Table 4.1 **Examples of preventive and protective controls**

Classification	Hazard	Hazard control example(s)
Preventive	Chemical spill	Don't step in it or touch it
	Chemical burn	Substitute a less corrosive chemical
	Jumping off loading dock	Take the stairs down
	Forklift emits carbon monoxide	Substitute with a battery-powered forklift
	CO_2 asphyxiation in a fermentor	Prohibit staff entering the vessel; use signs
	Blocked emergency exit	Keep exit clear; apply floor markings; perform emergency drills
	Drunken customer	Use de-escalation language; secure a safe ride home
	Contracting dangerous illness	Keep sick employees at home
Protective	Chemical spill	Use secondary containment; keep spill absorbent on hand
	Chemical burn	Enforce use of safety goggles and face shield; maintain eye-wash station
	Kettle boilover burn	Deploy foam sensor and antifoam agent; do not brew more wort than system design; spray wort surface with cold water
	Back injury from lifting kegs	Use pneumatic lifting device
	Outlets trip breaker when wet	Replace with wash-down rated enclosures
	Box cutter falling into mill rollers	Tether knife; install mesh guard above rollers
	CO_2 asphyxiation in walk-in cooler	Install wall-mounted CO_2 meter and gas line leak detector; enforce buddy system and having doors open while working inside
	Kegs falling onto forklift operator	Use protective cage around operator; load strapping; training
	Irritating dust from grain milling	Keep good ventilation and housekeeping; use respiratory protection
	Lightning during a beer festival	Take shelter inside building or vehicle
	Stress from overwork	Improve scheduling; hire more help; reduce overtime; provide employee counseling
	Fire in pub kitchen	Keep fire extinguisher on hand; install automated fire suppression system and fire alarm
	Blood cleanup after an injury	Use disinfectant and proper PPE; implement correct waste management

control fails while in use, cannot adequately reduce exposure to within permissible limits during normal operations, or is currently not feasible. A rendition of this typical hierarchy of controls is shown in figure 4.1. In this view, protective actions taken by employees are seen as vulnerable to the vagaries of employee behavior and extenuating circumstances (e.g., pressure to meet productivity targets, or discomfort caused by PPE). Thus, the employer will put hazard controls into place, design with safety in mind, provide necessary engineering controls and PPE, and train the workforce. Some opinions

from organized labor unions go as far as to suggest that even partially relying on a worker acting safely constitutes dereliction of safety and fiduciary responsibility on the part of the employer (Howe 2001; Robinson 1988).

Protective equipment and systems should, in theory, allow a worker to focus on production rather than worrying about their safety; their safety is largely seen to by the employer. Focusing mainly on these types of controls is to focus on hazard controls that are often capital-intensive. If an employer is not living up to their end of the bargain, the workforce may not

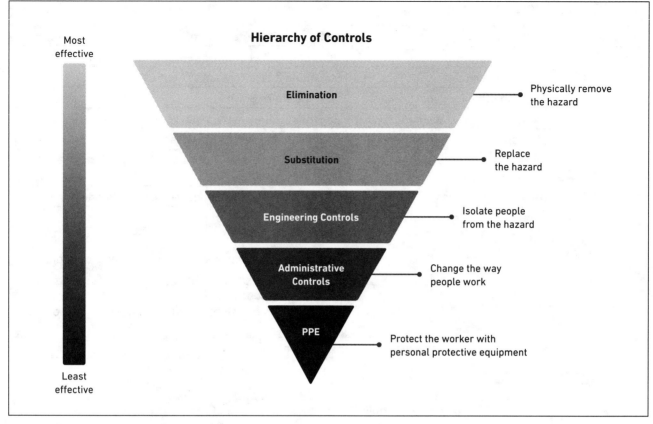

Figure 4.1. Traditional hierarchy of controls.

be prepared to protect themselves. An effective safety culture requires managers and employees to hold common certain beliefs, values, assumptions, feelings, and business systems in order for safety to really work. We look at this more in chapter 20.

One more criticism of the mainstream hierarchy of controls is how little value is given to the worker's own volition. This self-determination to remain safe is often called safe work practices, but I like to think of it as "thoughts and actions"—hazard controls that do not require equipment or other institutional controls.

A safety system where the worker is given a little more credit for avoiding hazards on their own can exist, but it relies heavily on the tenets of an effective safety culture. There are lots of behavioral pitfalls; we could even consider these to be behavioral hazards. A good safety culture has to overcome these behavioral hazards, just as it would with physical and chemical hazards in the workplace. Safe behavior is due to an individual's training, experience, and mentoring, but it will require management support, healthy communication, accountability, and a high level of participation.

Prevention Priority Hierarchy of Controls

An alternative view to the traditional hierarchy of controls, and one more suited to small businesses with an effective safety culture, is a model where the worker plays a more active role and is encouraged to use both employer-provided controls and their own volition. More and more, we see businesses focusing on developing a robust safety culture due to the realization that individual worker actions do have a profound effect on their own safety and that of their coworkers. This approach is often less expensive and puts more emphasis on safe work practices implemented by the employee. This is especially significant for smaller breweries that may not have the capital for sophisticated engineering controls.

In this view, there is less emphasis put on a hierarchy and more on the redundancy of controls. If the safety of a particular task relies mostly on PPE, so be it, but know that there is no additional control if the PPE fails.

The traditional hierarchy in figure 4.1 shows the ideal order of hazard controls. An alternate hierarchy diagram in figure 4.2 shows most of the same terms arranged from most preventive on the outside to

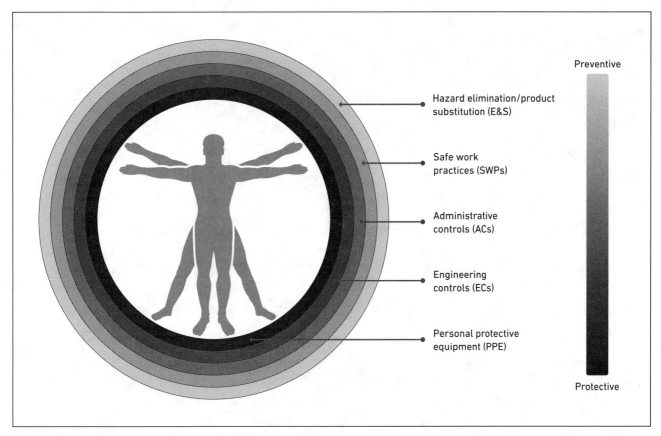

Figure 4.2. Hierarchy of controls on a prevention to protection scale.

protective on the inside. It puts more value on preventive thoughts and actions—namely, elimination, substitution, safe behavior by the worker, and preventive information provided through administrative controls. Engineering controls are further down the list than in the traditional hierarchy because they can still fail, they may only redirect a hazard, and they can be expensive. Prevention may be better in the ideal case, but often redundant layers of various approaches are necessary to create the safest work activity.

Safety by Design

There is no better time to eliminate hazards than when designing equipment or the layout of a brewery. Safety by design, also called prevention through design, is the commonsense idea to "design out" hazards before reaching day-to-day operations.

Starting in 2007, the National Institute of Safety and Health (NIOSH), a branch of the Centers for Disease Control and Prevention (CDC), has been working with partners in many trade sectors to learn about various industry hazards and make recommendations for prevention through design. Out

of this, NIOSH and the American Society of Safety Professionals (ASSP) have developed a consensus standard released in 2021 (American Society of Safety Professionals 2021).

The ASSP guidelines are pretty lofty and don't offer much in the way of specifics for brewers. But they do provide a good list of the types of devices and processes to think about during design, redesign, and retrofitting: new and existing work premises, structures, tools, facilities, equipment, machinery, products, substances, work processes, and the organization of work.

When laying out a new brewery space, or planning remodeling, it is vitally important to seek the opinions of experienced brewers and construction trade experts. Seriously consider hiring a brewery consultant to help identify future safety issues; if nothing else, the consultant can help review and interpret the mechanical, electrical, and plumbing drawings.

Architects can design a beautiful space, but left without expert production advice the results may not be safe or practical. Removing spent grain could be wholly unworkable. Business office staff may have to

PREVENTION THROUGH DESIGN EXAMPLES

Problem: A brewery manufacturer displayed a four-vessel brewhouse at a trade show. The brewery was excited at how compact the footprint was. From the tiny, squarish brew deck, all vessel manways opened toward the one place the operator could be standing. In case of a boilover, there was no way for the brewer to be out of the line of fire of scalding liquid.

Solution: A safety engineer explained to the manufacturer the inherent risks and explained the line of fire hazard. The manufacturer changed their brewhouse designs to allow a safe place for the brewer to stand.

Problem: An older building was being converted into a brewery. The architect put the business office upstairs, at the back of the building. Because the building was narrow, the staircase to the offices came down into the tank cellar, right between two rows of fermentors where there would be hoses, slipping hazards, and a trench drain. Office staff would be frequently walking through this crowded and hazardous workspace.

Solution: Plans for the cellar arrangement were modified and the staircase relocated to keep office personnel out of production areas.

Problem: A centrifuge room was being installed during a brewery expansion. There was only one place to put the centrifuges due to available floor space. Being very loud, the noise hazard would affect a large number of employees, likely triggering a full-blown hearing conservation program.

Solution: Acoustical-dampening tempered glass walls were included to enclose the centrifuges, isolating the noise and creating a focal point for brewery tour groups.

Problem: A new brewery was being built that would be making a lot of dry-hopped beers. Staff were not keen on climbing up and down stepladders, but the cellar alley was narrow. The budget did not allow for a catwalk to be installed.

Solution: The tank manufacturer added ladder hooks onto each vessel for a nominal additional cost. This allowed the cellar crew to use a fixed, straight ladder to be secured on the hooks, and the ladder reached higher than a stepladder, giving workers more to hold on to during dry hopping.

Problem: Production staff were usually using face shields and heavy nitrile gloves when working with cleaning and sanitizing chemicals, but they often set them back down on drums when still covered with chemical residues. One never knew what residues were on the PPE. The PPE needed to be washed after every use and stowed correctly.

Solution: A PPE pegboard was installed right over a nearby shop sink so workers would know where to find it. Because it was right by the sink, it became common practice to wash PPE after each use before hanging it on the pegs to dry.

walk through the brewery cellar to get to their desks. Glass walls that allow customers to view the brewhouse will condense steam and grow mold if a proper HVAC system isn't designed.

Equipment manufacturers can, and often will, build whatever brewing system you want, but it may have intrinsic safety issues that will forever put the workforce at risk. You may not know to ask for a kettle foam detector and interior lighting. Employing your electrician friend to wire the brewery because he'll trade services for beer may leave you with substantial commercial code violations and grave safety hazards.

HAZARD CONTROL STRATEGIES

There are four main hazard control strategies. Some are more preventive or "ideal," and others are more protective or "practical." All of these methods involve balancing cost, time required, worker attention required, hazard control effectiveness, and ease of implementation. The order the control strategies are discussed here is along the continuum of preventive to protective (refer back to fig. 4.2).

Thoughts and Actions

In a departure from typical hierarchy of controls discussions, I prefer to collect elimination, substitution, and self-actualized safe work practices under the heading "Thoughts and Actions." These concepts are all primarily preventive in nature. They may or may not need to employ other control strategies as part of their implementation, but, at the highest level, these approaches do not require any equipment other than a brain capable of critical thinking.

Elimination and Substitution

The common terms for this category of preventive hazard controls are elimination and substitution. *Elimination* answers the question, How can I remove the hazard altogether? *Substitution* answers the question, How can I redesign the activity to reduce the hazard?

In a brewery, a good example of elimination involves dry hopping. Suppose the current practice is for a solo worker to climb a stepladder, having to carry dry-hopping supplies up by themselves, and then perform the dry hopping from the stepladder. For this example, let's say the hazard we are trying to eliminate is the risk of falling from height.

Is it possible to dry hop from the floor level of the cellar? It is. Here are two possible ways of doing this: establish an air-free recirculation loop with a pump and ground floor fittings, or use a hop cannon-type device to deliver hops through existing hard piping. Notice we haven't eliminated all hazards; in fact, we may have added new hazards involving pressure, pumps, and electricity. But we have eliminated the falling from height hazard.

Using the same example activity, the worker could still climb up and dry hop through a top port, but we could substitute a safer way for the worker to get to that elevated work space. Switching out a mobile ladder stand with a fixed platform is a low-cost substitution. The platform is much steadier, has fall protection built in, and has room on the platform for the supplies necessary for dry hopping. Other substitutions include using a scissor lift or installing fixed catwalks with railings.

Consider two final thoughts on substitution and elimination. When taking this approach, make sure you are "trading up." That is to say, make sure the revised method is safer overall, not just different, from the prior method. And, secondly, notice that the tools we use to make substitutions or eliminations may be any type of hazard control available to us. For instance, if we switch to a rolling platform with stairs, we have used an engineering control to effect a substitution.

Safe Work Practices

Safe work practices is the third part of the thoughts and actions equation: the role of the worker in asserting their own avoidance of a hazard at any given moment. This is where we ask, How can I enable worker self-determination toward hazard control? The answers here include training, mentoring, and awareness of systemic workplace hazards. Notably, use of safe work practices has scientifically been shown to correlate to an effective workplace safety culture built on communication, expressed management values, progress accountability, workforce involvement, and an instilled safety skillset (competence). The relationship between safe behaviors, safe work practices, and peer-reinforcement of safety is discussed in the safety culture chapter (ch. 20).

Sticking with the dry hopping example, where we are trying to reduce or eliminate the risk of falling from height, let's say that dry hopping must go on and that there is nothing available but the same old stepladder. In this case, safe work practices advise the worker how to do the job as safely as possible. Stepladders are designed to be used only when the legs are fully splayed and locked open. The ladder must be set up perpendicular to the work, not parallel, since stepladders tip over more easily with side force than front-to-back force. To put it another way, the ladder needs to be aimed at the tank, not placed beside the tank. And the ladder is designed so that the top plate and top step below the top plate are not to be stood on. (More information on ladder safety can be found in chapter 10.)

In short, following safe work practices, the worker uses the tool the way it was designed to be used. They may have learned this in training, from observing coworkers, through skills transferred from another occupation, or by having had a previous mishap. The worker may also seek out the assistance of someone to help steady the stepladder and hand up supplies.

Safe work practices are the way we conduct ourselves in work activities. They rely on the worker being intent on being safe and not being derailed by any number of psychosocial phenomena, which we describe in chapter 18. This author believes that most individuals have a fundamental desire to be well and will exhibit safe work practices more regularly when workers are part of an effective safety culture. This culture is based on personal and social experience and the development of aligned beliefs, values, and assumptions about safety.

Administrative Controls

Administrative controls are the written and policy parts of hazard control. They are not as absolute as substitution and elimination, nor as fallible as PPE. They are as effective as employees' willingness to abide by them. Written guidance like an employee manual, standard operating procedure (SOP), safety data sheet (SDS), or lockout/tagout energy control procedure conveys the way to do the work safely. But written guidance can encompass more than this: written compliance programs, such as hazard communication, respiratory protection, emergency action plans, and forklift safety checklists, are administrative controls; so are signs, labels, placards, floor markings, and pipe labels.

Let's return to the dry hopping example once again. An administrative control for this activity would be ample labeling on the stepladder. It will state clearly, if only in small print, the duty rating (weight limit), how to use the ladder correctly, what American National Standards Institute (ANSI) standard it complies with, and possible consequences of misuse, like electrocution. Interestingly, while OSHA does have significant and detailed regulation on ladder specifications (29 C.F.R. § 1910), it does not require the labeling just described. Such labeling is placed there by manufacturers to inform proper use and limit liability (Hillenbrand 1983).

The brewery may have an SOP for performing dry hopping. Within this SOP could be specified safe procedures, including how to face the ladder toward the tank and obtaining a buddy worker to steady the ladder. The SOP constitutes another administrative control.

Like many written rules, administrative controls are subject to being bent or ignored. Administrative controls are good for reasons of OSHA compliance and legal defense, but they can be hard to enforce. Consequently, regular training and a healthy workplace safety culture is vital to the effectiveness of administrative controls.

Training will habituate behaviors that align with written guidance and the peer involvement of a strong culture reinforces them.

Engineering Controls

Engineering controls are typically among the most effective protective strategies. These are where we isolate or put distance between the worker and the hazard. Engineering controls may be passive or active. With passive controls, the worker is protected without having to think about the hazard. Equipment with moving parts, pinch points, high voltage, fall potential, etc. all need ways to limit a person's contact with the hazardous aspects of the device. Example controls include railings; a forklift operator cage; machine guarding around gears, pulleys, and belts; electrical circuit breakers; and ventilation systems activated by remote monitoring equipment. These are examples of protective appurtenances to equipment that are designed specifically to prevent human contact with the hazardous substance or agent.

Active controls require the worker to be conscious of and involved in controlling the hazard. Examples include using a CO_2 gas monitor or installing dock boards between the loading dock and a truck. With active controls, the worker has to recognize the hazard and use the equipment provided to improve workplace safety. There is a tie-back to safety training and preventive maintenance here as well.

Passive engineering controls for dry hopping include platform railings, pressure relief valves, and pressure gauges. Active controls for the same activity might include squeegeeing the floor to improve grip when standing on it or tying off the ladder in a way that keeps it from falling or slipping. As previously mentioned, a mechanical device for the worker to use that eliminates or substitutes the falling from height hazard (e.g., a hop cannon) would also qualify as an active control. There are other active controls associated with dry hopping that aren't associated with ladder use, like bleeding off CO_2 head pressure before opening the dry hop port.

Monitoring was discussed in chapter 3. Monitoring is also routinely used in the course of hazard control, sometimes because we are checking to make sure a hazard we identified has been effectively mitigated by our control choices. Monitoring devices either require the operator to understand the result and make corresponding operational changes (active control) or their signals have been programmed to analyze inputs and drive electrical or mechanical actions to take place (passive control).

Since we keep using the dry hopping example, you might be wondering what sort of monitoring could be used with climbing a stepladder. One way would be to review workplace security footage and identify ladder misuse. Another method, although fairly exotic, is to use a tribometer to measure the slickness of the cellar floor. Readings from the tribometer might be sufficient evidence to persuade management to invest in an alternative to the stepladder.

Personal Protective Equipment

While most safety professionals will argue that PPE is more than just the last line of defense, among workers themselves and the people who sell PPE you would think it is the best possible hazard control technology. A big problem with PPE is that it requires intelligent selection. With time and wear, PPE will fail. When it does fail, the hazard directly impacts the person relying on the PPE.

As a hazard control PPE can be very effective, but this requires selecting the proper equipment and using said equipment in accordance with the manufacturer's guidelines, including following the protocols for cleaning, inspection, and replacement. It is surprising to think that many employees using PPE don't even take the time to clean off chemical residues after use or inspect the equipment for damage. Again, regular training by an educated individual and a robust safety culture is a must. (More on PPE is found in chapter 9.)

HOW TO IMPROVE HAZARD CONTROL EFFECTIVENESS

As much as we might like to think that some things "just happen," this is rarely the case. A hazard exists because of human behavior, failure of equipment, choice of process, or extenuating circumstances like adverse weather. We know that things don't "just happen." Hazard assessments are a necessity and finding the best hazard controls for a given risk requires careful thought and planning.

Include Others

Hazard assessments conducted in a vacuum by one person often miss the true detail of how a task is performed and where the associated hazards are actually significant. Involving people who perform the task is vital. Task workers may have creative ideas about what

KETTLE BOILOVERS: CAUSES AND CONTROLS

Kettle boilovers are one of the most dangerous situations in a brewery. A boiling wort burn on more than 30% of the body has a better than average chance of killing the brewer.

Boilover Causes
- Brewing more volume than the system was designed for
- Overheating the kettle
- Failure to obtain hot break before putting heat on full
- Brewing a grist with a high percentage of high-protein grains
- Adding too large a charge of hops or additives too quickly
- Neglecting to use an antifoam agent
- Not having installed a foam sensor shutoff
- Standing in front of an open manway or standing with your back to an open manway
- Not having a cold-water hose handy to knock down foam

Boilover Hazard Controls
- Brew no more than the system was designed for
- Monitor the temperature of wort in the kettle
- Trim the heat just as the wort comes to its boil, then increase heat after hot break occurs
- Add hops and specialty ingredients cautiously while standing out of the line of fire
- Keep the manway hatch closed whenever possible
- Use an antifoam additive
- Install a foam detector in the kettle below the height of the bottom of the manway opening—the sensor will automatically shut down the heat source in the event of a pending boilover
- Do not stand in the line of fire
- Do not leave the kettle unattended during wort boiling
- Do not have untrained or unnecessary persons near the kettle while boiling wort

controls to design, their feasibility or practicality, and how much time they might cost or save during task execution. Certainly, involve others who know how to get the hazard assessment data into a workable SOP, but always have the original participants review it.

Gather Information

Once the hazards are known, get to work researching solutions. Search the internet for ideas and products, consult OSHA bulletins and industry colleagues, and, of course, use section 2 of this book. Beware of products that seem too good to be true and be sure to get references or further application data before laying down the money.

Prioritize Hazard Assessments

It takes a while to get the hang of hazard assessments. Start with tasks that you feel have a high severity or high probability. Climbing into a mash tun for cleaning without a lockout/tagout program is an example of a high-severity hazard. Tripping over a hose might be so frequent that it is essentially a constant hazard for those working in the cellar.

Throw in a few easy hazard controls to build your brewery's safety culture. This will also have the effect of reducing commonplace lesser hazards.

Use Multiple Controls for Optimal Safety

Use multiple hazard controls to create extra protection and redundancy. For instance, in the example of climbing into a mash tun, there are multiple engineering controls, administrative controls, and safe work practices that can be put in place. Use engineering controls like locking out the power disconnect to the rake motor; provide a waterproof written SOP or energy control procedure with lockout devices stored right on the brew deck and apply an adhesive sign to the mash tun reminding the operator to lockout; finally, reinforce safe work practices with regular training and safety meetings.

Systematize and Regularly Review

Schedule hazard assessments and track their completion and how the hierarchy of controls is specified. Periodically review each hazard control determination and decide if it is still the best safety precaution for the job. Sometimes new technologies can replace older hazard controls, or maybe the process steps are revised and the original hazard has been eliminated.

WARNING
FORKLIFT TRAFFIC

BEER

5

HAZARD ASSESSMENT

The employer has a legal responsibility to create a safe and healthful workplace (29 U.S.C. § 654). This is normally accomplished by understanding what hazards are present, then finding ways to eliminate or minimize those hazards. Once hazards are controlled, the workplace becomes a safer and more healthful place.

Hazards often cannot be fully eliminated, but they can be identified and contained enough that the work can be done as safely as possible. This relativism causes confusion for employers and workers alike. Both groups would like to know exactly what is expected of everyone in the performance of work. In other words, the idea of a safe workplace is a moving target across both time and industry sector. What passed as safe for a high steel worker in the 1930s would not be considered safe in modern times. And the hazards of high steel work are much greater than, say, those in an office setting.

The definition of safety will certainly continue to evolve as new detection and control methods are invented. Automation will continue to replace dangerous manual tasks. Past injury and illness data will inform our future safety perspective. And while we will never reach the nirvana of a zero accident and injury rate, we should always strive for that as the long-term goal.

Creating a safe workplace comes down to understanding each task, visualizing where the hazards lie in that task, determining to avert those hazards, and then performing the work in a way to disallow the hazards to the greatest degree practicable. The hazard assessment

HAZARD ASSESSMENT IN A NUTSHELL

- There's a hazard to doing this job: how can I avoid it, stop it, reroute it, destroy it, outsmart it?
- When I know how to do the job correctly and safely, I write it down in an SOP.
- Periodically, I review the process and see if we can make further improvements in safety, quality, or efficiency.

(HA) comprises this whole process. In its stripped-down basic version, the HA, also called the job hazard analysis (JHA), is essentially a strategic battle plan for how you will defeat your enemy, that is, workplace hazards.

VERSIONS OF HAZARD ASSESSMENT

There are different levels of sophistication when it comes to assessing and controlling hazards. Some procedures only get you as far as identifying the possible hazards. Some look backward, using past experience as a predictor of the future; with others, hazards are anticipated and planned for.

More advanced forms of HA include goal-setting and progress measurement systems that assess how well we have been doing and allow us to establish milestones for improvement in the future. Such assessments often rely heavily on *key performance indicators* (KPIs), which are quantifiable measures that gauge a company's performance against a set of targets, objectives, or industry peers. Two other progress-tracking systems are *objectives and key results* (OKRs) and the SMART system (*Specific, Measurable, Attainable, Reasonable, Time-bound*). While KPIs are often granular and set by teams within the company structure, OKRs and SMART goals typically express a broader company vision.

Key performance indicators need to be measurable in some way, even if the measurement is a simple "yes" or "no." In section 3 of this book, we explore ways to measure progress in making the workplace safer and systematizing an overarching safety program in chapters 21 and 22, respectively.

Table 5.1 lists some of the many processes used to identify risk. In fact, these processes are not limited to manufacturing safety or workplace HAs. They may be used in anything from financial management to hiring practices to cyber security, and beyond.

HAZARD ASSESSMENT PROCESS

Hazard assessment can be a simple process or it can be burdened with layers of bureaucracy and standardized tables. It can get so tedious that you will have SOPs telling you how to write SOPs. Let's not go there. A bare-naked HA has four steps:

1. Outline the task step by step.
2. Identify or speculate on what could go wrong at each step, i.e., what hazards are present.
3. Determine how you will eliminate, avoid, or control those hazards.
4. Document the safe task procedure, which includes the hazard controls you have selected. This will usually take the form of a standard operating procedure (SOP).

You can and should learn to use this stripped-down process on a routine basis. It doesn't involve any statistical methods and does not require specialized data. It *does* require an understanding of the task and it *does* require you to be open to modifying the way you perform your work.

In chapter 4, we mentioned how the best defense is a good offense. That's all a hazard assessment is: knowing what is coming at you and neutralizing it before it causes you harm. We compared defense to protective hazard controls and offense to preventive hazard controls. Just as a good offense is preferred over a good defense, prevention is better than protection in the world of hazard controls.

1. Outline the Task Step by Step

In our lean HA approach, step 1 is the easiest: outline the task step by step. This should be a bare-bones outline. Try to keep the number of steps to fewer than ten; five to

Table 5.1 **Hazard Assessment and Related Processes**

Process name	Methodology	Description
What-if analysis	Procedural	Brainstorming to identify what can go wrong during a task; part of hazard assessment
Checklist	Procedural	Listing of equipment and/or actions usually without additional instruction; can be part of standard operating procedure
Hazard assessment	Procedural	Identifying workplace hazards and assigning hazard controls; some focus only on PPE
Job hazard analysis	Procedural	JHA, synonymous with hazard assessment (HA)
Failure mode and effects analysis	Procedural and/or numerical	FMEA, or failure modes; a what-if analysis performed on a discrete component, action, or hazard; can be assigned numerical values for likelihood, severity, detectability, etc.
Benchmarking	Procedural and/or numerical	Measuring processes against those of organizations who are leaders in some aspect of operations, for example, safety
Risk assessment	Procedural	Often synonymous with HA or JHA
	Statistical	In environmental health, the chance of harmful effects to humans or ecological systems due to exposure to chemical hazard

Hazard Assessment Form

TASK:	HA DATE:
DEPT:	INITIALS:

STEP	DESCRIPTION	HAZARDS	CONTROLS	PPE	FMEA NO.

Figure 5.1. Hazard assessment worksheet. *Courtesy of Brewers Association.*

seven steps is a good target. You are not adding instructional detail here, just a top-level term or phrase for each step.

Typically, you will want to create a table that lists the steps of the task with room for the potential hazards and hazard controls on the same line. The Brewers Association has published a template, reproduced here (fig. 5.1). There are numerous examples freely available on the internet that can be found by searching for "hazard assessment template" or "job hazard analysis template." You will want the space to record any and all hazards and reasonable controls.

Everything in the HA and the subsequent SOP hinges on the outline step. Here is some helpful advice. First,

HAZARD ASSESSMENT FOR WORT BOILING

A hazard assessment can be described in just four steps. Let's walk through an example hazard assessment process for wort boiling.

Four Fundamental Steps
1. Outline the task step by step
2. Identify potential hazards for each step
3. Determine best hazard controls
4. Document the safe task procedure, which includes the hazard controls

Hazard Assessment Steps for Wort Boiling
1. **Outline the task step by step:**
   - Run 500 gallons sweet wort from lauter tun to boil kettle
   - Add steam to boil kettle cone when covered with wort
   - Add side jacket steam when 80% of wort is in
   - Trim side steam as wort is coming to the boil
   - Add foam control
   - Add first hops, observe hot break
   - Boil for 60 minutes, adding ingredients per recipe
   - Cut steam, knockout wort to whirlpool

2. **Identify potential hazards for each step:**
   - Overfilling boil kettle
   - Burns from hot wort
   - Loss of product from boilover
   - Damage to equipment from sticky wort residues

3. **Determine best hazard controls:**
   - Brew to a volume the boil kettle is designed for
   - Monitor wort temp
   - Control boil by trimming steam
   - Control foam with additive and/or foam sensor
   - Keep cold water hose at hand
   - Observe that the hot break has occurred before increasing steam
   - Add hops and other boil additions cautiously
   - Avoid manway line of fire
   - Keep manway closed unless adding ingredients
   - Do not leave boil kettle unattended while running steam
   - Ensure relevant staff undergo training and attend toolbox talks
   - Keep untrained people off brew deck
   - Deploy hazard signage

4. **Document the safe task procedure (SOP):**
   - Task name and purpose
   - Revision date and sign-off
   - Required equipment and supplies
   - Required PPE, lockout/tagout, etc.
   - List the steps for the task
   - Give instructions for each step
   - Include measurements for ensuring quality, i.e., time, temperature, gravity
   - Integrate hazard controls into each step's instructions
   - Include emergency action plan for foreseeable incidents
   - Reference related SOPs
   - Review periodically and revise as needed

involve a range of persons in the process. Get together a small working group that includes one or more workers who perform the task, a head brewer or production manager, and a representative from ownership or senior management. In a very small brewery, this may boil down to just a couple of people.

The inclusivity of involving a cross-section of personnel will give more detailed instruction in the SOP and increase worker buy-in. You also benefit from the experience of others. Sometimes it even happens that a participant who does not understand the process can push for greater clarity. This outline will become the backbone of your SOP later.

Now, if you already have an SOP, great. Hang onto it, but evaluate whether you went through the steps of identifying hazards and determining the best controls to put in place. Your SOP may lack the underlying health and safety documentation that provides the foundation for a robust SOP and brings it into OSHA compliance. If you have an amazing SOP with hazard control steps embedded into it, pass Go and move on to another task needing an HA.

2. Identify Potential Hazards

Chapter 3 of this book was about identifying and describing hazards. You and your team should by now be prepared to seek out and name the hazards in any task. In this step of the HA, use process knowledge, objective observations, and measurements to characterize each hazard belonging to each step of the process. For extra ideas about identifying and containing hazards, chapter 7, "Hazards are an Exchange of Energy," may help you discover more unusual hazards.

A good way to make sure you don't miss any possible hazards is to use the Brewers Association Hazard Placard (fig. 5.2), which displays pictograms for the most likely brewery hazards. Look at each pictogram and ask, "Could this be a hazard during this step?" We will also use these pictograms, and many others, throughout section 2 when describing various hazards.

Sometimes the same hazard occurs repeatedly throughout the task. A common brewery example is the risk of slips, trips, and falls. If your HA is about harvesting yeast from a fermentor cone, it is likely that the hazards of slipping on wet surfaces, tripping over hoses, and stepping into an open trench drain will persist throughout the whole procedure. Other times, a single step poses a difficult hazard challenge while the rest of the task is relatively safe.

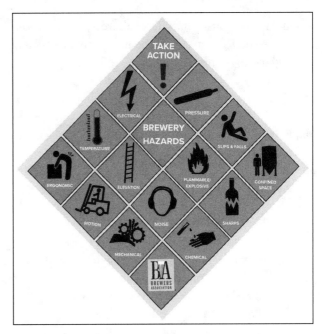

Figure 5.2. The Brewers Association Hazard Placard. *Courtesy of Brewers Association.*

Consider an automated keg cleaning and filling machine. Empty kegs are loaded onto the conveyor (minor lifting hazard), then the kegs go through an automated cleaning, purging, filling process behind acrylic and metal machine guards. But full kegs have to be moved from a waist-high conveyor to a pallet by manual labor. At another brewery, there is a similar keg line. Returned kegs with contents in them are lifted manually to the conveyor, then through the machine, where full kegs are mechanically brought to the floor. Same kind of job, each with one critical ergonomic hazard in just one step of the task. If you are already thinking about using a mechanical or pneumatic lifting device for this last step, then you are right on track for step 3, determining hazard controls.

3. Determine Best Hazard Controls

Determining the best hazard controls can be the most challenging step in the HA process. Remember that PPE is not the only hazard control! Elimination and substitution remain the best controls; but, where other preventive and protective controls are necessary, the employer should consider that a truly effective safety culture is one where the worker exercises their own agency through safe working practices and is encouraged to use both employer-provided controls and their own volition. Refer back to chapter 4 for added detail. Your brewery's approach may differ from another's based on how robust your safety culture is, what training resources you have, and the amount of capital available for engineering controls.

Different kinds of controls have different advantages and disadvantages. One control might be highly effective but costly. Another might meet OSHA recordkeeping requirements but it does not make the workplace any safer in practice. Still another works well, but only when the worker takes charge of their own well-being. Here are some key themes to remember:

- Prevention is best, if you are able to apply it.
  - Elimination of hazard by any means
  - Substitution for a less hazardous process
- Preventive "Thoughts and Actions," with behavior driven by workplace safety culture
  - Thoughtful action by the worker to avoid a hazard
  - Administrative controls directing the worker how to act
- Protection, where the hazard still exists but we isolate it from workers
  - Engineering controls are devices that manage hazards
  - Personal protective equipment; important, but best used in conjunction with other controls

4. Document the Safe Task Procedure

Once you have completed the first three steps of the HA, the last step will seem easier and give you a great sense of accomplishment. You are going to use what you documented on your hazard assessment form and build it out into an instructional document: the standard operating procedure (SOP).

The SOP is the "recipe" for how to perform a certain task, the same way, every time, and in a safe manner. Using SOPs enhances quality, product consistency, productivity, profitability, time management, and safety. The SOPs form a vital link in safety compliance documentation, food safety, cost of goods sold, and many other important business processes.

In a nutshell, the SOP is your original task outline, fleshed out with a list of the required equipment and supplies and instructional details for each step of the task, all with safe work instructions included. The internet has many SOP worksheets and templates to choose from. An example from the Brewers Association is given in figure 5.3.

Be efficient in creating SOPs by using a template that suits your business. The template can be identified with your company name or logo, numbered in any way, printed on any color paper. Often, SOPs are laminated and kept right at the location in the plant where the activity is taking place. In other words, there is no standard SOP because there is no general requirement to have them.[1] Standard operating procedures were developed for process control and are widely used in all types of businesses for that reason, not because OSHA mandates them.

Watch out for overly verbose or lengthy SOPs. Too long, and no one will read them. Consider breaking complex tasks into sub-tasks with separate SOPs. It is fine to use phrases, leave out unnecessary words, or add checkboxes. Just like a cooking recipe that reads "cook until soft, add herbs, plate," your SOP might read "buddy system for stepladder, open and locked, three points of contact."

DOCUMENTATION BEST PRACTICES

Performing HAs and developing SOPs takes time and effort. This is especially true when a team of coworkers takes charge of the process. Creating templates early in the company's evolution will lend consistency, while still promoting the contribution of many voices. When an outsider or a senior employee is the sole author of these documents, employees may sometimes feel disenfranchised and may be non-compliant. As a reminder, a robust approach for preparing these vital documents includes involvement from persons regularly performing the tasks, experienced staff, engineers, and safety specialists, as well as the guiding hand of management.

Hazard assessment and SOP documentation needs to be periodically reviewed and updated—these are living documents that are subject to changes in process, policy, and the availability of new hazard control products. The SOPs will serve as important training documents for new employees and those assigned to new duties in the business, so it is worthwhile making sure they reflect current best practice in terms of process and hazard control. Also bear in mind that valuable improvements in SOPs can come from new hires who bring their own experience with them. Just be mindful you don't adopt a worse or more unsafe procedure for the one you already had.

Although SOPs aren't generally required by OSHA regulations or other consensus standards, they can act

---

[1]   Special exceptions include the Energy Control Procedure under lockout/tagout rules and Confined Space Entry Permits (*see* ch. 19).

Standard Operating Procedure (SOP) Form

TASK: _____	SOP NO: _____ REVISION DATE:_____
DEPT: _____	INITIALS: _____

1) Purpose

This SOP describes Brewery _____'s procedure for safe and effective _____.

2) Scope

This SOP is limited to _____.

3) Responsibilities

It is the responsibility of every Brewery _____ employee to maintain the highest standards of safety, quality, and sustainability. Any Brewery _____ employee who performs the task described herein will be first trained and approved for the use of the equipment and materials specified, according to this SOP. Any changes to this procedure shall be first approved by _____.

4) Equipment and Materials

__ _____   __ _____

__ _____   __ _____

__ _____   __ _____

5) Procedure
1) _____
    a. _____
    b. _____

2) _____
    a. _____
    b. _____
    c. _____

Figure 5.3. Standard operating procedure worksheet. *Courtesy of Brewers Association.*

as important documentation. If there is a quality problem with a batch, it is often recommended to go back to the SOP and review it with employees involved in producing that batch. You may discover that a step was missed or that the SOP needs improvement to clarify a step. An SOP that contains hazard control actions is invaluable for your operation's safety culture, not only because it helps workers perform the task safely, but because it is written evidence that a hazard assessment has been carried out and hazards are being controlled to the extent possible.

Keep SOPs locally available. A busy worker is far more likely to review an SOP when the document is right at hand. Standard operating procedures can also be simplified into process checklists. Workers can use checklists to help them perform the task faster while maintaining quality and safety. Checklists can also be used to create batch documentation for ongoing quality control.

SECTION II
PROCESSES

If you think about starting any activity you want to be good at—brewing, playing the guitar, going downhill on a mountain bike—you know you'll have to make an effort to learn about the parts involved. Take bombing down a mountain on a bicycle: in selecting equipment, you will naturally think about what the activity is going to be like and what hazards it presents, then choose your equipment and riding method accordingly.

Let's start. You know you will be accelerating going downhill, so you'll need excellent brakes. The surface will vary: the trail could be sandy, then rocky; there may be tree roots in your path. You will need tires with lots of grip, just the right air pressure, and the ability to withstand hard impacts. You will want shock absorbers in the fork and frame to keep you from being bounced right off the bike. A helmet is a must—if you do crash, a helmet could save your life. Maybe there are other elements, like specific pedals, shoes, bar grips, gloves, shin pads, and so on, that you think you will need. Without realizing it, you've just done a hazard assessment.

SECTION II ORGANIZATION

The chapters in section 2 are to help you identify unsafe situations and select safe solutions for activities you may encounter in beverage manufacturing. Before we get into specific hazards and methods to control them, the first three chapters of this section provide some key definitions and address overarching themes that apply broadly to all hazard assessments. Chapter 6 begins by looking at how we can describe the nature of injuries and disease, and the routes by which hazards can bring these about when workers are exposed. Chapter 7 shows how all hazards can be thought of as a damaging transfer of energy to the human body, equipment, products, or the environment. Nearly all hazards follow predictable physical laws that describe potential, kinetic, or molecular energy. The exception is those issues involving human behavior, either psychological, like anxiety or autism, or psychosocial, where two or more people are having a problematic interaction.

Systemic Hazards

Usually when we think about a workplace hazard, we associate a certain hazard with a specific situation. For example, if you are moving pallets of grain with a forklift to a storage rack, there is a risk of damaging the pallet or grain sacks (among other risks). This is called a *situational* hazard. Tripping is a situational hazard associated with hose transfers. Electrocution is a situational hazard associated with wiring in new equipment.

But let's say you haven't been certified in forklift operation. In fact, no one at your company has ever been trained on how to safely operate a forklift. Furthermore, there is no company policy on who can use the equipment, there is no preventive maintenance schedule, and the seat belt doesn't work either. To make matters worse, the shop floor where the forklift is operated lacks marked pedestrian lanes and safety devices like bollards, dock plates, and corner mirrors. It can be seen that there is a general failure to operate and maintain forklifts safely and appropriately. This safety issue is not related to just to one task or situation, it is throughout the entire operation. This overarching negligence is a *systemic* failure and creates a systemic hazardous condition.

Chapters 8 and 9 discuss systemic hazardous situations in the workplace and provide general suggestions for improvement. While this concept can be applied to any number of business types and activities, we present six categories of operations that commonly occur in breweries. Where safety regulations or design criteria exist, they are summarized in tables.

Situational Hazards

Situational hazards are grouped in chapters 10 through 18 by their nature: physical and ergonomic, chemical, biological, and psychosocial. Chapter 19 covers confined space activities and control of hazardous energy (lockout/tagout), for the simple reason that each of these complex activities common to daily brewery operations is a perfect example of a combined hazard, where multiple situational hazards must be controlled simultaneously.

Within each situational hazard chapter, each type of hazard is identified with a box that shows one or more hazard pictograms, the term for the hazard, the type of energy the hazard involves, where in the manufacturing process you will likely encounter this hazard, and the typical outcomes that result (*see* "Caught Between" example). This is the kind of information you would collect in step two of the hazard assessment process: What hazards exist at a certain point in the performance of a task?

The information that follows describes what prevention and protection tools are commonly used to control that type of hazard. Therefore, this part of each chapter is like step three in hazard assessment: identify hazard controls for the known or anticipated hazard.

Some hazard discussions involve specialized terms, ratings, or are closely tied to another type of hazard. When this is the case, the hazard outcome discussion will include this extra information. If there is an OSHA compliance component to the subject, you will also find a table summary of the requirements and references to regulations and supporting documents.

This layout is designed to benefit you when you are following the hazard assessment process. Recognizing a potential hazard is usually not difficult: there is boiling wort in this vessel, the dry-hop port is ten feet off the floor, the forklift comes around a blind corner, that coupling could fly off, etc. The part that is often more difficult is knowing how to confront the hazard once you know what it is. This is what section 2 is all about. As such, you should be familiar with its contents and layout, and use it often as a resource in hazard assessment and work activity planning.

OTHER USES OF THIS SECTION

Section 2 has many other applications. For a safety meeting, or toolbox talk, you could pick a single hazard, identify where in your processes the hazard may occur, address hazard controls, and discuss possible improvements to be made.

If your company has a safety committee, pick a hazard and then go through the plant to identify where that hazard exists and decide whether existing controls are adequate. This process often discovers uncontrolled or poorly controlled hazards.

You may use section 2 to identify appropriate warning and security systems, such as signs, barriers, locks,

Caught Between

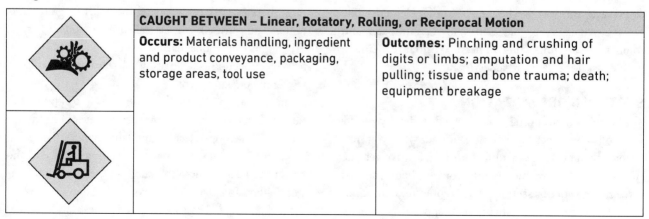

	CAUGHT BETWEEN – Linear, Rotatory, Rolling, or Reciprocal Motion	
	Occurs: Materials handling, ingredient and product conveyance, packaging, storage areas, tool use	**Outcomes:** Pinching and crushing of digits or limbs; amputation and hair pulling; tissue and bone trauma; death; equipment breakage

monitors, and alarms. Maybe thinking about your brewery tour and the proximity to certain hazards will lead you to create a safer itinerary for visitors.

You may want to create a reading list or training regimen for workers in a certain department. What are all the typical hazards encountered by a canning line operator or a warehouse person or a cook? For that, flip to the appendix, "Job Activity Hazards Index," and search the chart for the job activity to find the most important sections on relevant hazards and their controls. This chart can also be helpful when creating orientation or training agendas according to a worker's expected duties.

6

HOW HAZARDS CAUSE INJURY

Safety has a parlance that allows us to conveniently speak about the nature of hazards and how they manifest through injury or illness. We first saw this in chapter 3, where it was pointed out that hazard and outcome are preferred over incident and accident, respectively (p. 25). Some hazards result in an immediate outcome, whereas other times the results are delayed. An injury might occur where contact was made or the damage may show up in another part of the body. An eye injury may be caused by a physical agent like a flying metal fragment, or by a chemical agent like a splash from an acid. So, where there are terms that relate to more than one type of hazard (e.g., a physical, chemical, or biological hazard), we have defined them in this discussion to avoid repetition.

ACUTE VERSUS CHRONIC INJURY

We discussed the nature of acute and chronic injuries in chapter 3 (p. 26), but let's have a quick recap. Acute injuries occur as a result of a single incident or exposure. Normally, acute injuries do not result in disease, but the physical damage may remain long after. You can touch a hot steam pipe and be instantly burned, but your skin will take time to heal; similarly, a full keg you are lifting slips from your grip and smashes your foot. Something like acid cleaner might permanently damage your vision if it splashes in your eyes.

Chronic injuries and diseases result from repeated, low-level exposures to a hazard. Repetitive lifting can lead to lifelong back pain. Frequent loud noise exposure will lead to permanent hearing loss.

In general, individuals are more concerned about acute exposures than chronic ones. Acute injuries are real-time and tangible. Chronic exposures increase the *probability* of long-term injury or disease, but do not guarantee it.

DISEASE

Disease is not always easy to define as it covers so many conditions. A disease can be a communicable negative health condition, like hepatitis caused by viral infection, or Lyme disease where the bacteria responsible is transmitted through tick bites. But a disease can also be a non-communicable ailment like cardiovascular disease, cancer, substance addiction, or a chronic musculoskeletal disorder.

In safety circles, the term *disease* is broadly used to encompass negative health conditions caused by infectious, genetic, physical, or environmental agents. We can use the following definition of disease to capture how disease is discussed in this book:

> *Disease –*
> *A negative health consequence that:*
> *a) affects one or more parts,*
> *functions, or processes of the body,*
> *b) has definable signs or symptoms, and*
> *c) will progress if not treated.*

LOCAL VERSUS SYSTEMIC INJURY

Acute and chronic injuries and disease can act upon the part of the body where the exposure occurred or can show up elsewhere in the body. When the injury manifests at the initial site of exposure, we say the outcome is a *local* injury, for example, a burn, a cut, or an object lodged in the eye. If we know a hazard acts locally, that points us toward using engineering controls and PPE. If we need to use our hands for a job involving a corrosive chemical, then we know we should protect the hands.

When the outcome of a hazard manifests elsewhere in the body, that is a *systemic* injury. Infection resulting from exposure to contaminated body fluids may damage specific body organs. Certain persistent pollutants accumulate in the body's fat tissues, while some heavy metals can affect the skeletal or neurological systems. Quite often, the controls for these sorts of hazards involve elimination, substitution, and training to ensure established protocols are followed. Engineering controls and PPE may also be valuable.

The word *systemic* is used in an entirely different way in safety parlance to describe systemic hazardous situations, as we saw in the introduction to section 2 (p. 55). This complicates matters—it is important not to confuse the two uses. (Systemic hazards are covered in chapter 8.)

Table 6.1 **Examples of injury occurrence and symptom location**

Manifestation	Duration	
	Acute	**Chronic**
Local	Debris particle in eye causes irritation	Frequent contact between bare hands and caustic substance causes dermatitis
Systemic	Ingestion of peanut-flavored beer by allergic individual causes anaphylaxis	Long-term alcohol use damages liver and pancreas

Using the four basic terms—acute, chronic, local, and systemic—we can be pretty descriptive about an injury or illness, yet not have to be overly technical. Note that acute injuries are not always local and chronic injuries not always systemic. Furthermore, some injuries can manifest in multiple ways from contact with the same

hazard. A lead bullet can cause an acute-local outcome in the form of a bullet wound. If not removed, lead fragments from the bullet can be a source of lead toxicity, resulting over time in chronic-systemic outcomes.

Why is it so important to think of hazards this way? When we consider the physical and chemical properties of hazards together with the physiological effects, we get a more accurate picture of the overall behavior of the hazard. Knowing the hazard's behavior helps us select effective controls, as well as informing monitoring and testing regimes when looking for symptoms should a worker be exposed.

ROUTES OF ENTRY

Routes of entry are how a hazardous substance or agent gets into the body, where it may then do local or systemic damage. We most often think of exposure to chemical hazards when discussing routes of entry, but this applies to physical and biological hazards too: exposure to radiation, inhalation of simple asphyxiants, and exposure to contagious agents are several such examples.

From a safety perspective, the routes of entry are typically respiration (inhalation), ingestion, and dermal (or eye) contact or penetration. Some substances are most hazardous by a specific route of entry, while others may cause injury through any route. Understanding the route(s) of entry for a particular hazard allows us to select the most effective PPE and engineering controls. With a corrosive cleaning chemical that affects the skin and eyes, we know we should be using chemical protective gloves and eye protection. For bothersome dust, we might select a respirator, improve the area ventilation, or reengineer the equipment to emit less dust.

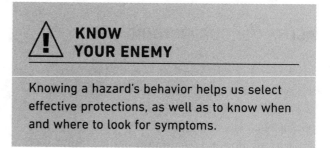

⚠ **KNOW YOUR ENEMY**

Knowing a hazard's behavior helps us select effective protections, as well as to know when and where to look for symptoms.

Respiratory Route of Entry

Respiratory System Overview
In anatomy, the respiratory system is divided into the upper respiratory tract and lower respiratory tract.

The upper respiratory tract includes the nostrils and nasal cavity, the sinuses, the oral cavity, the pharynx (throat; includes the entrances to both the esophagus and the larynx), and the larynx (voice box). The larynx is the gateway to the lower respiratory tract and has a moveable flap (the epiglottis) that acts to route air and food into the windpipe and esophagus, respectively. The nasal cavity and sinuses help filter, warm, and moisten air during inhalation.

The lower respiratory tract comprises the trachea (windpipe), bronchi, bronchioles, alveoli, and the lungs themselves, which are the organs housing most of the bronchi and all the bronchioles and alveoli. In the lower tract, the trachea forks to the two lungs through large air pipes called primary bronchi. Each primary bronchus forks, tree-like, into smaller bronchi that enter the lungs, and then to even smaller branches called bronchioles, which ultimately end at the grape-like clusters of alveolar sacs, tiny pouches (sacs) made up of individual alveoli covered with capillary blood vessels; this arrangement allows the rapid two-way exchange of gases between inhaled air and the bloodstream.

The diaphragm is a large muscle that lies outside of the respiratory tracts. It is a vital part of the respiratory system because it causes inhalation and exhalation by rhythmically contracting and relaxing. It also divides the torso, forming the floor of the chest cavity (thoracic cavity) and the roof of the abdominal cavity.

Without adequate oxygen we can only survive for a few minutes. The lungs are the organs that get oxygen into our bloodstream to supply all the cells of the body, and they are the major route by which waste gases, mainly carbon dioxide generated by cellular metabolism, are expelled from the body. This gas exchange takes place across the boundaries of the approximately 480 million alveoli that are in an average pair of lungs. Larger, healthier lungs have more, but the alveoli remain consistently sized between individuals (Ochs et al. 2003). This multitude of alveoli represents a significant absorptive surface area, on the order of 30 to 180 m^2, depending on study conditions. Current respiratory aerosol absorption models generally use 78 m^2 for the lungs, about the size of half a tennis court. This vast area, only a cell's thickness from the bloodstream, helps explain how oxygen, toxic gases, and substances like nicotine can be absorbed so quickly into the body when inhaled.

How Hazards Enter the Respiratory System

The respiratory route has defense mechanisms, including mucus and cilia. Many surface areas of the respiratory tract above the bronchioles include cells that produce mucus, a watery, sticky substance that traps particles and bacteria entering the tract. Cilia are fine, hair-like structures also found lining the same areas of the respiratory tract. Particulates caught in the mucus are carried upward by the coordinated oscillations of the cilia, which propels the contaminated mucus until it can be dumped down the esophagus or expelled by spitting, coughing, or sneezing. Despite the effectiveness of this system, there are many contaminants that do get through, including asphyxiating gases, solvent vapors, narcotic substances, airborne viruses, mold spores, and microscopic irritating particles.

In general, solid and liquid particles larger than about 10 micrometers (μm) are mostly caught in the upper respiratory tract, while constituents smaller than 10 μm advance into the lower respiratory tract but are caught before reaching the alveolar surface (Ferrari, Carugno, and Bollati 2019). The sum of particulates 10 μm and smaller is called the PM_{10} and is often used in expressing particulate air pollution levels. Total particulates smaller than 2.5 μm are called the $PM_{2.5}$, or the fine fraction. Particles sized between 2.5 and 10 μm are referred to as the coarse fraction (World Health Organization 2003, 7).

Fine fraction particles, gases, vapors, viruses, and bacteria smaller than 2.5 μm may make it all the way to the alveoli. In the absence of cilia and mucus, the alveoli are protected by special white blood cells called phagocytes. Phagocytes recognize invasive material, including fungal cells, bacteria, and mineral dusts. Phagocytosis is the process whereby a phagocyte envelopes foreign matter and attempts to destroy it using enzymes and other biochemical processes. It works well to prevent infections, but does little for contaminants like silica dust or asbestos fibers.

To be a respiratory hazard, the substance must be able to become airborne or already be in the air mixture. Industrial hygienists use a variety of terms to classify these physical forms in more specific ways than we use in common parlance. This is because each of the terms says something specific about the physical form and how it was formed, which, in turn, leads to proper selection of monitoring and respiratory protection strategies. Table 6.2 summarizes these terms, but we have already seen them outlined in more detail in table 3.2 (pp. 28–29) with some examples of where they can occur in a brewery. Figure 6.1 shows the size ranges of examples that could be encountered in a brewery setting.

Table 6.2 **Summary of physical forms of respiratory hazards**

Physical form	Class		Definition
Aerosols	Any aerosol		A general term for a suspension of fine solid or liquid particles in a gas or air
	Solid aerosols	Dust	Particulate generated by abrasive or destructive processes: • Suspended atmospheric dust, 0.003-1 µm diameter • Settling dust, 1-100 µm • Heavy dust, >100 µm
		Fume	Particulate resulting from vapor condensation
		Soot	Fine, black, carbonaceous particles produced by incomplete combustion of an organic material
		Fiber	A particle with an aspect ratio of at least 3:1 length to width
		Viruses	Nonliving infectious agent that can only replicate inside infected cells
		Bacteria	Single-cell prokaryotic microorganisms, some of which cause infections and disease in humans
		Fungi	A broad taxonomic group that includes yeast, mold, mushrooms, mildew and rusts
	Liquid aerosols	Mist	Suspended microscopic liquid droplets generated by mechanical force such as spraying or splashing
		Fog	Suspended microscopic liquid droplets generated by vapor condensation
Gases		Gas	A gas-phase substance, which may be pure or combined with air, originating from a substance normally a gas at ambient conditions
		Vapor	A gas-phase substance, usually in a mixture with air, emanating from a substance normally in a condensed (liquid or solid) state at ambient conditions
Complex mixtures		Smoke	Mixture of visible and colorless byproducts of incomplete combustion, including soot, hydrocarbons, ash, and gases and vapors
		Smog	A combination of smoke and fog, and possibly other airborne pollutants, that forms in the atmosphere, typically over urban areas

Brewery Dusts and Mists

There are many airborne substances produced in the course of normal operations in a brewery, including grain milling dust, powdered or granular brewing additives, floor dust, nuisance mold spores, and environmental pollen. During construction or remodeling activities, cement dust, sawdust, welding fume, and fiberglass insulation will also be present. Any of these substances can be hazardous, affecting the upper respiratory system and/or the eyes and resulting in irritation, coughing, sneezing, mucus production, and allergic reactions.

Irritating substances often induce the production of excess mucus. Coughing or swallowing this mucus can either be helpful, say in the case of malt dust, which can be digested in the stomach, or can cause further illness if a toxic substance enters the body through the small intestine. Breathing rodent poison powder or lead paint dust are examples.

When a material does not have an established permissible exposure limit (PEL) under OSHA regulations and does not cause serious health outcomes, it is called inert. Industrial hygienists often call these nuisance materials, because they can still cause worker discomfort or cleanliness problems in the workplace. Grain milling dust is such a material, and as such is assigned a generic OSHA exposure limit intended to avoid irritation. However, different individuals respond differently to dust from milled grain, some being sensitive or allergic well below established limits.

Personal protective equipment that offers respiratory protection from nuisance or hazardous substances

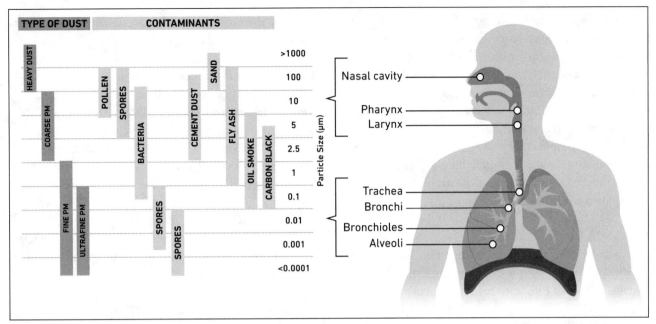

Figure 6.1. Common brewery air contaminants and other familiar substances are shown with their typical size range and the region(s) of the respiratory system they target. *Illustrationn © Getty/paveugra*

(dusts, mists, fibers, and fumes) incorporate filters that are designed to capture materials down to a certain particle size. These range from disposable dust masks to sophisticated respirators that pump filtered air into the breathing zone. The equipment may be rated by its particulate removal efficiency or by a protection factor that considers how tightly the unit fits the face and how effectively it can separate inhalation and exhalation valve operation. Where respirators are a suggested hazard control in later chapters, more specifics for respirator selection will be given.

Chapter 9, which looks at PPE, contains a section on respiratory protection and the associated regulations. Respiratory protection is a very sophisticated topic with significant regulation, monitoring science, and consensus standards. Often the best approach is to employ elimination, substitution, and engineering control solutions instead of going down the path of respiratory protection using PPE.

Simple Asphyxiants

Brewery substances that can penetrate to the alveolar surfaces include ultrafine particulates, the smallest biological molecules, and gases and vapors. Asphyxiation can occur from relatively non-toxic gases because they displace oxygen to the extent that we cannot breathe. These *simple asphyxiants* are sometimes considered physical hazards, sometimes chemical. Brewery gases

that can do this include carbon dioxide, carbon monoxide, nitrogen, argon, propane, and natural gas. Too much oxygen is also a problem, as it can cause lightheadedness and dramatically increases the intensity of a fire, if one were to occur. In this book, we discuss brewery gases with the toxic asphyxiants in chapter 16, since the hazard controls are so similar.

Chemical Asphyxiants

When an asphyxiant penetrates to the bloodstream and acts locally in the blood or systemically in some organ, we call them *chemical asphyxiants*. Exposure to these agents requires more than getting some fresh air, because the toxic agent cannot be displaced or removed this way. In breweries operating propane-powered forklifts, gas boilers, or direct-fire kettles, there is going to be some production of carbon monoxide (CO). Whether the CO amounts to a hazard will depend largely on the efficiency of the combustion device and the presence of adequate ventilation.

Exposure to CO results in the CO entering the bloodstream and displacing oxygen from hemoglobin, lowering the oxygen-carrying capacity of the blood and starving cells of oxygen. The best treatment for this exposure is placing the worker in a hyperbaric chamber with excess oxygen. Unless CO poisoning can be resolved quickly, permanent brain and organ damage or death are likely outcomes.

Immune Response to Respiratory Hazards

The body's immune system might recognize a substance as foreign and attack it with white blood cells or antibodies. Immune-mediated inflammation of the airways is known as bronchoconstriction. This narrowing of the airways makes it difficult to breathe. Within the lungs, defensive white blood cells called alveolar macrophages can try to latch onto and breakdown foreign substances. If contaminants make it into the bloodstream, another class of white blood cells called neutrophils will try to break down and eliminate the hazard.

Ingestion Route of Entry

One usually does not intend to ingest an irritating or poisonous material, but it does happen. Typically, poor hygiene or lack of PPE leads to accidental ingestion.

Chemical cleaner, sanitizer, welding fume, pest control product, or similar hazardous substance can be desposited on a worker's hands and, through the act of eating, drinking, smoking, nail biting, etc., the contaminant is ingested. Worker hygiene is discussed as part of housekeeping (ch. 8) and PPE (ch. 9).

Dermal Absorption and Skin Penetration Routes of Entry

The surface area of the skin is around 2 m^2 (22 sq. ft.) and makes up about 16% of a person's body mass, making it the largest organ of the body (Zimmerman 2018).

Unlike the alveolar surfaces in the lungs, the skin is many cells thick. Skin is categorized into two layers, the epidermis and underlying dermis. The epidermis consists primarily of specialized cells that make the protein keratin, which is strong, flexible, and water resistant. The bottom layer of the epidermis is where new cells form, which then migrate toward the surface to replace the top layer of hardened dead cells that are constantly being sloughed off.

The dermis, the thickest layer, consists largely of cells that make up connective tissue and is supplied with nerves, blood vessels, sweat glands, and lymphatic vessels. This layer senses pain and heat and is where significant injury occurs during a brewery burn. Regulation of blood flow and sweating in the dermis helps regulate body temperature in adverse hot or cold situations.

Lying below and partly interwoven with the dermis is the hypodermis. Not technically part of the skin, the hypodermis is mostly fatty tissue known as subcutaneous fat. This layer insulates the body and helps protect internal organs by cushioning.

Chemical Absorption and Reaction

Despite being water-repellent and made of many cellular layers, skin is especially vulnerable to contact with corrosives, oxidizers, solvents, and pesticides. From a prevention standpoint, conducting work with such hazardous agents is safest when there is no possibility of direct contact between the chemical and the skin. This might be done with an enclosed system, like a chemical dosing pump and plumbing. If there is a chance of direct contact, PPE is needed. Specific substances, outcomes, and controls are discussed in chapter 16.

Open Wounds, Contusions, and Burns

Open wounds in the skin can be caused by laceration, dermatitis, blistering, burns, or impact contusions from a variety of physical and chemical hazards. Common outcomes from these injuries include blood loss, infection, and the admission of foreign agents into the bloodstream.

In a brewery, burns due to hot liquids or direct contact with hot surfaces are most common and can cause significant skin injury. (Frostbite can also result in the same effects but is much less likely to occur in a brewery.) Lacerations, a deep cut or tear in the skin, can be caused by knife accidents, flying objects, or contact with objects that tear the skin, while a common skin irritation or dermatitis can be caused by exposure to oxidizing sanitizers.

An impact contusion, which can be caused by an incident such as being hit by a flying object or falling hard against a surface like a metal staircase, is an especially difficult injury to treat. It results from a severe impact that craters the tissue rather than cutting it open. Impact contusions often cannot be treated with suturing. Open wounds can be the result of many kinds of physical or chemical injuries.

Preventing these types of skin damage can be achieved with a full range of controls, including safe work practices, elimination, engineering controls, and PPE. Lacerations, burns, and impact contusions are discussed as part of physical hazards in chapters 10 and 14 and dermatitis is discussed under chemical hazards in chapter 16.

7

HAZARDS ARE AN EXCHANGE OF ENERGY

With practice we become more adept—you could say intuitive or sensitive—about knowing there is a hazardous condition nearby and how to be safe. There's that feeling you get when you realize you don't have a seat belt on. It is a subconscious thought in your head that something is different; neurons flash, and an idea boils up into your conscious mind: "Oh, seatbelt, right."

But the process isn't as simple as recognizing that you need to put on the seatbelt. The seatbelt is the hazard control—without even realizing it, you have gone through the first three steps of a hazard assessment and instantly arrived at the hazard control. You have a task, say, to get from point A to point B. The task requires a car. Driving in a car is inherently dangerous because of speed, mass, vehicle design, road conditions, your actions, and the actions of others. A seatbelt will reduce the chance of serious injury or death in the event of a collision. But all you thought was, "Oh, seatbelt, right."

This subliminal process is possible because of the habituation we develop with driving. We may be in and out of a vehicle several times a day and have ample opportunity to develop the habit of wearing a seatbelt, even as a passenger. In a brewery setting, we can habituate wearing eye protection if the necessity presents itself frequently enough. You will get to the point where every time you enter the packaging hall a little voice in your head says, "Oh, safety glasses, right."

Relying only on habituation has some pitfalls. We can develop unsafe habits just as easily as safe ones, maybe more easily. There are those among us who routinely stick bare hands into diluted sanitizing solutions. Over time, these practices can lead us to stick our hands into anything, which might result in a serious injury or long-term bodily harm. If we develop a bad habit, we will have to make a conscious effort to rewrite our mental code so that safe practices, like wearing gloves, becomes the normal condition.

Sometimes the hazard just isn't encountered frequently enough that we can develop that alert response. Let's say you operate a forklift only on rare occasions. Will your intuition remind you about the walk-around inspection, the seatbelt, the horn, looking in all directions, keeping your hands inside the cage and away from the mast, travel speed, or the load height, balance, and tilt? Probably not.

You will need to make up for lack of habituation with a more methodical recall of what you learned during your training or certification, which may have taken place some time ago. You will need to pay conscious attention to the hazards around you, identifying them by the energy they possess or are exhibiting. You create safety in the moment with this conscious awareness. The one universal thing that all hazards have in common is that the conveyance of their energy can and will cause harm.

ENERGY—THE BASICS

In the absence of developed good habits and reflexes, we have to consciously look for hazards and anticipate the harm or undesired outcome that could result. A scientific way to approach hazard identification is to recognize that all hazards pose the risk of a deleterious transfer of energy to the human body, equipment, products, or the environment.

Energy can be categorized as either *potential* energy (E_p) or *kinetic* energy (E_K), but kinetic energy is always responsible for injury and damage. Potential energy is formed when work is done against a force such that energy is stored due to, for example, deformation (say, tightening a spring) or position (say, lifting an object off the ground). The stored energy is not doing any work at the moment—it has the potential to be converted to kinetic energy.

Kinetic energy is the energy of things in motion. Kinetic energy does work. Most often we think of the motion of objects: a keg falling from above, a moving forklift, rollers in a grain mill, etc. This subset of things in motion is often called mechanical energy, but this term is also used to refer to potential energy that resides in mechanical devices (e.g., a wound spring, or pressure in a gas cylinder). Sticking to the terms *potential* and *kinetic* can help avoid confusion. There also many kinds of kinetic energy that are invisible: gravity, energy waves, radiation, magnetism, electrons, and molecular bonds.

In nature, we do not see one form of potential energy transforming to another type of potential energy without a kinetic energy intermediary. Even the most unnoticeable neurological processes involve biochemical reactions that can change thoughts, feelings, and behaviors, potentially with an adverse outcome.

Often, there is some intermediate actor made of matter that the energy is put into. Then that energized matter does the work or creates the harm. Other times, energy is transferred from one entity to another, to another, and so on down the line until it reaches a stopping point.

Frequently, an object will possess both potential and kinetic energy at the same time. Look at the first frame in figure 7.1. Notice the ball on the left that is lifted and then dropped: when it is being held in the lifted position, the ball has potential energy; when it is dropped and is moving toward the other balls, its kinetic energy increases with the amount of vertical drop, while the potential energy weakens correspondingly until it comes to rest after the impact with the next ball.

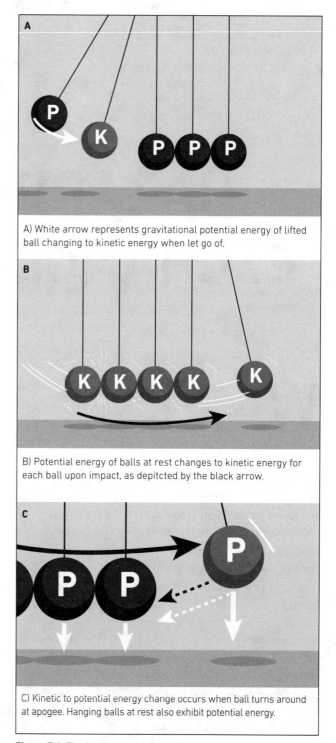

A) White arrow represents gravitational potential energy of lifted ball changing to kinetic energy when let go of.

B) Potential energy of balls at rest changes to kinetic energy for each ball upon impact, as depitcted by the black arrow.

C) Kinetic to potential energy change occurs when ball turns around at apogee. Hanging balls at rest also exhibit potential energy.

Figure 7.1. The familiar toy called a Newton's cradle exemplifies the processes of $E_P \rightarrow E_K$ (a), $E_K \rightarrow E_K$ (b), and $E_K \rightarrow E_P$ (c).

Picture a brewer on a ladder adding dry hops through the top of a fermentor. Let's say the brewer didn't have an SOP and forgot to release the head pressure in the vessel. Upon removing the tri-clamp blank end "coin" from the top fitting, the E_p from the contents being under pressure changes to E_K in the form of rapid air flow, causing the coin to fly through the air. The change of vessel pressure causes potential energy in the carbonated beer to lose CO_2 kinetically as it rapidly leaves solution. The rapid phase change of dissolved CO_2 instantly becoming gaseous CO_2 (in the form of bubbles) causes the surrounding beer to quickly expand and shoot out of the opening in another moment of kinetic energy. Finally, this energetic mass of gas and liquid transfers its kinetic energy to the brewer. The brewer is pushed over and falls from the ladder: this transforms the brewer's potential energy, which is due to his mass having earlier been moved off the ground to the top of the ladder, into kinetic energy as he moves (falls) back toward the Earth.

EXAMPLE ENERGY TRANSFORMATIONS IN A FOOTBALL PUNT

Kinetic to Potential

Moving ball lands in punter's hands above the ground

Punter's muscles lift ball higher

Potential to Kinetic

Punter's hands and ball drop

Leg muscles contract to move leg through the kick

Kinetic to Kinetic

Falling ball meets moving foot

Ball absorbs leg energy and flies

Kinetic to Potential

Punter sits on bench

FORMS OF ENERGY

Now we have two hazard paradigms. In chapter 3, we listed four hazard classes: physical, chemical, biological, and psychosocial (*see* table 3.1). This characterization is based on the mode in which the hazard acts, but also just as much the sort of outcomes that typically result.

In this chapter, we look at hazards from the viewpoint of the form of energy they possess. Potential and kinetic energies can be classified further into nine scientifically determinant forms of energy (US Energy Information Administration 2020). These are described in table 7.1, along with a tenth form of hazardous energy that does not adhere strictly to the laws of physics: human behavioral energy. Human behavioral energy is a manifestation of neurological processes, which themselves are transformations involving chemical and electrochemical energies that can be further translated into the kinetic effects of muscles in motion. On a chemical level, yes, natural laws are kept. How this human energy is directed and the behaviors that it may cause are not entirely consistent or predictable.

Just remember, energy causes harm by inducing some form of matter into an encounter.[1]

- A falling (linear kinetic energy) tool (matter) hits us on the head (outcome).
- A reactive (chemical energy) cleaner (matter) burns our eyes (outcome).
- An angry (behavioral/psychosocial energy) customer (matter) causes us stress (outcome).

Habituation to the safe way of doing things is a powerful tool when it comes to safety, but it works best when we have ample opportunities for reinforcement. When we encounter a task and its hazards infrequently, habituation is unlikely to be effective. For less common hazards, or when conducting a hazard assessment, we will need a logical process to discover what hazards are there, how they do harm, and how we can best intercede in that harm with the right hazard controls. Understanding the energy processes taking place provides valuable insight into specifying what controls will be effective.

---

[1] Electromagnetic radiation is the exception, because the energy itself can be absorbed directly without intermediate transfer to other matter.

Table 7.1 **Examples of energy to outcome processes**

Potential energies		Storage mechanism	Outcome example(s) upon exposure when converted to kinetic
Chemical	Chemical	Energy within atomic and molecular bonds	Caustic chemical reaction converting skin fats to soap
	Biochemical	Disease-causing agent	Toxins are introduced into the body; disease agent replicates using bodily resources
	Electrochemical	Capacitors and batteries	Electric shock to worker as current flows due to redox reaction between anode and cathode
Mechanical	Tension	Springs, elastics	Bungie cord eye injury
	Pressure	Pneumatics (gas systems)	Keg rupture during cleaning
		Hydraulics (liquid systems)	Forklift mast freefall
Gravitational		Objects[a]	Keg falling from stacked pallets
Nuclear[b]		Subatomic particles in nuclei of atoms	Heating reactor water

Table 7.1 **Examples of energy to outcome processes** (cont.)

Kinetic energies		Transfer mechanism	Outcome example(s) upon exposure
Motion	Linear	A form of translational motion where an object moves in one direction along a straight line; includes tension	Forklift striking object; cask spile being ejected; worker falling from ladder
	Rotatory	Wheels, pulleys, gears, etc. that rotate in a circular fashion around an axis	Rotating mash rake kills worker; hand crushed in bottle crowner
	Vibratory	Motors, engines, conveyors	Equipment failure from imbalanced bearing
	Rolling	Wheeled vehicles use a combination of rotatory and translational motion	Un-chocked truck rolls away
	Reciprocating	Machines that move back and forth or up and down in repeating cycles	Hand pinched while reaching into carton erector
	Projectile	Flying objects, pressure release	Tri-clamp coin flies off during dry hopping; overpressurized cans explode
	Oscillatory	Regular motion back and forth between two points, e.g., electromagnetic waves, or fluctuating voltage and current in alternating current	Worker electrocuted by contacting exposed wire connected to alternating current
	Flow	Pressure moving gases, liquids, or solids; density differences of materials	Wort lost by opening valve, worker engulfed in grain silo
Electrical		Electrons move through a conductive material	Overloaded power strip melts wiring
Thermal		Kinetic energy of molecules increases as they move faster	Boiling wort causes burn
Sound		Acoustic waves propagating with repeating pressure variations through air or other medium	Hearing loss from centrifuge noise
Radiant[c]	Ionizing	X-rays, gamma rays	Genetic mutation; cancer
	Non-ionizing	Visible light, ultraviolet, infrared, microwaves, radio waves	Tissue burns; eyesight damage; skin cancer

Human		Transfer mechanism	Outcome example(s) upon exposure
Behavioral		Neurochemical processes that affect behavior, reflexes, reasoning, and emotion	Anger, violence, depression; substance use disorder

[a] Energy increases with object mass and height.

[b] Not relevant to beverage manufacturing.

[c] Transfer mechanisms for radiant energy, also called electromagnetic radiation, are massless photons.

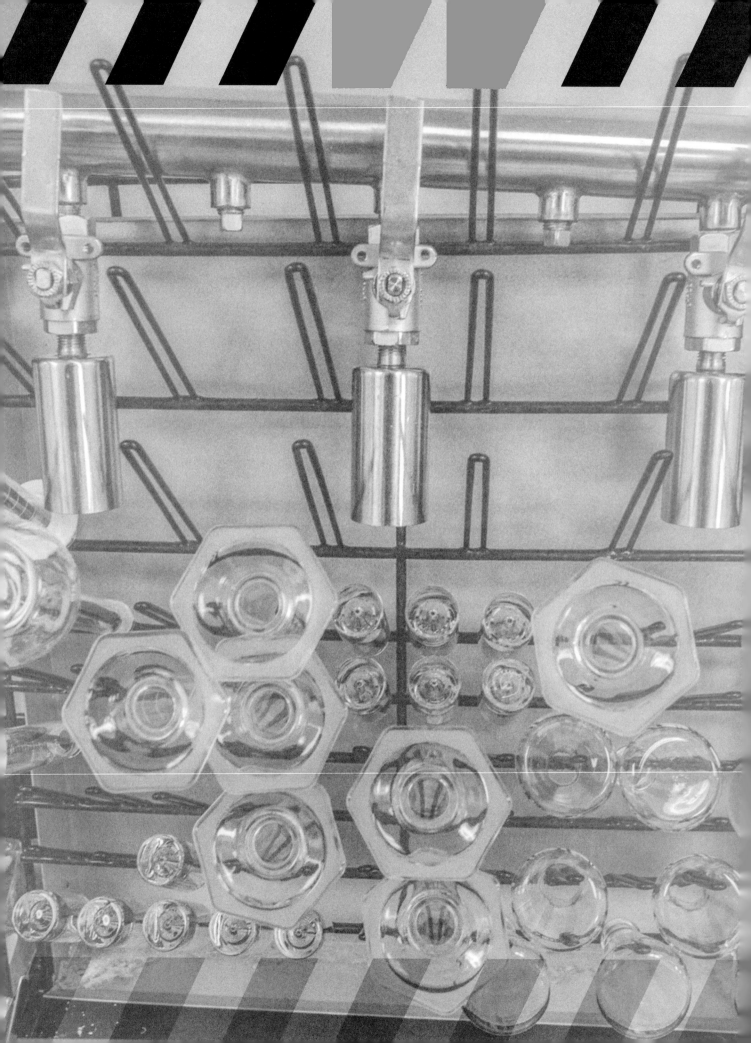

8

SYSTEMIC WORKPLACE CONDITIONS

Before venturing into discreet hazards that have a tangible basis in specific tasks, there are some overarching themes that can pervade an entire operation. Systemic risks are widespread contributing factors that can lead to a hazardous incident but are not specific to a singular task. They are generic, endemic, on-going safety shortcomings that occur in all parts of business activities (National Safety Council n.d.[a]).

This chapter deals with these broad, systemic areas of concern. Systemic conditions are not typically improved by focusing on a specific task. Instead, improvement comes through ownership and employees working together to build a culture that values the good condition of the premises, buildings, and equipment. This includes safety training, technical information, signage, and protective equipment being in constant use and updated as needed. Also, the social framework of the workplace must be conducive to functional communication, collaboration, and recognition of individual diversity. Safety is inclusive and shared by all employees under an umbrella of ownership support.

Intention to Create a Safe Workplace

COMPANY SAFETY PROGRAM – **Systemic Plant Condition**
Occurs: Plant-wide
Outcomes: Increased accident and injury rates, disenfranchisement, legal and financial problems

The main tenet of an effective safety culture is that the employer intentionally and sincerely transmits the message that safety is a core business value that requires everyone's participation. The employer must regularly highlight the importance of safety, support efforts to identify and manage hazards, and provide necessary resources, like training, engineering controls, and PPE.

 ILLNESS AND INJURY PREVENTION PLAN

Your company should develop a written company safety and health program to create a structure that management can use to proactively improve safety conditions in the workplace. OSHA usually uses the term *injury and illness prevention plan* to describe such programs, variously abbreviated by companies and institutions as I2P2, IIP program, or IIPP. IIPP plans are outlined in Table 22.1.

OSHA's "Safe + Sound" initiative is designed to encourage all companies, regardless of size, to have a company safety and health program. The depth and sophistication of the program will increase with the size of the company and the severity and variety of workplace hazards. Safe + Sound emphasizes three components: management leadership, worker involvement, and the hazard assessment process (US Department of Labor n.d.[b]).

The message that safety is a business essential should be clearly conveyed from initial hiring onward. Job descriptions should outline universal safety responsibilities for all employees, and list further position-specific safety requirements (see sidebars). The same universal responsibilities can be translated into the employee handbook as well. It is a good idea to have all employees sign an acknowledgement of their safety responsibilities. Employees should be given a chance to ask for clarification of any requirements. Non-native speakers or those with a reading disability may benefit from assisted interpretation.

For most brewery businesses, the safety and health program starts simply and evolves with time. Establishing expectations in writing and exemplifying

UNIVERSAL SAFETY RESPONSIBILITIES

The universal responsibilities might include the following:

- Safety is a top priority for our company, its employees, our customers, contractors, and suppliers.
- The company is required by law to create a safe and healthful workplace.
- The company has a safety and health program; key points are included in the employee handbook.
- As work conditions or hazard assessments determine, safety requirements are subject to change.
- Following safe procedures and using the engineering controls and personal protective equipment provided is part of your job.
- You will not report for work under the influence of any drug or alcohol. You will not partake of any drug or alcohol during your shift, unless it is a permitted sensory evaluation of product, subject to a daily volume limit of [specified volume] ounces.
- Communicating about safety with coworkers and managers is part of your job.
- The company has procedures and/or documents to report near misses, observed hazards, health concerns, injuries and illnesses; you are protected from termination by law for reporting unsafe conditions.
- We may enlist you to help identify hazards in your work area(s), suggest safe ways of performing work, participate on the Safety Committee, walk-through on inspections, or otherwise provide ideas concerning safety for yourself and others.
- You will do your part to keep the workplace clean, orderly, and sanitary.
- You will not engage in any horseplay, practical jokes, social media stunts, or other antics that could result in injury or show improper or unsafe conduct.
- You will not post images or narratives of incidents, accidents, injuries, fighting, disorderly conduct, or stunts to any public forum, including social media platforms, news periodicals, news networks, blogs, vlogs, or other media outlets.
- You will be mindful of your fellow workers and will communicate and behave with respect for their cultural, racial, gender, religious, or political identity. Harassment, aggression, or other threatening behavior is considered unsafe and will result in disciplinary action up to and including termination.
- You are expected to learn and follow company emergency action protocols in case of injury, fire, spill, natural disaster, or terrorism; you will do your part to keep exits clear at all times.
- We will inform you in the event of major workplace changes, new equipment, or process changes that could involve new health or safety hazards.
- If you are required to receive safety training, the company will compensate you for the time it takes; you are expected to make all reasonable efforts to attend training as required and scheduled.

⊶ POSITION-SPECIFIC SAFETY REQUIREMENTS

For specific job positions, there may be safety requirements that should be included in addition to the universal responsibilities. Position-specific safety requirements might include the following examples:

- You have a right to know about hazardous chemicals required for your work. The company maintains and makes available safety data sheets on all chemical products. You will receive any necessary training on the safe handling and storage of chemical products.
- When specific personal protective equipment (PPE) is required according to hazards inherent in the task, specified in a hazard assessment or standard operating procedure, or required by your direct supervisor, said PPE shall be worn, cleaned, and stowed properly. Unless otherwise stipulated, the company will provide all required PPE.
- Some ingredients in use at the brewery are known food allergens. If you have any known food allergies, it is your responsibility to inform your supervisor. If you are unable to work in the proximity of our ingredients, you may be reassigned.
- [If there is a PPE allowance.] The company will reimburse up to $[limit] annually for the purchase of hard-toed work shoes/boots.
- [If there is a prescription eyewear allowance.] The company will reimburse up to $[limit] every other year for the purchase of prescription safety glasses.
- You shall not operate any forklift or other powered lift without first obtaining forklift training certification satisfying the company's approval. If you are not certified, the company or its contractor will provide training.
- You shall adhere to workplace safety notices, floor and pipe markings, and designated pedestrian pathways.
- You will be working around confined spaces and shall not enter them unless you are specifically trained, equipped, and adhere to written procedures for a particular vessel.
- [If a respiratory protection program is in effect.] If required to wear a fitted respirator as part of this job position, the company will provide required fit testing, respirator fitness exam with doctor's opinion, respirators, filters, and training in proper respirator use and care; if you are unable to qualify for respirator usage, you may be reassigned. This does not apply to voluntary wearing of disposal dust masks.
- [If a hearing conservation program is in effect.] If required to wear hearing protection as a part of this job position, the company will provide audiometric testing, hearing protection device(s), and training in noise reduction and hearing protection use.

them at the senior management level will greatly assist in developing an effective safety culture. Culture is discussed in detail in chapter 20 and safety and health management systems in chapter 22.

Housekeeping

HOUSEKEEPING – Systemic Plant Condition
Occurs: Plant-wide
Outcomes: Slips, trips, falls, puncture injury; fire, explosion; infection, proliferation of vermin; exposure to unknown substances; equipment damage.

Housekeeping is a term borrowed from home life and applied to work. Good housekeeping implies a place free of clutter, dirt, and refuse. It means that everything is stowed in the place where it belongs and not left about as a hazard or where it can be damaged.

We call housekeeping a systemic safety issue because it is not related to just certain tasks. It is a concern in all parts of the business: manufacturing, warehousing, retail; inside and outdoors; in public areas and private ones. Every workplace task imaginable benefits from good housekeeping.

Good housekeeping is not busywork. It serves two important purposes. It reduces the opportunities for slips and trips, the accumulation of fire hazards, and complications during emergency actions. It also sets a tone, psychologically speaking, that the business takes organization seriously.

If orderliness is so important, it follows that the company values safety, employee well-being, quality, and efficiency.

HOUSEKEEPING TELLS A STORY

A brewer once told me about having an OSHA inspection. The compliance officer made a comment about the amount of clutter he saw. The inspector didn't cite the business for it, but the shame the brewer felt drove him to action. It took many long workdays and weekends to get things into good shape. Years later, the brewery now prides itself on its tidiness.

I have audited many manufacturing facilities. The first thing I pay attention to is the housekeeping. Good housekeeping usually signals good safety and environmental practices. Normally, a business with poor housekeeping cannot quickly clean the place for visitors, as the previous example illustrates. This is one reason why OSHA does not typically give advance notice of an inspection.

WALKING AND WORKING SURFACES

All the places we might put our feet during the workday are called walking and working surfaces, sometimes abbreviated to WWS. These surfaces could be ground level, they could be on an elevated brew deck, or they could be on the building's roof. Walking and working surfaces are given a detailed treatment in chapter 10. For the purposes of housekeeping, a core value should be maintaining floors, aisles, stairs, and ramps free of obstacles.

HOSES AND CORDS

Trips and falls are among the most common cause of injury in breweries. These can be caused by slippery surfaces or tripping over hoses or cords in the path of the person. Reducing the chance of these injuries comes down to accepting that it is better to squeegee the floor or put away the transfer hose between tasks than to just work around the hazards. Workers often argue that it is a waste of time to put something away that you'll need again in a little while, but how much time is lost with a trip to the emergency room or sick leave for a key employee?

TOOLS AND PARTS

Leaving portable tools lying about is a common housekeeping issue. This can result in tripping accidents as well as damage to the equipment. Small parts that are involved in a repair or new installation can be easily lost. Keeping these bits on a tray or in a can or bucket can improve job quality and workplace neatness.

Pay special attention to not leave tools on stairs. This is a common occurrence, probably because it avoids having to bend all the way to the floor to pick up the item. Tripping and falling on stairs can cause serious injury and even death.

Most breweries have some sort of maintenance area or tool bench. Keep these areas clean like the rest of the brewery. Workbenches and other flat surfaces have a way of collecting improperly stowed tools and small parts and fasteners. Use bins, drawers, shelving, and hooks on pegboard to manage all these small tools and parts. The same goes for areas where clamps, gaskets, and valves are stored for brewery and cellar use.

SANITARY CONSIDERATIONS

Keeping process equipment, contact surfaces, and bathrooms clean and regularly sanitized limits the risk of workers and customers being exposed to illness-causing microorganisms and residual chemicals. Clean production equipment has a direct benefit on your product quality.

Cleaning and sanitizing products are designed for particular applications. A one-size-fits-all approach to cleaning products usually results in some surfaces not being adequately cleaned.

In general, an effective cleaning operation has four influencing factors: chemical action, mechanical action, temperature, and time. Together, these form the "Sinner circle," named in 1959 after German detergent chemist, Dr. Herbert Sinner (Henkel Corporation n.d.). A principle of the Sinner circle is that when one pie slice is weaker, the others must make up for it; only with a full-strength circle is the cleaning considered effective. One way to remember the parts of the Sinner circle is the acronym CHAT: chemical, heat, agitation, time.

Let's say that you would like to substitute a sodium percarbonate granular cleaning product in place of strong caustic for clean-in-place (CIP) tank cleaning after fermentation. Since the chemical action of percarbonates is weaker than caustic for organic brewing deposits, you will need to compensate by increasing time and temperature, assuming the mechanical action of the CIP system is a constant. Time and temperature both have production costs. Increasing temperature could harm gaskets or hoses and may represent a higher scalding risk. The Sinner circle helps you think about everything that goes into cleaning when looking for product substitutions.

Cleaning and sanitizing occurs in a series of logical steps. The early steps remove dirt, films, stains—both organic and inorganic—and are followed by rinsing. If a final surface mostly free of microorganisms is desired, sanitizing chemicals can follow the cleaning and rinsing steps. Sanitizer effectiveness is greatly reduced when used on a surface that has not been pre-cleaned.

Chemists use the expression "like dissolves like." In other words, ionic cleaners such as acids and bases are *hydrophilic* (Gk: "water-loving") and remove water-based soils. Organic cleaners like citrus oil-based cleaners and organic solvents are *lipophilic* (Gk: "lipid[fat]-loving") and dissolve oily soils. Detergents are a class of cleaners in which the molecules are hydrophilic on one end and lipophilic on the other, allowing the detergent to homogenize emulsions of watery and oily soils.

Likewise, food preparation areas, glassware, and draught dispense lines need constant attention to reduce disease transmission and avoid product quality issues. In commercial kitchens, poor housekeeping can lead to slippery floors and can attract vermin. The buildup of grease from frying and range cooking can make firefighting especially challenging.

COMBUSTIBLE MATERIALS

Housekeeping is not limited to clutter, trip hazards, and obstacles. Things that can catch fire and be difficult to put out often track back to housekeeping issues. Breweries handle a lot of corrugated cardboard and paper packaging materials. Fiber materials should be consolidated and kept in safe locations until used or recycled. Examples of safe locations include outdoor covered storage and pallet racks or available floor space beneath sprinklers. Keep combustibles away from chemicals, moisture, sources of ignition, grain mills, electrical panels, and boilers or furnaces. Assuring good access for fire service personnel is critical, should these materials ever catch fire.

Grain dust is a classic housekeeping issue for breweries. Accumulations of grain dust can be slippery, cause quality problems, and, in the right situations, increase the chance for a grain dust explosion. Grain dust also attracts vermin. Because malt husks house natural spoilage bacteria, grain dust reaching your fermentation processes will likely affect product quality.

WHEN NOT TO PERFORM HOUSEKEEPING

There are situations where the method chosen as part of housekeeping may result in hazardous exposure. In the case of crystalline silica, which can be a constituent of diatomaceous earth (DE), the biggest hazard is inhaling silica suspended in the air. Silica exposure can cause a life-shortening respiratory disease called silicosis. If you intend to clean up after working with DE, a push broom would

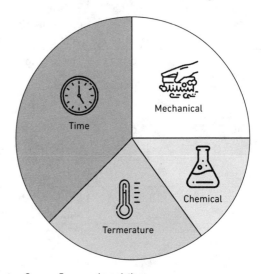

Figure 8.1. The Sinner circle shows the relative contribution of each cleaning factor. *Source: Brewers Association.*

Table 8.1 **Occupational Safety and Health Standards (29 C.F.R. Part 1910) that mention housekeeping**

Short title of standard	Title 29 citation	Key requirements
Exit routes, emergency action plans, and fire prevention plans	1910 Subpart E, Appendix (4)	It is the intent of this standard to assure that hazardous accumulations of combustible waste materials are controlled so that a fast-developing fire, rapid spread of toxic smoke, or an explosion will not occur.
Flammable liquids	1910.106(e)(9)	(i) General. Maintenance and operating practices shall be in accordance with established procedures which will tend to control leakage and prevent the accidental escape of flammable liquids. Spills shall be cleaned up promptly.
		(ii) Access. Adequate aisles shall be maintained for unobstructed movement of personnel and so that fire protection equipment can be brought to bear on any part of flammable liquid storage, use, or any unit physical operation.
		(iii) Waste and residue. Combustible waste material and residues in a building or unit operating area shall be kept to a minimum, stored in covered metal receptacles and disposed of daily.
		(iv) Clear zone. Ground area around buildings and operating areas shall be kept free of tall grass, weeds, trash, or other unnecessary combustible materials.
Sanitation	1910.141(a)(3)	(i) All places of employment shall be kept clean to the extent that the nature of the work allows.
		(ii) The floor of every workroom shall be maintained, so far as practicable, in a dry condition. Where wet processes are used, drainage shall be maintained and false floors, platforms, mats, or other dry standing places shall be provided, where practicable, or appropriate waterproof footgear shall be provided.
		(iii) To facilitate cleaning, every floor, working place, and passageway shall be kept free from protruding nails, splinters, loose boards, and unnecessary holes and openings.
Handling materials— general	1910.176(c)	Storage areas shall be kept free from accumulation of materials that constitute hazards from tripping, fire, explosion, or pest harborage. Vegetation control will be exercised when necessary.
Grain handling facilities[a]	1910.272(c)	The employer shall develop and implement a written housekeeping program that establishes the frequency and method(s) determined best to reduce accumulations of fugitive grain dust on ledges, floors, equipment, and other exposed surfaces.
Electrical work practices	1910.333(c)(9)	Where live parts present an electrical contact hazard, employees may not perform housekeeping duties at such close distances to the parts that there is a possibility of contact, unless adequate safeguards (such as insulating equipment or barriers) are provided.
Bloodborne pathogens	1910.1030(d)(4)	Employers shall ensure that the worksite is maintained in a clean and sanitary condition.

Table 8.1 **Occupational Safety and Health Standards (29 C.F.R. Part 1910) that mention housekeeping** (cont.)

Short title of standard	Title 29 citation	Key requirements
Respirable crystalline silica	1910.1053(h)	(1) The employer shall not allow dry sweeping or dry brushing where such activity could contribute to employee exposure to respirable crystalline silica unless wet sweeping, HEPA-filtered vacuuming or other methods that minimize the likelihood of exposure are not feasible.
		(2) The employer shall not allow compressed air to be used to clean clothing or surfaces where such activity could contribute to employee exposure to respirable crystalline silica.
Laboratory standards	1910.1450 (Appendix A)	Proper housekeeping includes appropriate labeling and storage of chemicals, safe and regular cleaning of the facility, and proper arrangement of laboratory equipment.

National consensus standards and recommendations from other professional organizations	
Organization	*Relevant code or standard*
National Fire Protection Association[b]	NFPA 91: *Standard for Exhaust Systems for Air Conveying of Vapors, Gases, Mists, and Particulate Solids.*
	NFPA 70®: *National Electrical Code®.*

[a] More detail in section on engulfment and grain handling in chapter 11 of this book.
[b] "Codes & Standards," National Fire Protection Association, accessed February 16, 2023, https://www.nfpa.org/Codes-and-Standards.

not be a good idea because you will put more silica into the air where it could be inhaled. In this case, you would look at other control methods, like gently washing the surface down or using a HEPA filter-equipped shop vacuum.

Cleaning in and around electrical panels, disconnects, and outlets can cause injury or fire. If grain dust, corrosion, or other fouling is observed inside distribution panels, circuit breaker panels, or disconnects, employ a qualified electrician to do the work. Infrared imaging devices can show overheated connections without direct circuit contact.

APPLICABLE STANDARDS

OSHA does not have a single standard dedicated to housekeeping, like they do with, say, hearing conservation. However, there are many standards that address the importance of housekeeping that can be applicable to breweries. These are summarized in table 8.1.

Emergency Planning and Signage

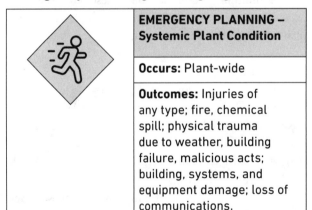

	EMERGENCY PLANNING – Systemic Plant Condition
	Occurs: Plant-wide
	Outcomes: Injuries of any type; fire, chemical spill; physical trauma due to weather, building failure, malicious acts; building, systems, and equipment damage; loss of communications.

EMERGENCY ACTION PLANNING

An effective response to any type of workplace emergency relies on four steps. To begin with, staff should consider what sorts of emergencies they believe could occur, and what sorts of outcomes could result. Essentially, this is using the hazard assessment (HA) process for a group of events that require immediate action. Since one cannot perform an HA in a timely and effective manner while an emergency is unfolding, assessment takes place in advance of potential emergencies. Once emergency possibilities are identified, they are prioritized according to their risk and the company's resources.

The next step is to make a plan on how to respond. The first part of planning is the pre-plan. This is where key resources, equipment, and staff are identified. Next is the operations or event plan. This is the stage where operational guidance is needed. In keeping with the HA process, the operations plan can be thought of as an SOP or a checklist. It will describe how the job is done—in this case how the response should occur—and will outline responsibilities for on-site personnel and also name emergency services, contractors, utility companies, and, potentially, media contacts. The final step in planning is to develop a recovery plan, that is, the process for normalizing everything after the incident. This step might include contacts for contractors or off-site facilities. Once an emergency plan is developed, it should be reviewed periodically to make sure the contact information is up to date and the planned procedures are still the best course of action.

Since OSHA's purview is only over employees, its regulations only apply to emergencies as they involve employees. Thinking more broadly, an emergency at a brewery could involve employees, contractors, customers, or the general public. Realizing the OSHA requirements are the minimum required, adding these other classes of affected persons will make for a more complete emergency plan without much additional effort.

Emergency Action Plan

The requirements under OSHA related to emergency action planning are straightforward and easy to accomplish. The required emergency action plan in 29 C.F.R. § 1910.38 is a written plan for workplaces with more than 10 employees. It may be unwritten and given orally with 10 or fewer employees.

The instructions for notifying emergency services, what exit routes to use, and where to assemble are most often included on a floor plan diagram placed in work areas. It should be clearly labeled with a title such as "Emergency Services and Evacuation Routes" or similar.

Employees with special roles need to be listed by name or position and best contact method. They may not be on the site when the emergency occurs and will need to be notified. Those special roles are anyone who needs to shut down critical systems, anyone qualified to provide rescue or render first aid, and anyone of a supervisory nature who can explain the emergency action plan to employees. At least a subset of employees must be trained in assisting others to evacuate.

EMERGENCY PLANNING FORMULA

Emergencies can include fires, explosions, chemical spills, injuries requiring medical treatment, adverse weather and natural disasters, site security, robbery, or terrorism. All of these benefit from the same formula for hazard analysis and planning:

- Assess possible emergencies
- Plan and document the response strategy
  - Pre-plan
  - Operations or event plan
  - Recovery plan
- Identify resources for each step of the response
  - Designate key persons and contact information
  - Identify non-company resources
- Conduct periodic plan review

A useful exercise is to take an example from the news. Maybe a tornado touched down or there was a major power blackout; even something like a train derailment or the local beach experiencing a red tide can be used. Get together your safety committee and go through the emergency planning process based on data from media coverage of the event. See what you could have done to limit the losses from the emergency. How could alarms, warnings, training, response equipment, and contractors have improved the outcome?

Once you have a feel for it, apply the same formula to emergencies you think could directly jeopardize the brewery. They will likely fall into classes of evacuation, fire and explosion, hazardous spill, injuries or illnesses, hostile persons/terrorism, and adverse weather or natural disasters.

Lastly, the employer must notify all employees of the emergency action plan and make sure employees understand it. This is most often done during new employee orientation. The emergency action plan can easily be included in the employee handbook.

Fire Prevention Plan

OSHA requirements for the fire prevention plan are found in 29 C.F.R. § 1910.39 and are closely analogous to the emergency action plan. Ten or fewer employees may receive the information orally; if you have more than 10 employees then the fire prevention plan must be written. As the name suggests, this plan is geared toward prevention of fires, with only limited information on how fires might be controlled.

The fire prevention plan will list major fire hazards, handling and storage procedures for hazardous materials, ignition sources and ignition control, and type(s) of fire protection equipment necessary to manage each major hazard. The plan lists procedures to control the accumulation of combustible materials and also procedures for maintaining controls of ignition sources.

Two special classes of employees are those responsible for maintaining fire prevention controls and fire handling equipment, and those who are responsible for controlling fuel sources. Both classes of employee must be listed by name or job title in the fire prevention plan.

Spill Prevention Plan

Chemical spills are another hazard that can be hard to avoid but whose outcomes are much improved by having a spill prevention plan in place. That plan should include necessary spill control and cleanup supplies and protective equipment. Employees should be trained in using equipment and supplies before a crisis exists.

OSHA and the US Environmental Protection Agency (EPA) both have several spill prevention or control standards but, technically speaking, they apply to hazard waste, oil materials, or specific toxic chemicals called out individually. Most breweries would be hard-pressed to find a specific citation. There is a general call for hazard control training in the construction standard in 29 C.F.R. § 1926.21(b), but an equivalent requirement is not evident in the general industry regulations.

A progressive safety program in a brewery would create a spill prevention plan equivalent to the emergency action and fire prevention plans. Alternatively, spill prevention and response procedures might be covered in SOPs for activities involving a risk of chemical spill. Spill training is a valuable investment that makes a difference.

Medical Services and First Aid

OSHA's rules on medical services and first aid (29 C.F.R. § 1910.151) are scant. The employer should take stock

SPILL TRAINING MAKES A DIFFERENCE

I have a sister who was a metallurgist in a testing laboratory. She and her colleagues had to work with hydrofluoric acid, which has some alarming properties. For one, it dissolves glass and has to be kept in Teflon® containers. For another, it reacts strongly with calcium, which is prevalent in the human body as it functions as a messenger between human nerve cells and is a major element in our skeleton and teeth.

If hydrofluoric acid is spilled on the skin, it may take hours to be detected, at which point it has penetrated tissue to the extent that nerve signals are short-circuited (this manifests as extreme pain) and, possibly, one's bones begin to erode. First aid treatment involves flooding the affected tissue with calcium gluconate, a calcium-rich substance that is injected subcutaneously over and over. This is a substance that requires very specific handling and specialized first aid supplies.

My sister asked me what she should do to prepare her lab for a hydrofluoric acid accident. I told her to assemble a hydrofluoric acid spill kit with a special neutralizing agent, specific heavy-duty PPE, and other cleanup supplies. Then I suggested that one slow work day, a Friday afternoon or whatever, she should spill some water on the floor and use the spill kit to practice responding to an incident. She did all that.

More than a year later, my sister called me to tell me about a new employee in the lab who was instructed to make up a dilution of hydrofluoric acid. He did this in a manner common for hydrochloric, nitric, or sulfuric acid. He used an empty two-liter glass acid bottle called a 9-pounder. Within 30 seconds of mixing the acid into water in the jug, the bottle dissolved and broke apart. Alerted by the white fumes given off, the chemist got the failing bottle over the sink in the nick of time. All the residues were cleaned up using the pre-made spill kit and no one suffered an exposure. Training works and practicing drills works, whether for evacuation, fire, hazardous chemical spills, natural disasters, or hostilities.

of the hazards in their workplace and their proximity to trained emergency medical services (EMS). In paragraph (a), the employer is required to have ready access to medical personnel for consultation. The presumption is that the employer identifies a hospital or occupational clinic where they can seek advice and send injured or ill employees.

Paragraph (b) states, that if such a facility is not "in near proximity," a person or persons will be trained in administering first aid and first aid supplies will be at the ready. The final paragraph, §1910.151(c), requires that suitable equipment for flushing the eyes or body be located "within the work area for immediate emergency use" when corrosive materials are in use.

For a brewery, the employer needs to determine if they are going to rely on nearby EMS or provide first aid training and supplies. Any brewery will require flushing facilities due to the frequent use of strong acids and bases in brewery and draught line cleaning operations.

PREPARING FOR EXTRAORDINARY EVENTS

Unusual and exceptional incidents are typically unexpected and hard to prepare for. While you may not be able to predict or prevent an event that is out of the ordinary, you can plan generic actions to improve the business's response in terms of both employee and customer safety and the protection of company assets.

Firstly, evaluate the workforce and find personnel resources. Who knows chemical safety or first aid and CPR? Who knows how to shut down the boiler or other key systems? Who has connections to municipality departments (e.g., police, fire, the mayor's office) or the bank? Identify these individuals and put them in a planning role based on their experience. You will be planning for evacuating employees and customers to a safer place, while also obtaining emergency assistance and possibly contractor support.

Your business's occupancy permit will be based on meeting minimum standards for exit signs and exits, emergency lighting, and fire protection, but it likely ends there. Beyond those, planning for unusual events is left to the business. Emergency alarms, exit routes, and rallying points should be determined and, when there is not an emergency, practiced periodically. Conduct fire drills and define sheltering to be used in cases of adverse weather or persons with malicious intent.

Importantly, practicing drills helps identify opportunities for improving your emergency plan. Expect things to go wrong during drills and use these mistakes to improve response planning. Emergency response staff from local police, fire, medical, and social services are typically very happy to help with response planning and even drills.

There is often a need for extra private resources in the case of extraordinary events. A business might need security personnel, information technology experts, tradespeople and laborers, construction supplies, and emergency equipment.

Building Infrastructure

Consider uncommon, catastrophic events that could happen to the workplace itself. These include fire, explosion, or structural failure due to something within the building. Wiring not up to code could overheat and start a fire. A gas leak goes unnoticed over the weekend, contacts a source of ignition, and the building explodes. Workers stack heavy objects on the walk-in cooler ceiling and it caves in.

You can never reduce the chance of these incidents happening to absolute zero, but they will become much less likely if proper building, fire, and electrical codes are followed. All utilities work—electrical, mechanical, HVAC, structural, and plumbing—needs to be designed and installed by contractors who understand commercial requirements. Do not use a residential electrician because he's your drinking buddy and he'll wire your brewery in exchange for beer.

In the craft beverage industry, many businesses are situated in older buildings. The structures were built to different standards that are often not compliant with current standards. The utilities and structure should be upgraded to current design standards, even if the municipality grandfathers the premises and says it does not require upgrading.

Natural Disasters

There are many sorts of natural disasters. It is a long list, including wildfires, tornados, hurricanes, monsoons, earthquakes, volcanos, sinkholes, floods, mudslides, blizzards, avalanches, and lightning strikes. Most of these can easily result in utility outages. A disease epidemic or prevalence of a disease-carrying vector also fall into the natural disaster category. Examples include the COVID-19 pandemic and mosquito-borne West Nile virus.

The hazard controls we use for natural disasters are usually administrative controls and engineering controls. Planning and preparing for emergency action, documenting and practicing the plan, and training the workforce to execute the plan is the administrative controls part.

Depending on the disaster one is planning for, pre-planning meetings with local emergency services will make their critical response more effective. Contracts with service providers can be put into place. These might include contracts for boarding up the building, making emergency repairs, or performing post-disaster cleanup. Plans can also be made with equipment rental businesses for pumps, generators, and waste bins.

Enacting protective measures is the engineering controls part. Before a disaster, these might include installing backup power, water, or gas supplies. Engineering controls during or after an event could include structural stabilization, building dikes, flood water pumping, volcanic ash removal, or, in the case of an epidemic, ventilation and vaccination.

Hostile Persons

There is always a chance of encountering a hostile person who intends to act on their feelings. Essentially, these are individuals or groups with malevolent intent that affect the business, staff, or customers. In the context of a brewery or taproom, this could involve a customer or former employee who is intoxicated, angry, or suffering from an untreated mental illness. Such an encounter could include weapons and a premeditated intent to do harm.

Keeping an incident log that describes the event or events, the participants, vehicle plate numbers, and actions taken by the business can help to identify problematic individuals, while providing a level of liability control for the business. Staff training can be very helpful for managing such incidents. Local police departments are often happy to conduct this sort of training for free. The training should include when to engage and when to dial emergency services, de-escalation techniques, sheltering and security, alarms, exits, and rallying points.

An act of terrorism or war is also the result of persons acting with hostile intent, albeit these are more like random hazardous events. Nevertheless, basic emergency planning is well-advised. Preventing these kinds of events is next to impossible, and insurance policies will have fine print about excluding most situations from coverage. We can still plan for those most likely to happen according to our local situation, and conduct emergency planning and evacuation training and prepare necessary procedures, equipment, and supplies ahead of time.

HAZARD SIGNAGE

Hazard identification is used both preventively to suggest safe work practices, and protectively to warn of existing hazards. It takes many forms: illuminated exit signs, workplace hazard signs, placards, floor and pipe markings, and posters. Collectively, we can call this *signage*.

Some types of signage are legally required and have exact specifications of color and dimension, while others can have any design the business feels will be effective. In either case, signs should be consistent, visible, and durable according to their location.

Some workplaces put little to no thought into signs beyond what the fire marshal requires for exits or occupancy. Others have gone overboard and there are signs everywhere. Too many signs is referred to as "sign pollution." People begin to ignore signs when they become ubiquitous. What follows here are the recommended bare essentials.

Emergency Exits

The local fire marshal or building codes superintendent will make sure that exits, fire extinguishers and prevention systems, and first aid stations, showers, and eyewashes have been identified with acceptable signs. Additional signs can be required on a case-by-case basis for bathrooms, automated external defibrillators (AEDs), tornado shelters, and the like.

OSHA stipulates in 29 C.F.R. § 1910.37(b) that each exit route be adequately lighted. This means that an employee with normal vision could see their way along the exit route. Each exit must be clearly marked with an exit sign. This sign must be back lighted or phosphorescent to give off light sufficient to be seen in the dark.

If a door or passage along an exit route could be mistaken for an exit, a sign indicating that it is not an exit should be installed. If the door goes into a specific area, like a restroom or a closet, it is best to use a sign to indicate the room's purpose.

Hazard Areas

OSHA prescribes a number of different accident prevention signs and tags. These standardized administrative controls identify hazards and, in some cases, provide instruction on safe behaviors. These rules are found in 29 C.F.R. § 1910.145. OSHA also accepts signs that conform with the ANSI/NEMA Z535 consensus standards.[1] The ISO standard *Graphical symbols — Safety colours and*

---

[1]   ANSI, American National Standards Institute; NEMA, National Electrical Manufacturers Association

safety signs (ISO 3864) provides a uniform international consensus standard for signage.

Signs and tags have specific coloration and a signal word. The words "danger," "warning," and "caution" are signal words. Signs and tags may also have a major message, which can be either text or a pictograph. They convey specific meanings and employers are required to instruct employees on the meaning and required actions associated with these signs.

The danger sign is colored in red and black on a white background. It conveys immediate danger and that special precautions are necessary. The caution sign is written in black on a yellow background and warns against potential hazards or to caution against unsafe practices. In between them in severity, although not prescribed in the standard, is the warning sign, which is colored in black on an orange background. Like the danger sign, it conveys a hazard that could result in serious injury or harm, though not necessarily immediate.

OSHA also defines the instruction sign. This type of sign is used where there is a need for general instructions and suggestions relative to safety measures. A green panel overlays a white background area with black writing. Another type of sign that has grown out of these standard formats is the notice sign. This sign is intended to publicize a policy, such as "all visitors must check in" or "authorized personnel only." The notice sign looks a lot like the instruction sign but uses a blue panel on a white background.

Fire extinguisher signs, in particular, do not have a specified format in § 1910.145. However, it is recommended they use red coloring, provide the words "Fire Extinguisher," and any pictograph or major message that would be helpful.

Process Signage and Floor Markings

It is advisable to label process piping, containers, and traffic lanes with hazard information. Fixed process pipes in a brewery commonly convey steam, hot water, chemical cleaning solutions, or compressed air or other gases. Process labeling can also indicate the direction of flow if this is important to know.

In the case of surface contact burn hazards, such as with steam pipes or brew kettles, it may be helpful to indicate "burn hazard" clearly. Signage should complement the use of pipe insulation, machine guarding from hot surfaces, and other engineering controls.

Figure 8.2. Proper safety signage uses similar components to emphasize the safety message.

Traffic control markings can be used to delineate forklift or pedestrian traffic lanes. Bollards, mirrors, and warning signs placed on walls will enhance forklift safety, protecting both people and equipment. Electrical panels, drench showers, and eye wash stations are emergency systems that need to be kept free of all obstacles. A combination of signage and floor markings is often used. With showers and eye washes, remember the victim seeking to use them may have limited vision or be in distress, so visibility should be a key consideration.

Labor Law Poster and Public Safety Postings

In 29 C.F.R. § 1903.2(a)(1) OSHA requires a workplace poster entitled "Job Safety and Health: It's the Law!" It can be downloaded from https://www.osha.gov for free in many languages. The poster informs workers of their general rights to a safe workplace, access to their medical records, and protection from whistleblowing, among other rights. In jurisdictions that run a state plan, the state agency may have a modified version

Figure 8.3. Examples of common signs and tags consistent with OSHA standards.

of this poster that covers the same information plus state-specific rights and requirements. The poster must be posted where employee notices are typically put and cannot be covered or defaced.

For a commercial space to be opened to the public, the local public safety authority, usually a fire marshal or municipal public safety official, will issue an occupancy permit. This usually takes the form of a certificate that must be displayed on an interior wall of the premises where it can be readily found. A business will be given a maximum occupancy load based on square footage of assembly space, the number and type of occupiable spaces (standing, bar seating, table seating), the bathroom facilities, and the type of permitted assembly. It is up to the business to assure that the occupancy is not exceeded or else it faces disciplinary consequences from the municipality.

Exceeding occupancy can lead to structural instability, sprinkler system inadequacy, or injuries during evacuation. Boilers and sprinkler system standpipes are normally required to be inspected by a licensed authority and each will have a signed certificate or tag to be mounted on or beside the equipment.

Beyond safety notices, there can be requirements to post a liquor license, food service establishment permit, brewing permit, business hours, sales tax certificate of authority, and any local ordinances.

APPLICABLE STANDARDS

OSHA has numerous standards that apply to emergencies. The four most important are emergency action planning, fire prevention, first aid, and signage. These are summarized in table 8.2. Chemical emergencies are discussed in chapter 16.

Table 8.2 **Standards pertaining to emergency planning**

Short title of standard	Title 29 citation	Key requirements
Emergency action plans	1910.38(b)	Written plan required for more than 10 employees; orally communicated plan allowed for 10 or fewer employees
	1910.38(c)	Plan must include procedures for reporting a fire or other emergency; procedures for emergency evacuation; procedures to be followed by employees who remain to operate critical plant operations before they evacuate; procedures to account for all employees after evacuation; procedures to be followed by employees performing rescue or medical duties; and the name or job title of every employee who may be contacted by employees who need more information about the plan or an explanation of their duties under the plan.
	1910.38(d)	An employer must have and maintain an employee alarm system.
	1910.38(e)	Employer must designate and train employees to assist in a safe and orderly evacuation of other employees.
	1910 Subpart E, Appendix (4)	It is the intent of this standard to assure that hazardous accumulations of combustible waste materials are controlled so that a fast-developing fire, rapid spread of toxic smoke, or an explosion will not occur.
Fire prevention plan	1910.39(b)	Written plan required for more than 10 employees; orally communicated plan allowed for 10 or fewer employees.
	1910.39(c)	Plan must include a list of all major fire hazards, ignition sources, and control equipment; accumulation controls; procedures for maintenance and inspection of safeguards; list employees responsible for maintaining equipment to prevent or control sources of ignition or fires; list of employees responsible for the control of fuel source hazards.

Table 8.2 **Standards pertaining to emergency planning** (cont.)

Emergency exits	1910.37(b)	Lighting and marking must be adequate.
	1910.37(d)	Exit routes maintained during construction.
	1910.37(e)	Must have a distinctive, operable employee alarm system, unless employees can detect fire or emergencies without one. See also 29 C.F.R. § 1910.165.
First aid	1910.151(a)	Ensure the ready availability of medical personnel
	1910.151(b)	Either have an infirmary, clinic, or hospital in near proximity to the workplace, or a person or persons adequately trained to render first aid; adequate first aid supplies readily available.
Accident prevention	1910.145	Specifications for danger, warning, caution, instruction signs and tags.

National consensus standards and recommendations from other professional organizations	
Organization	*Relevant code or standard*
National Fire Protection Association[a]	NFPA 101®: *Life Safety Code*®. 2021 ed. Methods to protect people based on building construction, protection, and occupancy features that minimize the effects of fire and related hazards.
	NFPA 13: *Standard for the Installation of Sprinkler Systems*. 2022 ed. Occupancy classes for design, installation, and water supply requirements for sprinkler systems.
American National Standards Institute[b]	ANSI Z535 Series. Consensus standards for colors, signs, symbols, labels, tags, barricade tapes.
International Organization for Standardization[c]	ISO 3864: *Safety colours and safety signs* International consensus standard for colors, graphical symbols for signs, labels and markings.

[a] "Codes & Standards," National Fire Protection Association, accessed February 16, 2023, https://www.nfpa.org/Codes-and-Standards.
[b] American National Standards Institute [homepage], accessed February 16, 2023, https://ansi.org.
[c] "Standards," International Organization for Standardization, accessed February 16, 2023, https://www.iso.org/standards.html.

Repair and Maintenance

REPAIR AND MAINTENANCE – Systemic Plant Condition	
Occurs: Plant-wide	**Outcomes:** Injuries of any type; fire, chemical spill; physical trauma due to equipment or building failure; building, systems, equipment damage; loss of communications.

GENERAL

Safety includes avoiding damage to the facility and its equipment and mechanical systems. Unsafe equipment can lead to worker injury or illness. The failure of equipment itself is bad for business due to work shutdowns, equipment repair, or product damage.

It can be estimated that, for every six incidents of equipment or property damage in the workplace, we can expect a worker injury (Phimister et al. 2003, 446). We can learn two important things from this: where there is equipment damage there can also be worker injury, and that damage to property occurs at a higher rate than injuries.

Repair and maintenance of equipment is essential for creating a safe workplace. There are many areas where OSHA specifically calls out repair and maintenance but many other areas where the question of maintenance is left to the employer.

BUILDING SYSTEMS

The building system can be thought of as the building structure itself and every piece of equipment that is either built-in (permanent) or is a component of operations. The building structure itself requires maintenance, as does everything inside it. The following are examples of building structures and built-in components:

- Floors, walls, windows, doors, and roof(s)
- Utility service connections to the outside
- HVAC, plumbing, electrical, and natural gas systems
- Fixed and portable fire protection systems, alarms, and security systems
- Boiler, compressor, hot water heater, and refrigeration systems
- Brewing equipment, tankage, packaging line, and coolers
- Materials storage systems (e.g., silos, super-sacks, totes, pallet racks)
- Materials handling equipment
- Draught beer serving systems

- Shop and maintenance operations
- Laboratory areas
- Office areas
- Retail areas
- Outdoor areas and grounds (separate from building, but still requiring maintenance)

There are two main concerns pertaining to repair and maintenance. First, keeping equipment and systems in good working order improves safety, quality, production up-time, and conserves energy. Second, OSHA has a substantial body of regulations requiring regular inspection, maintenance, and repair, and necessitating documentation of the same.

A brewery with a good maintenance program will have developed schedules for routine inspection, schedules for preventive maintenance, and will keep on hand expected replacement parts and tools to shorten the turnaround on repairs. As a matter of policy, damaged equipment is taken out of service upon the first signs of weakened integrity, excessive wear, physical damage, or obsolescence, or when scheduled for preventive maintenance.

Factors that go into determining the frequency and type of inspections and maintenance include both technical and human factors. The cost economy of preventive maintenance is maximized when the frequency is high enough to repair systems before they fail, but not so often as to waste labor or process delays with inspections. One has to select meaningful measurements that will indicate a maintenance need. Sometimes this can be scheduled according to the manufacturer's instructions. For example, nylon parts in a canning line might be replaced every 100,000 cans, because wear is expected to get critical at around 120,000 cans. Alternatively, the measure can simply be by a time estimate, say, weekly, monthly, or annually.

The human factors include whether management has expressed its support for preventive maintenance and provided adequate time in the workday schedule to accomplish inspections. In an effective safety culture,

Table 8.3 **Suggested inspection and maintenance frequency for a brewery**

System/Component	Inspection/Maintenance and Frequency
Doors, windows, locks	Inspect daily as a matter of opening and closing duties
Roofing, building fabric	Clear debris, check for damage; record quarterly or annually
Gutters, drains	Clear debris, check for damage; record quarterly or annually
Utility connections	Inspect/record quarterly or annually
Outside systems[a]	Inspect/record weekly or monthly
Video security	Check proper function daily
Grounds	Trim trees/shrubs as needed; mow, rake, sweep, shovel, plow, as needed; use chemical fertilizers, pesticides and de-icers responsibly
Ingredient warehousing	Integrated pest control monthly
Electrical, distribution	Keep breaker panels and disconnects easily accessible; check daily
Electrical, point of use	Inspect/repair any damaged housings, receptacles; inspect quarterly
HVAC system	Inspect ducts for dust buildup, check fittings, wash cooling fins as needed; semi-annually or annually
Air conditioning unit	Preventive service; semi-annually or annually
Lighting and exit signs	Inspect quarterly; keep spare lamps on hand; replace bulbs safely if working at height
Waste plumbing[b]	Maintain as needed to keep traps clear and wetted
Eye wash or drench shower	Test quarterly and maintain records
Portable extinguishers	Inspect quarterly; licensed certification annually
Fixed fire prevention systems	Inspect quarterly; licensed certification according to jurisdiction
Fire alarms, other alarms	Test quarterly
Exit signs, emergency lighting	Inspect and test quarterly
First aid supplies	Restock monthly or quarterly
Glycol chiller	Inspect glycol level monthly; wash cooling fins as needed; keep spare solenoid valves and thermistors on hand
Power-pack chiller	Inspect glycol level and temperature monthly; replace pump head as needed
Boiler	Blowdown daily or weekly; add treatment according to field test results; inspect quarterly; licensed certification according to jurisdiction
Water treatment system	Inspect, backflush, and/or restock consumables according to manufacturer's recommendations
Air compressor	Inspect tank pressure and secondary gauge (if present), bleed condensate, empty oil trap weekly or monthly
Bulk gas systems	Inspect tank pressure and secondary gauge (if present), bleed condensate or trap, if present
Draught beer system	Clean with caustic every two weeks; clean with acid quarterly
PPE, chemical gloves	Inspect with each use; retire from use if damaged
PPE, reusable gear[c]	Clean, inspect with each use; retire from use if damaged; mark for repair or replacement
Powered industrial trucks	Inspection every shift
Tools, power cords, ladders	Inspect on each use; retire from use if damaged; mark for repair or replacement

[a] Outside systems can include chillers, transformers, grain silos, bulk gases, fixed ladders, security fences, and lighting.

[b] Waste plumbing can include toilets, sinks, floor drains, trench drains, floor integrity; if there is a kitchen, then a grease trap too.

[c] Reusable PPE can include face shields, heavy gloves, aprons, chemical boots, respirators, hearing protection, and fall protection harnesses and lanyards.

preventive maintenance is the normal state of things. In a less robust safety culture, repairs are made on a reactionary basis when equipment fails.

Early on, every brewery should make a list of systems and components that will benefit from regularly scheduled inspection and/or maintenance. Whether frequencies are established by manufacturers or standards organizations, or even just based on a best guess, ultimately, it is up to the end user to establish a schedule. Table 8.3 lists examples of the sorts of actions to be taken and their suggested frequency.

APPLICABLE STANDARDS

OSHA recognizes that maintenance is a necessary component of almost every safety and health program or IIPP. Consequently, there are a great number of Title 29 standards listed in table 8.4. For the most part, the details and scheduling for inspection and maintenance is left to the employer.

Table 8.4 Occupational Safety and Health Standards (29 C.F.R. Part 1910) with repair and maintenance requirements

Short title of standard	Title 29 citation	Key requirements
Walking/working	1910.22	Inspect and maintain all walking and working surfaces
Emergency exits	1910.37	Exit routes maintained during construction, repairs, and alterations
Fire prevention plan	1910.39	Equipment in the fire prevention plan must be maintained; see also Appendix to 29 C.F.R. Part 1910 Subpart E
Powered platforms	1910.66	Equipment to be maintained and inspected
Manlifts	1910.68	Equipment to be maintained and inspected
Ventilation	1910.94	Equipment to be maintained and inspected
PPE	1910.132	Inspect, maintain, and clean PPE
Respirators	1910.134	Training required on inspection, maintenance, and cleaning of respirators
Lockout/tagout (LOTO)	1910.147	LOTO to be used when servicing or maintaining machines and equipment in which unexpected energization or start-up could occur
Portable fire extinguishers	1910.157	Extinguishers to be regularly inspected and maintained; see also Appendix A to 29 C.F.R. Part 1910 Subpart L §§ 1910.157–.165)
Employee alarm systems	1910.165	Alarms to be regularly tested and maintained
Fixed fire suppression equipment	1910.159, .160	Maintain all fixed fire systems, e.g., sprinklers and kitchen hood systems, in an operable condition; see also Appendix B to 29 C.F.R. Part 1910 Subpart L §§ 1910.157–.165
Fire detection systems	1910.164	Maintain all systems in an operable condition
Powered industrial trucks	1910.178	Includes forklifts; equipment to be maintained and inspected
Hand and portable powered tools	1910.243, .244	Maintain all tools in an operable condition and with associated guarding
Resistance welding	1910.255	Inspect, maintain, and keep certification record of equipment
Grain handling facilities	1910.272	Training required to cover preventive maintenance and lockout/tagout procedures
Electric equipment	1910.306	Required clearances for electrical equipment maintenance

CAUTION

DUST MASK
MUST BE
WORN

SmartSign.com • 800-952-1457 • 52-0721

9

PERSONAL PROTECTIVE EQUIPMENT

While it is often said that personal protective equipment (PPE) is the last line of defense, it is nevertheless still very important. When PPE fails there is usually no other control between the hazard and our body, resulting in harmful exposure and injury. Selecting the best PPE, and regularly inspecting, cleaning, and replacing it, are essential aspects of this form of hazard control that are often overlooked.

Personal Protective Equipment

PERSONAL PROTECTIVE EQUIPMENT (PPE) – Chemical, thermal, linear, rotary, projectile, sonic	
Occurs: Plant-wide	
Outcomes: Dermal and subdermal chemical or thermal burns, lacerations; injury to eyes, head, hands, feet, hearing; electric shock, electrocution	

GENERAL REQUIREMENTS

Selection of PPE is best done by considering the expected hazards and the human factors of use. Hazards might be chemical, physical, electrical, or biological. Human factors include visibility, dexterity, comfort, ease of use, and even fashion. The employer is responsible for providing most PPE, as well as giving training on its proper use.

While some PPE is disposable, much of it is designed to be worn repeatedly. Non-disposable PPE should be inspected before and after each use because PPE takes a beating and certainly wears out. This wear can be seen as soil buildup, cracking, splitting, broken seams, discoloration, changes in texture or flexibility, or other evidence of deterioration. If eye protection or face shields become scratched or fogged, they should be replaced. Gloves, boots, aprons, and hearing protectors that show signs of decay or cannot be effectively cleaned need replacing.

EYE PROTECTION

The eyes are delicate anatomical structures and can easily be injured by chemical splashes, irritating dusts or mists, flying objects, biological agents, and heat, ultraviolet, and infrared energy. Many consider sight to be the most important of the five senses and the one that is most disabling to live without. Even so, the National Institute for Occupational Safety and Health (NIOSH) estimates there are 2,000 eye injuries requiring medical treatment daily in the US, with about 5% of those injuries leading to lost work days (National Institute for Occupational Safety and Health 2013). It is estimated that over 90% of these workplace injuries could be reduced in severity or prevented by simply wearing appropriate eye protection (American Optometric Association n.d.).

Understanding the energy of eye and face hazards informs how we choose protective equipment. Selecting

PPE is based on understanding the mode of the hazard one is trying to avoid. The primary types of damaging forces are chemical, airborne irritants, flying objects, blunt force trauma, and optical radiation.

Under the general requirements given in 29 C.F.R. § 1910.133(a), OSHA requires the use of "appropriate eye or face protection," meaning the employer needs to conduct a hazard assessment and make a determination. Side shields are required if there is a risk of flying objects. Workers requiring prescription eyewear can either be fitted with prescription safety eyewear or wear protection over their regular glasses, so long as the function of the prescription lenses is not affected. The minimum level of protective shade for welding goggles or masks is also detailed.

Chemical Exposure Protection

Acids and bases are common in the brewery, notably in wort production, water treatment, and cleaning operations. These chemical agents, particularly acids with a pH less than 4 or bases with a pH greater than 10, will combine with moisture in the eyes to create a corrosive mixture that will damage eye tissue (Lim, Ah-Kee, and Collins 2014). Bases are generally worse than acids, penetrate further into tissue, and bring about an immune response that decomposes eye proteins as an undesired side effect. These injuries can be very painful and result in temporary or permanent vision loss. Lasting damage can include a hazy or opaque cornea, as well as vascular and epithelial injury (Singh et al. 2013).

The simplest PPE for chemical splashes is either chemical splash goggles or safety glasses with side splash shields. Safety glasses are not entirely effective on their own and may only stop some of the splashed material. Combine safety glasses with face shields to increase effectiveness and protection against splashed chemicals. Goggles provide much better coverage but are prone to fogging up and being worn too tightly, which can result in headaches. Beyond PPE, having eye wash facilities close by the work area is required when corrosives are in use.

Foreign Matter and Physical Trauma

Physical injury to the eyeball can occur from the presence of dust, smoke, welding fume, or flying objects from tools or machinery. Grain milling dust and flying objects from pressurized or motorized systems are two brewery activities that are known to cause eye injuries.

SAVED BY THE LENS

Wearing glasses since childhood has worked out well for me. Because of some early experiences, I chose to never have vision correction surgery or wear contact lenses.

In junior high during shop class, a grinding wheel came apart and shot ceramic bits into my face. My glasses saved my eyesight. In college, I was responsible for firing a black powder cannon whenever the football team scored. One time, when I went to light the cannon, a rogue wind gusted and blew the flame right down the vent, bypassing the fuse. The cannon fired with my face at point blank distance above the vent. I spent the rest of the day in the emergency department having cinders removed from my face. Other cinders were embedded in my glasses, but I was fine with that.

Later in life, I worked at an environmental testing lab. A coworker put a 1-liter flask on a Bunsen burner just before the mid-morning coffee break. It contained a water sample and concentrated sulfuric acid. The coworker must have been hungry for his daily bagel and he forgot to swirl the liquids to mix them. Being denser than water and having a much higher boiling point, the concentrated acid settled into a bottom layer beneath the water and gradually became superheated. My coworker and I got back from our break just in time for the flask to explode and rain hot sulfuric acid on us both. We got under the shower and stripped down to our safety glasses.

Working around other people can expose you to hazards you hadn't thought about and that can affect you in the blink of an eye. That is why it is a good idea to wear safety glasses everywhere and anytime in the production process.

Foreign matter in the eye can scratch the surface of the eyeball in what is called corneal abrasion. These are usually painful injuries that cause prolific tears to be produced. Remove contact lenses right away. Do not rub the eyes, but rinse them with a buffered eyewash solution, if available, or fresh water. Lifting and pulling the upper lid over the surface of the eye sometimes frees foreign matter, as does tilting towards the nose and crying to free it.

If the foreign matter is safely removed, either by crying or by intentional flushing of the eye, corneal abrasion often heals within a few days. It is still a good policy to be examined by an ophthalmologist to assure all contaminants are gone.

The foreign matter could be biological in nature. Infectious materials, like bodily fluids, can transmit disease-causing bacteria, viruses, protozoa, or fungi to the ocular fluid and then into the body. Allergens can illicit an immune response called allergic conjunctivitis in the eyes, resulting in red, itchy eyes, inflamed capillaries, and swelling from fluid retention. Some allergens can cause allergic sensitivity that persists long after exposure. If working around bodily fluids, such as when cleaning up after a bloody injury or a restroom mess, avoid eye contamination by wearing eye or face covering and using good personal hygiene practices.

Another type of physical eye injury occurs with trauma from forceful impact with an object. This could result in puncturing of the eyeball, retinal detachment, or fracture and caving in of the eye socket (orbital fracture). In the brewery, these injuries could occur from a fall from height, vehicular impact, failure of a pressurized system, or a physical assault. Eye puncture or orbital fracture require immediate attention by a qualified medical practitioner. Retinal detachments can be caused by a blow to the head, the detachment occurring at the back of the eyeball. A retinal detachment may be suspected if the victim experiences flashes of light, floaters, or loss of vision. The presence of retinal detachment should be assessed by a qualified medical practitioner if such symptoms occur after an accident.

Eye Strain

Intense or prolonged visual activities can injure the eyes. Eye strain can result from long periods of looking at computer screens or focusing on fine manual tasks. Symptoms of eye strain include itchy or watery eyes, blurry vision, eye muscle soreness, and headache. Building in task breaks and optimizing workstation lighting can help reduce eye strain. These types of eye injuries do not usually have a PPE hazard control.

Hazardous Light

Concentrated light energy can be found in some breweries. Laser equipment in breweries is used for date coding, cutting films, scoring boxes, and kiss-cutting labels. Laser fill monitors perform poorly with foaming beverages but may be used in flat beverage bottling. Lasers in packaging applications range from ultraviolet (UV) light to visible light to infrared (IR) and beyond. If the lasers are set up for marking or cutting plastics, fiber, or foils, they will usually be CO_2 lasers that produce light in the long IR range.

Laser equipment can cause permanent eye damage within milliseconds. Knowing the wavelength of the laser system allows the user to specify the type of filtering safety lenses required to protect the eyes. Consult the equipment manufacturer to understand the class of laser and its wavelength of operation.

Aside from lasers, UV light hazards in a brewery can be found in water disinfection systems and ink curing. Ultraviolet light must not be admitted to the eye.

Welding emits hazardous UV light in addition to visible light. When a welding arc is viewed without protection it can result in a condition known as arc eye or welder's flash. Arc eye includes simultaneous inflammation of the cornea and conjunctiva (a condition known as keratoconjunctivitis) and reddening of the skin (erythema) and can be thought of as akin to sunburn of the eyeball.

Welding should never be performed without an appropriate face shield, which filters light and protects the face from splattered molten metal. Many modern welding helmets now have auto-darkening filters for which the response time and degree of shading can be pre-set.

Vision Correction

For workers who require vision correction, prescription safety glasses are widely available and can be configured with nearly any lens type. Lenses can be made of tempered glass or high impact plastic. Prescription safety glasses have the benefit of high compliance, since most people who wear glasses will be wearing them during work anyway.

HEAD PROTECTION

Whenever there is the potential to be hit with a falling object or contact a live electrical conductor, head protection is required by OSHA. Hard hats should meet or exceed the ANSI Z89.1 consensus standard (29 C.F.R. § 1910.135(b)). Metal hard hats should never be worn around electrical hazards. Qualified head protection will carry a label or embossed mark stating it meets ANSI Z89.1 requirements.

The wearing of hard hats is not routine in breweries, but hard hats should be available to workers and worn for any task where falling objects could occur. Examples include for support personnel who are stabilizing ladders or handing up supplies to an elevated worker. Another aspect of head protection is to cover or restrain long hair to prevent it from getting caught in machinery.

Hard hats have an internal suspension system that keeps the hard shell out of direct contact with the skull. In the event of an impact, the hard shell absorbs most of the energy and the suspension keeps the shell from transferring most of that energy to the skull. Hard hats that have been struck should be thoroughly inspected and replaced if there is any sign of structural damage to the shell.

Bump caps are lightweight head protection that may not meet the ANSI standard. They have little to no suspension and may have only a layer of foam cushioning. They are commonly used by workers who are in short workspaces, such as beneath a brew deck, to minimize the effects of bumping one's head on low pipes and such. Bump caps now come as an insert that can be installed inside a baseball cap. They are unobtrusive and comfortable.

FOOT AND LEG PROTECTION

Whenever there is a danger of foot injury due to falling or rolling objects, objects that can pierce the sole, or an electric shock hazard, the employer must ensure that the worker's feet are protected. If other controls do not adequately protect the worker, appropriate work shoes or boots are used. Notice that the employer is not required to provide such footwear, only ensure that it is in use. However, many employers provide a footwear stipend for employees.

The applicable consensus standards for occupational footwear are either ANSI Z41-1991 or ASTM F2412-05 and F2413-05. However, any footwear that the employer demonstrates meets or exceeds these ANSI or ASTM standards is allowable (29 C.F.R. § 1910.136(b)).

Foot protection can be thought of as an equipment system designed to meet the needs of comfort,

activity, and ergonomics combined with the necessity for providing protection from other workplace hazards. These additional hazards include contact with corrosive or poisonous materials, risk of fire or hazardous debris, slippery surfaces, puncture hazards, and hot substances. Protection can be built entirely into the footwear or it can be an accessory such as toe guards, rubber pullovers, gaiters, or leggings.

Common foot hazards in breweries include slippery surfaces on floors, ladders, and stairs; impacts from falling objects; chemical exposure; and burns from boiling wort or other hot liquids. Workers on the brew deck and cellar typically wear over-the-calf polyvinyl chloride (PVC) rubber boots with reinforced toes and shanks. With the pantleg drawn over the outside, the boot protects against slips, falling objects, chemical burns, and hot wort. If the pantleg is tucked inside the boot, liquids can be conveyed into the boot, dramatically worsening the injury.

Another common approach is to wear close-toed approved work shoes or boots that provide physical protection but not chemical or thermal protection. This type of footwear is useful in warehousing, dryside packaging, kitchen, and front of house. Caution against using them when working around harsh chemicals and thermal hazards. Open-toed footwear should not be permitted in any production or warehousing area.

For legwear, if workers are exposed to manual material handling or chemical or thermal hazards, long work pants made of synthetic fibers or a synthetic-cotton blend are best. Avoid denim jeans, as they can be hot and retain moisture. Shorts, skirts, or kilts should not be worn anywhere in manufacturing where scrapes, cuts, impacts, burns, or chemical contact are even remotely possible.

HAND PROTECTION

Nowhere in PPE standards is the employer given more flexibility than in hand protection. Due to the wide variety of hazards, manual dexterity requirements, length of use, and materials of construction, there is a nearly infinite variety of gloves on the market. The OSHA rules on PPE for hand protection do not recognize any consensus standards, but they do require the employer to assess hazards and select appropriate protection.

Hand hazards in the brewery and front of house include sharp objects, abrasions, hazardous chemicals,

Table 9.1 **Glove types for common brewery tasks**

Glove type	Brewery task(s)
Leather or synthetic work glove, mechanic's glove, or performance glove	Handling of freight, packaged ingredients, packaging materials, maintenance activities
Knit glove with or without applied texture or rubberized coating	Warehouse work, cold room, inventory handling
Disposable nitrile, 3–8 mil thickness	Lab work, sampling, working strength chemical use, routine cellar tasks, bodily fluids cleanup, messy or oily cleanup and maintenance activities
Disposable vinyl or nitrile, 2–6 mil	Food handling
Heavy duty PVC- or neoprene-coated knit glove or flock-lined glove with gauntlet[a]	Wort production, hot clean-in-place, steam
Chemical-resistant nitrile, neoprene-coated latex,[b] neoprene, or butyl, 10–28 mil	Concentrated chemical dispensing, clean-in-place tasks, dishwashing
Latex, neoprene-coated latex,[b] 18–28 mil	Dishwashing

[a] Not typically described by thickness, but generally heavy duty and often sold by the single pair.
[b] Some individuals experience latex sensitivity.

and high and low extremes of temperature. Some considerations for the employer include:

- Nature of the hazard(s)
- Required dexterity, tactility, lifespan (also called performance factors)
- Lifetime of the glove
  - Single use, disposable
  - Multiple use, washable
- Cost and availability of the glove

Employers should make qualified glove selection decisions based on these factors. It is helpful to consult with glove manufacturers or review chemical compatibility charts. Safety data sheets occasionally provide specific glove recommendations and can be used as a second source of advice. Table 9.1 lists the types of gloves most common in the brewing industry paired with typical applications.

HEARING PROTECTION

Hearing protection is regulated in a separate part of the Occupational Safety and Health Standards, falling under Occupational noise exposure, (29 C.F.R. § 1910.95). Hearing protection is required when the workplace noise levels exceed 85 decibels on the A scale (i.e., 85 dBA) on an eight-hour time weighted average (TWA). When required by the standard, the employer is responsible for providing hearing protectors. High noise levels can cause irreversible gradual hearing loss or audio range impairment. Noise also causes physical and psychological stress.

Protectors for shielding or deadening sound may be earplugs or earmuffs. A pair of protectors is given a noise reduction rating (NRR), which indicates the relative amount of protection provided, usually in a range between 0 and 30, although an NRR can be higher than 30. The higher the NRR, the more protection is provided. OSHA provides a series of calculations for applying NRRs and for calculating acceptable noise exposure intensities and periods based on conditions. Additional discussion on hearing protection is provided in chapter 10.

Earplugs can be disposable or reusable. Earplugs are typically made from foam rubber or silicone. Reusable earplugs should be inspected and cleaned regularly—replace any that cannot be cleaned. Inserting plugs with clean fingers is important. Contaminants can adhere to earplugs and be transferred to delicate tissues in the ear canal. Earmuffs are reusable over-the-ear devices.

Loud activities in a brewery can include compressors, pumps, centrifuges, and packaging lines. One complication caused by the wearing of hearing protection is that normal workplace communication becomes more difficult. Look for ways to reduce workplace noise, thereby reducing the need to have hearing protection.

RESPIRATORY PROTECTION

Adverse health effects can be brought on by breathing air contaminated with harmful dust, fog, fume, mist, gas, smoke, vapor, spray, or other aerosol. Two extenuating factors lead most breweries to control respiratory hazards with elimination, substitution, engineering controls, and safe work practices. The first factor is the prevalence of facial hair, which makes tight-fitting respirators ineffective. The second factor is the complexities of the respiratory protection program requirements in 29 C.F.R. § 1910.134(c). Along with hearing conservation, respiratory protection program requirements are among the most complicated of OSHA's PPE standards.

When exposure to respiratory hazards exceeds the permissible exposure limit, eight-hour time-weighted average (PEL-TWA), the employer can use respirators or any other appropriate hazard controls to lower the inhaled exposure to the worker to below the PEL-TWA. Whenever respirators are used as the means of exposure reduction, in part or in whole, a respiratory protection program is required. Respirators generally cover the nose and mouth or the entire face with a tight skin-to-respirator seal. Disposable dust masks are not respirators.

Breweries with a respiratory protection program must conduct respiratory fitness screenings through an occupational clinic. The brewery must also perform quantitative fit testing for each individual assigned to wear a half-face or full-face respirator. Medical records must be maintained for 30 years and meet confidentiality requirements. These stringent requirements usually provide sufficient motivation to reduce or eliminate respiratory hazards by other means.

Voluntary Protection

When workers are not being exposed to any respiratory hazards above the PEL, their breathing may still be bothered by airborne materials. When this occurs, the airborne material is called a nuisance. In breweries, nuisance materials can occur with fugitive dust from grain milling, as well as during housekeeping activities like sweeping.

In situations where nuisance materials are present, employees may choose to wear disposable dust masks for their own comfort. Dust masks—designed to catch particles, not vapors—can be provided by the employer or the employee, but the employer cannot require their use without triggering the full requirements of a respiratory protection program. Disposable dusts masks offer limited protection and should not be used against toxic substances. Their effectiveness is greatly reduced when used over a beard or other facial features that prevent direct contact with the skin.

If any other type of respirator is worn voluntarily, the employer must make sure the employee is fit to wear the device. This is normally done through pulmonary function testing. The employer must also train the employee in proper inspection, cleaning, and maintenance of the respirator. Other parts of the respiratory protection program are not required, however.

Respiratory Protection Program

If exposure to one or more substances in the workplace exceeds their respective PELs, then the full respiratory protection program will come into play. The program is also required if the employer requires respirators to be worn for any other reason. The written respiratory protection program includes the following:

- Procedures for respirator selection and use
- Medical evaluation of employees who are required to use respirators
- Fit testing for each employee required to use respirators
- Procedures for respirator use in routine and reasonably foreseen emergencies
- Procedures and schedules for cleaning, disinfecting, storing, inspecting, repairing, discarding, and otherwise maintaining respirators
- Procedures for ensuring adequate air quality, quantity, and flow of breathing air for atmosphere-supplying respirators (rarely required for a brewery)
- Training in the respiratory hazards to which employees are potentially exposed during routine work and emergencies
- Training in donning and doffing respirators, their limitations, and maintenance
- Procedures for respiratory protection program review for effectiveness

For the majority of breweries, a full respiratory protection program is rarely applicable because exposures above the PEL do not occur and the employer does not require any type of respirator or dust mask. However, to document that the business does not need a written program, medical evaluation, and fit testing, the brewery may benefit from conducting exposure assessments with a qualified industrial hygienist.

PROTECTION FROM OTHER TYPES OF BODILY INJURY

There are hazards that occur less frequently in a brewery setting that will require specialized PPE or protective shielding. This may include protection against radiant heat, ionizing radiation, ultraviolet light, boiling liquid splashes, bodily fluids resulting from an injury, and stabilization of overuse injuries using ergonomic appliances. These are discussed on a case-by-case basis in chapters 10–17.

SENSIBLE WORK CLOTHES

When not protecting against chemical incursion, work clothing should be selected to provide protection against abrasion, protruding objects, slippery floors, and blistering of feet and hands. Clothing should not be so loose that fabric can catch in moving machinery or interfere with work. To be clear, the wearing of shorts, cargo pants, sneakers, or sandals should be prohibited.

After meeting these needs, clothing should be comfortable and appropriate for the ambient conditions. Brewhouse activities can be warm and humid. Lightweight synthetic blend long pants and work shirts should remain comfortable and not become weighed down with excessive moisture. Some workers prefer a coverall with light shirt underneath. Avoiding chafing and rashes is discussed in chapter 10 (p. 105).

Warehousing may be cold and require insulating layers, a warm jacket or fleece, and insulated gloves and a hat. Planning work activities to reduce the frequency with which workers have to change clothes will tend to reduce chafing and also improve productivity.

Having satisfied these criteria, clothing can be chosen to meet the aesthetic needs of the workplace or company culture. Just as good housekeeping sets a tone for safety in the workplace, so does a uniform or a workplace dress code.

APPLICABLE STANDARDS

OSHA has a primary standard for PPE in 29 C.F.R. §§ 1910.132–.138, and § 1910.140. Personal protective equipment and various kinds of protective shielding are called for in numerous other national consensus standards. Table 9.2 lists the most common standards relating to PPE, including national consensus standards from other professional organizations recommended by OSHA or incorporated by reference into the Title 29 standards.

Table 9.2 **Standards related to personal protective equipment**

Short title of standard	Title 29 citation	Key requirements
Compliance duties owed	1910.9	Employer responsible for providing PPE.
General requirements	1910.132	Employer responsible for hazard assessment, selection, providing PPE, and training in proper use.
Eye and face protection	1910.133	General requirements, protection from eye hazards, employer ensures employee use of PPE, accommodation for vision prescription, use of filters/shields for light hazards.[a]
Safe use of lasers		National consensus standards for the safe use of lasers that operate at wavelengths between 180 nm and 1000 μm: classification of lasers (1 to 4) based on biological harm (ANSI Z136.1-2014); safe use of lasers in manufacturing (ANSI Z136.9-2013).
Bloodborne pathogens	1910.1030(b)(3)	PPE required for bloodborne pathogen exposure protection.
Hearing protection	1910.95	Requirements for hearing conservation program when time-weighted average sound level of 85 dB is exceeded; voluntary below this level.
Respiratory protection	1910.134	Requirements for respiratory protection program if permissible exposure limit for any hazardous chemical or listed substance is exceeded; voluntary below this level, so long as employer does not require respirator. (See also ASTM F3407.)
Head protection	1910.135	General requirements, protection from falling objects and electrical hazards.[b]
Foot protection	1910.136	Employer ensures use of qualified footwear but is not required to provide it.[c]
Electrical protective equipment	1910.137	Rubber insulating equipment requirements.[d]
Hand protection	1910.138	Employer responsible for hazard assessment, selection, providing PPE, and training in proper use.

Table 9.2 **Standards related to personal protective equipment** (cont.)

National consensus standards from other professional organizations	
Organization	*Relevant code or standard*
American National Standards Institute[e]	ANSI Z41-1999 American National Standard for Personal Protection - Protective Footwear
	ANSI Z87.1-2003 Occupational and Educational Personal Eye and Face Protection Devices
	ANSI Z89.1-2009 American National Standard for Industrial Head Protection
	ANSI Z136.1-2014 American National Standard for Safe Use of Lasers
	ANSI Z136.9-2013 American National Standard for Safe Use of Lasers in Manufacturing Environments
American Society for Testing and Materials[f]	ASTM D120-09: Standard Specification for Rubber Insulating Gloves
	ASTM D178-01: Standard Specification for Rubber Insulating Matting
	ASTM D1048-12: Standard Specification for Rubber Insulating Blankets
	ASTM D1049-98: Standard Specification for Rubber Insulating Covers
	ASTM D1050-05: Standard Specification for Rubber Insulating Line Hose
	ASTM D1051-08: Standard Specification for Rubber Insulating Sleeves
	ASTM F819-10: Standard Terminology Relating to Electrical Protective Equipment for Workers
	ASTM F1236-96: Standard Guide for Visual Inspection of Electrical Protective Rubber Products
	ASTM F2412-05: Standard Test Methods for Foot Protection
	ASTM F2413-05: Standard Specification for Performance Requirements for Protective (Safety) Toe Cap Footwear
	ASTM F3407-21: Standard Test Method for Respirator Fit Capability for Negative-Pressure Half-Facepiece Particulate Respirators

[a] Incorporates by reference ANSI Z87.1-2003.
[b] Incorporates by reference ANSI Z89.1-2009.
[c] Incorporates by reference ASTM F2412-05 ("F-2412-2005"), ASTM F2413-05 ("F-2413-2005"), ANSI Z41-1999, and ANSI Z41-1991.
[d] Note that § 1910.137(a)(3)(ii)(B) refers to the following ASTM consensus standards related to electrical personal protection: ASTM D120-09, ASTM D178-01, ASTM D1048-12, ASTM D1049-98, ASTM D1050-05, ASTM D1051-08, ASTM F1236-96, and ASTM F819-10.
[e] American National Standards Institute [homepage], accessed February 16, 2023, https://ansi.org.
[f] "Standards & Publications," American Society for Testing and Materials, accessed February 16, 2023, https://www.astm.org/products-services/standards-and-publications.html.

10
ENVIRONMENTAL, ERGONOMIC, AND TRIP/FALL HAZARDS

Noise and Hearing Protection

NOISE – Sound Pressure Waves
Occurs: Packaging lines, centrifuges, pumps, generators, compressors; workplace music and personal listening devices
Outcomes: Short- and long-term hearing loss, reduced communication effectiveness

OCCUPATIONAL NOISE

Hazardous sound levels, conventionally called *noise*, are vibrations in air that transmit energy to the body's hearing apparatus. When these vibrations last too long or are of too great an intensity, permanent hearing loss results. Hearing loss can occur from damage to any part of the hearing apparatus, from the eardrum at the end of the ear canal to the tiny hairs in the cochlea deep in the inner ear.

Brewery activities range from quiet and contemplative to raucous and distracting. Noise results primarily from two sources: process equipment and entertainment. Applying OSHA noise limits and hearing conservation standards to workplace noise caused by equipment is relatively straightforward, but applying them to entertainment noise is not.

OSHA regulates allowable noise much the same way as it does with exposure to chemical hazards, by applying a permissible exposure limit (PEL) on an eight-hour daily average of exposure called the time-weighted average (TWA). OSHA may also require audiometric testing of workers' hearing over time.

OSHA's noise standard does not contemplate personal listening devices, loud workplace music, or the effects of noise on non-employees exposed in taprooms, live performances, and festivals. Nor does it consider damaging noise outside of the workplace due to hobbies and domestic activities (hot rods, chainsaws, firearms, fireworks, and loud music).

Onset of Hearing Loss

Hearing loss comes on gradually, imperceptibly. We may make assumptions about hearing loss, such as it being an inevitable circumstance of aging. Hearing loss can also result from scuba diving, head injury, buildup of ear wax, and having had certain diseases such as measles or meningitis. These facts make it difficult to determine where workplace noise injury truly occurs and they can also make it challenging for workers to take hearing conservation seriously.

Hearing loss due to occupational noise is typically permanent, progressive, and irreversible. The degree of hearing loss is most often measured with audiometric testing (discussed below).

Brewery Noise Sources and Controls

Loud equipment common to most breweries includes centrifuges, pumps, air compressors, and forklift warning systems. Breweries may have one or more continuous batch centrifuges, which can effectively "scream" during use. Pumps can also produce shrill noises.

Packaging breweries will be noisier due to the banging of containers, operation of multiple conveyors, and discharge of compressed air. Glass bottle packaging is generally noisier than aluminum can packaging.

There are other noises in many production and retail spaces, namely the broadcasting of loud music and the wearing of personal listening devices. Music played in the production space will be able to be measured during workplace monitoring using a noise meter. It is important that conditions are representative during sound monitoring and that the total environmental noise is measured.

Any physical structure that can impede the progression of sound waves will reduce worker exposure. For fixed equipment like compressors and centrifuges, isolating the equipment in a separate room with concrete or soundproof glass can be very effective. Keep in mind that communication with such a room may require specifically designed public address systems or alarms. Quieter equipment can be substituted for older, noisier devices.

Hearing protectors are a widely used noise control. They can be passive or active. Passive protectors, which include earplugs and earmuffs, simply block sound. Earplugs and earmuffs should be marked with a noise reduction rating (NRR) that follows the requirements of ANSI S3.19-1974: *Method for the Measurement of Real-Ear Protection of Hearing Protectors and Physical Attenuation of Earmuffs*, the national consensus standard for hearing protection devices. Active protectors include sound dampening headphones. Devices should be tested in the workplace to determine how well verbal communication and machine malfunction noises can be heard while still offering adequate protection.

Earplugs come in many varieties, including single-use, reusable, and connected by a lanyard or hoop-shaped clip; the plugs may be foam or silicone, and may be a high-visibility color. The major selection considerations include comfort and fit, whether the worker will have dirty hands when putting in the plugs, and whether others need to visually identify that their coworkers are wearing their protection. Contamination of any sort on the fingers may cause an exposure to the worker through contact with the ear canal, especially with foam rubber earplugs where the worker will compress the plug by rolling between their fingers before installing.

Administrative controls include placing warning signage in noisy areas and prohibiting the use of personal listening devices. Headphones for personal listening devices interfere with the accurate measurement of sound exposure because they produce surplus sound levels that are not measured with a sound level meter. Listening devices also reduce peer-to-peer communication in the workplace and interfere with detection of brewery sounds such as pumps, liquid transfers, pressure relief valves, gas monitoring alarms, forklift backup alarms, and cries for help.

HEARING CONSERVATION PROGRAM

Employers are often confused over whether monitoring or a hearing conservation program is required. In a nutshell, if any worker is exposed above the specified *action level*, a hearing conservation program is required. The action level specified by OSHA is 85 dBA on an eight-hour TWA. How has OSHA arrived at this action level?

In accordance with ANSI standards, OSHA has defined this action level as an equivalent noise dose of 50% of the PEL-TWA. The PEL-TWA is set at 90 decibels on the A-weighted scale, that is, PEL-TWA equals 90 dBA. This is the "criterion sound level." The 85 dBA action level is due to the fact that OSHA treats 5 dBA as the amount by which the sound level changes that either halves or doubles the effective exposure time. For example, an increase in sound level of 5 dBA doubles the dose. So, the action level specified by OSHA is 85 dBA on an eight-hour TWA; in other words, half the dose of the 90 dBA PEL-TWA. (The decibel scale is logarithmic, which is why you cannot simply say 20 dB is double the sound level of 10 dB.)

Because revising OSHA standards requires congressional approval, sometimes other bodies can provide safer guidelines with less red tape. This is true with occupational noise. The National Institute for Occupational Safety and Health (NIOSH) has established a more conservative consensus standard, which it terms recommended exposure limits, or RELs. Unlike OSHA, NIOSH treats an increase of 3 dBA as double the dose (National Institute for Occupational Safety and Health 1998, 25). You can see in table 10.1 how this produces a more conservative exposure limit for workers. Notice that as the exposure is a smaller portion of the worker's day so the allowable sound level increases. This is because we are interested in the average exposure over the course of a typical day.

Table 10.1 **Permissible noise exposure limits according to OSHA and NIOSH**

Exposure duration (hr./day)	SOUND LEVEL (slow response dBA)	
	OSHA	NIOSH
8	90	85
6	92	86.5
4	95	88
3	97	89.5
2	100	91
1.5	102	92.5
1	105	94
0.5	110	97
0.25	115	100

When the 8-hour TWA exceeds 85 dBA, the action limit threshold is crossed and OSHA requires a complete hearing conservation program. A hearing conservation program requires workplace noise monitoring, worker hearing exams, and the provision of hearing protection and training, all in addition to whatever engineering, administrative, and PPE controls are in use. OSHA has produced a helpful booklet to further explain the hearing conservation program (US Department of Labor 2002).

An eight-hour TWA below 85 dBA does not require the employer to implement a hearing conservation program; however, the employer may make hearing protection available for use on a voluntary basis. This is somewhat analogous to OSHA's respiratory protection rules, where a full program is required above the PEL and voluntary protection is allowed below it (*see* p. 96). It is best for the employer to conduct adequate and accurate noise monitoring of actual conditions to understand where they fall on the compliance requirement spectrum.

Brewery Noise Monitoring

OSHA's principle of workplace noise monitoring is to measure the ambient sound levels in the immediate vicinity of the worker's ears. Measurements are typically made with a noise meter calibrated to read sound pressure levels in the A-weighted frequency bands in a slow response mode and report the time-weighted average over the monitoring period in dBA (i.e., A-weighted decibels).

If the noise meter reports octave band analysis results instead, OSHA provides a chart for converting octave band results to dBA. OSHA accepts that this conversion may result in a different result than a simultaneously obtained dBA analysis. In any case, determining if a complete hearing conservation program is required is based on direct or calculated dBA values.

Measurements are interpreted without regard to hearing protection. These external noise measurements do not actually measure what volume levels are reaching the eardrum or if hearing injury has been sustained. They are a tool to indirectly assess whether unacceptable hearing damage may be occurring.

OSHA requires that the noise measurements include all continuous, intermittent, and impulsive noise within an 80 dB to 130 dB range. Sampling must represent the employee's work activities. Two common approaches to this are personal monitoring and area monitoring.

With personal monitoring, a datalogging sound level meter is worn by the worker. It has a microphone that is clipped onto the worker's collar. The device operates continuously throughout the day, logging sound levels and integrating them according to duration. At the completion of monitoring, the worker's TWA exposure, along with maximum and average readings, can be downloaded. Personal monitoring is best suited to workers who move between areas with different sound levels and spend variable amounts of time in each.

Area monitoring is mostly used when the affected workers are not moving between different noise areas. Noise levels are measured at worker stations within each area and the workplace is then mapped to show the typical noise in each. By knowing how long a worker spends in each area, the TWA can be calculated as a sum of fractional exposures using a calculation provided in the standard (20 CFR § 1910.95(b)).

Employers have the responsibility and flexibility to determine when and where the monitoring is to take place and that it is truly representative of the worker's potential exposure. Monitoring must be repeated whenever workplace conditions change, workers are reassigned, or new equipment is operated.

Noise meters can be purchased, rented, or be provided by an industrial hygienist (IH). The IH may be employed as a state-level OSHA On-Site Consultation representative, or as a commercial consultant paid for by the brewery or its insurance company. While operating the device may be straightforward for a technically minded brewery employee, interpreting the results will benefit from having a trained or certified IH involved.

Audiometric Testing

How sound energy affects the worker depends on the magnitude and duration of exposure, the age of the worker, whether they are wearing a personal listening device, and the extent to which they have already experienced some hearing loss. In other words, measuring sound levels measures the environment, not the worker. It allows an estimate of the dose the worker has received.

The occupational noise exposure standard provides for a way to measure the effects of the noise on the individual through audiometric testing (29 C.F.R. §1910.95(g)). Unfortunately, it measures hearing loss after the worker has suffered an exposure. Audiometric testing involves the employee being supplied sounds of various frequencies and volumes in a soundproof booth. The results can be plotted as an audiogram showing ranges of frequency and volume hearing loss, called shifts. The standard requires a baseline audiogram be conducted within six months of the identified need for a hearing conservation program, and annually thereafter.

APPLICABLE STANDARDS

OSHA's standard for occupational noise exposure in 29 C.F.R. § 1910.95 is the primary regulation applying to hearing conservation (table 10.2). The reader is encouraged to become more familiar with the requirements of the hearing conservation program by reviewing appendixes A through I in the standard. These include information on sound level measurement and calculations, audiometric testing, and additional resources. Appendix G, while labeled as non-mandatory, contains supplemental information that is very useful.

Table 10.2 **Occupational Safety and Health Standards pertaining to noise and hearing protection**

Short title of standard	Title 29 citation	Key requirements
Occupational noise exposure	1910.95(c)	Hearing conservation program requirements
	1910.95(d)	Workplace monitoring
	1910.95(g–h)	Audiometric testing program and requirements
	1910.95(i–j)	Hearing protectors
	1910.95(k–m)	Training and recordkeeping

National consensus standards and recommendations from other professional organizations		
Organization	*Citation*	*Criteria document*
National Institute for Occupational Safety and Health	DHHS (NIOSH) Pub. No. 98–126	Recommended noise exposure limits lower than OSHA standard

Lacerations, Abrasions, Punctures, Avulsions, and Amputations

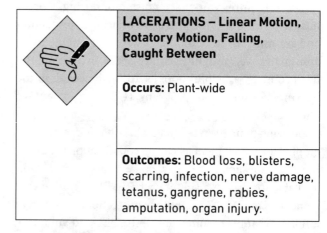

	LACERATIONS – Linear Motion, Rotatory Motion, Falling, Caught Between
	Occurs: Plant-wide
	Outcomes: Blood loss, blisters, scarring, infection, nerve damage, tetanus, gangrene, rabies, amputation, organ injury.

TYPES OF TISSUE INJURIES

Tissue injuries vary greatly in their severity and treatment requirements. They may be minor, such as a shallow cut or a splinter, or require immediate medical attention due to blood loss, threat of infection, or additional trauma such as crushing or burning. Injuries are generally considered minor if they are limited to the dermis and hypodermis (p. 64), but once they involve cut or torn major blood vessels, muscles, tendons, ligaments, bone fractures, or internal organs they should be treated as emergencies and 9-1-1 called.

Tissue injuries may occur as a result of tool use, sharp protrusions and surfaces, falls from the same level, falls from height, flying objects, rotating machinery, kitchen and bar tasks, and many other situations.

Lacerations

A tearing of one's tissues, technically called a laceration, is a hazard found in production, packaging, bar service, and kitchens. Lacerations are commonly caused by manual mistakes or improper tool use with cutting implements or by handling sharp materials, like broken glass or sheet metal waste.

When a laceration is minor, we often call it a cut. Cuts do not result in excessive blood loss and are often managed with first aid. When they are more serious, lacerations require medical attention, which usually involves wound irrigation and either suturing or the use of compressive bandaging.

Abrasions and Rashes

Abrasion occurs when the skin is rubbed against a rough surface. Abrasions can result in sensitive skin, redness, chafing, blisters, or infection. Abrasions are common on the hand or foot from continued contact with a tool handle or ill-fitting shoes. It also occurs where clothing or tool belts rub against the skin or where the body has scraped a rough surface. Continuously damp or moist areas of skin can be particularly vulnerable to chafing, where the skin becomes worn and sore and susceptible to infection.

A rash is a skin irritation that may have similar symptoms as abrasion but might be caused by contact with an irritating chemical or an allergen. Heat rash, also called prickly heat, is a painful rash caused by occluded sweat ducts in the skin and is discussed later in chapter 14.

Prevention of abrasion injuries includes wearing clean, properly fitting work clothes and socks. Rough surfaces on tool handles, ladders, and other regularly touched areas can be smoothed or refinished. For those in packaging lines, extended contact with paperboard products does cause abrasion and dry skin. For protection against hand and finger abrasion, choose correctly fitting gloves,

keep hands clean between tasks, and use moisturizing creams if the skin become dry or brittle. Most abrasion injuries can be treated with first aid: cleansing the surface, possibly using an antibiotic salve, and protecting from further abrasion during the healing process.

Chafing is a frequent occurrence for workers in the humid brewing environment. Chafing often occurs in the groin, armpits, where rubber boots contact the calf, or between the toes where damp skin is subjected to excessive rubbing or contact. Keeping these vulnerable areas clean daily and drying them before donning underwear or socks is important, and workers should not share towels, clothing, or chemical boots.

While chafing itself is painful, the breakdown of the top layers of the skin makes these moist regions susceptible to yeast (fungal) infection by *Tinea pedis* (athlete's foot), *Candida balanitis* or *Tinea cruris* (jock itch), or *Candida albicans* (thrush and vaginal yeast infection). These fungal infections respond well to topical antifungal treatments usually available over the counter or by prescription. In more severe cases, an oral regimen may be prescribed by a practitioner. While these infections are not usually classified as sexually transmitted diseases, there is evidence of crossover.

Punctures

Punctures occur when an object, usually pointed, penetrates the dermis and possibly deeper, but results in a wound that does not readily bleed. One effect of bleeding is that it helps carry contaminants outside of the wound. With punctures this does not happen or happens only to a small degree, so contaminants remain inside the body where they can cause infection. Examples include stepping on a nail, getting a severe splinter, being bitten by a dog, or wounded by a bullet. (Insect bites and stings are covered in chapter 17.)

Other than minor splinters that can be managed with first aid, a puncture means a qualified medical practitioner should be consulted. The wound may need to be cleared of residual material and countermeasures administered, such as a tetanus injection or antibiotics.

Avulsions

Avulsions are generally serious injuries where skin and tissue have been torn away forcibly from the body. An avulsion can look like part of the body is missing. They occur during accidents involving significant forces and are often seen in car or bicycle accidents and in bullet injuries. They usually result in rapid blood loss and require an emergency response—call 9-1-1.

An avulsion fracture is a separate type of injury. It occurs when a tendon or ligament tears away a small piece of a bone. These are considered to be musculo-skeletal disorders.

In a brewery setting, an avulsion injury can occur if a person's limb, hair, or clothing is caught or tangled in packaging machinery or an unprotected gear or pulley system. Avulsions may also occur where workers suffer severe trauma, such as with a forklift or vehicle accident.

Amputation

Amputation is defined separately from an avulsion. OSHA rules define amputation in 29 C.F.R. § 1904.39(b)(11) as "the traumatic loss of a limb or other external body part."

> *Amputations include a part, such as a limb or appendage, that has been severed, cut off, amputated (either completely or partially); fingertip amputations with or without bone loss; medical amputations resulting from irreparable damage; amputations of body parts that have since been reattached. Amputations do not include avulsions, enucleations, deglov-ings, scalpings, severed ears, or broken or chipped teeth.*

It is important the employer understands what constitutes an amputation because amputations require OSHA be notified within 24 hours of an amputation occurring (*see* fig. 2.2 on p. 17).

Fortunately, amputation hazards are rare in a brewery setting. Possibilities might include remodeling or maintenance involving a circular or reciprocating saw, or reaching into a moving mash rake or malt mill with a bare hand. While rare, possibilities like this should not be overlooked in a hazard assessment.

SYSTEMIC LACERATION HAZARDS

In the broad context of possible activities—brewing, packaging, distribution, bar service, food preparation, and maintenance—there are many possible cut and puncture hazards. Rather than attempt to list every scenario, three situations are described. Those persons

RESPONDING TO AN AMPUTATION

It is important to contact emergency medical services immediately if a workplace incident results in amputation. Remember these few key steps:
1. Apply pressure to slow bleeding and elevate the injury, if possible.
2. Dial 9-1-1.
3. Keep the patient conscious and calm, to the extent possible.
4. If the detached body part is safely accessible, wrap it in sterile cloth or plastic wrap and place in a cooler for EMS to deliver to the hospital. Some detached body parts can successfully be reattached.
5. The employer must report the injury to OSHA within 24 hours of the incident occurring.

responsible for performing hazard assessments can consider the principles discussed here and apply them to their brewery's particular setting and activities.

Unprotected Surfaces, Edges, and Protrusions

Periodic investigative walks throughout the premises can reveal protruding objects, sharp edges, splintering wood, exposed fasteners, sheet metal edges, and many other laceration, puncture, and abrasion hazards. Identifying such hazards is a good hazard assessment exercise for all workers. A near miss reporting system will be helpful in identifying hazards and ensuring they are remedied.

It may be common for workers in a particular part of the operation to know about certain hazards and avoid them out of habit. But other persons not familiar with the work area are liable to fall victim to the same hazards. The best time to remedy dangerous hazards is when they are first identified.

During construction or renovation activities, maintain good housekeeping and waste management. Pick up lost fasteners with a magnetic sweeper and mark protrusion hazards with fluorescent tape, signs, or, in the case of rebar or utility stub-outs, with a bright plastic cap designed for the purpose. Demolition tasks are prone to injure workers with fastener punctures, lacerations from metal edging, ceiling grids, wiring, metal lath, and such. Wearing heavy leather or synthetic work gloves, safety glasses, and head and foot protection should be required for such activities.

Cutting Tools and Knives

Shears, utility knives, pocketknives, razor blades, kitchen cutlery, hand tools, and office equipment represent a wide range of cutting hazards. When the wrong device is chosen for a job or the worker is not trained in its proper use, injuries are more likely.

Lacerations in brewery operations most often occur while cutting open grain sacks, opening or breaking down cardboard boxes, and performing electrical, construction, or maintenance tasks. Dropping a knife into a grain mill is not unheard of, so a lanyard is recommended for these situations.

Common knife injuries result from cutting an object or material held in place on top of the leg, cutting the hand that is holding the work, or dropping a knife onto the foot. Poor placement of the hand or other body part while using a screwdriver, wrench, or powered hand tool can result in laceration, puncture, avulsion, abrasion, or amputation.

The best safe work practice controls to avoid these types of injuries are to know how to properly use the tool, to position body parts out of the way of slipping with the tool, and by performing work deliberately. Selecting the right tool for the job translates to fewer incidents of the tool slipping off the work or requiring the use of excessive force. Personal protective equipment may include mechanic's gloves, leather gloves, or, in exceptional cases, Kevlar® or chainmail gloves. Some activities may benefit from a leather apron, such as during metal fabrication.

If a kitchen is part of your brewery operation, there will be many opportunities for laceration. Knives should be kept where they are secure from falling onto the legs and feet of cooks and do not protrude from countertops. Knives should only be washed by hand and not placed into dish sinks. Knife skills should be discussed and practiced, including knife selection, hand and food positioning, knife action, and sharpening. The adage that a sharp knife is a safe knife is true. Sharp knives require less force, produce better results, are less prone to slipping, and if a laceration does occur, it results in a cleaner cut that will heal better.

Kitchens may have graters, mandolins, and motorized slicers. Always use guards if provided and do not try to get the last bit of food sliced through these devices.

CUTS LIKE A KNIFE

Choose the right tool for the job
- Retractable or foldable knives should be used for boxes and bags
- Spring-loaded, auto-closing knives are preferred
- Use brightly colored knives
- Add a lanyard if a knife is used over machinery or at height
- Longer blade knives should be used for cordage and fabric
- Know the proper tools for cutting in the kitchen, bar, and shop

Practice safe cutting techniques
- Learn how to use each cutting device
- Cut away from your body and the bodies of others
- Cut things on a solid surface, free of clutter, in good lighting
- Don't rush or take shortcuts
- Don't use a device with a loose blade
- Don't use a device that's missing its factory-installed guard
- Don't use a cutting blade as a screwdriver, pry bar, or chisel
- Wear mechanic's gloves or cut-resistant gloves

Look after your tools
- Close the blade or stow the tool after each use
- Keep blades sharp and replace when dulled
- Keep blades free of sticky residue
- Store cutting devices in a sheath or drawer

Learn to sharpen blades and be everyone's friend
- Whetstones
- Diamond
- Carbide
- Ceramic
- Butcher's steel
- Leather strop

Rather, finish cutting them manually on a cutting board or use the scraps for other cooking uses. Lacerations can occur equally during tool use and during cleaning. Wash these tools by hand right after use rather than immersing in a sink of dishwater. If they are washed in a dishwashing machine, be sure to secure sharp edges facing downward. It will help to place a heavier object over them so they are not thrown around during washing. Closable cutlery boxes are available for commercial dishwashers.

Broken Glass

It is not uncommon for glasses to break in front of house or in the bar sink. It is important to deal with broken glass promptly to reduce the chance for injury to others, but don't rush it. Take the time to use tools, like a dust-pan and brush, to round up shards. Place them in a spare cardboard box or plastic pail, not a plastic garbage bag.

If a glass is broken in a bar sink, don heavy dishwashing gloves and carefully reach in to remove the drain plug. Again, use a tool or a protected hand to pick up glass for disposal. It can be helpful to shine a flashlight on the affected area to make sure pieces haven't been missed.

Keep appropriate tools nearby and make sure staff know where they are. For broken glass pickup, have a broom, dustpan, brush, and labeled box or pail. Additional tools might include a squeegee, flashlight, or shop vacuum, depending on the setting. Glass will be broken at some point, so planning for it will make even the first cleanup less difficult. Other locations in a brewery where broken glass occurs are laboratories, sensory testing areas, and, of course, bottle packaging.

Keep first aid supplies stocked and available near food and service operations. Adhesive bandage strips should be blue or otherwise brightly colored so as not to be confused with any comestibles.

APPLICABLE STANDARDS

In general, lacerations, punctures, and related injuries are not called out in their own occupational health and safety standard. A search of 29 C.F.R. § 1910 reveals 15 mentions of "puncture," for example, that are spread across various standards. The most relevant standards are given in table 10.3.

Table 10.3 **Occupational Safety and Health Standards (29 C.F.R. Part 1910) pertaining to lacerations, punctures, etc.**

Short title of standard	Title 29 citation	Key requirements
First aid	1910.151(a), -(b)	Readily available medical personnel or training and supplies for rendering first aid[a]
Sanitation	1910.141(a) (3)	(iii) Every floor, working place, and passageway kept free from protruding nails, splinters, loose boards, and unnecessary holes and openings
Fall protection	1910.29(b)(6)	Guardrails smooth to prevent lacerations, punctures, or snags
Ladders	1910.23(b)(7)	Surfaces free of laceration and puncture hazards

[a] First aid training and supplies must be "adequate" (29 C.F.R. § 1910.151(b)). Check with the Red Cross or similar accredited provider for training.

Musculoskeletal Disorders

	MUSCULOSKELETAL DISORDERS – Motion of the Body: Lifting, Bending, Pushing, Collision/Pinching, Climbing, Reaching, Slipping, Twisting, Overexertion, or Repetitive Motion	
	Occurs: Plant-wide	**Outcomes:** Pain, reduced mobility and strength, edema (swelling), chronic debilitation of limbs or back.

This section deals with tissue injuries that are beneath the surface of the skin. Musculoskeletal disorders are "injuries or disorders of the muscles, nerves, tendons, joints, cartilage, and spinal discs" (Centers for Disease Control and Prevention 2020a). They may be the result of an acute (one-time) event or they may be due to chronic exposure to a stress or repetitive motion that causes injury over time. A worker may have a chronic injury before they become employed at the brewery and be more likely to sustain injury while on the job. Overexertion, ill-health, awkward posture, poor work practices, and lack of engineering controls are some of the most prevalent contributing factors.

Despite the current lack of ergonomic standards by OSHA, musculoskeletal disorders are cited by the US Bureau of Labor Statistics and other workers' compensation studies as being the number one type of injury in manufacturing, both in terms of frequency and severity. The back is the most frequently injured body part, followed by the limbs. Musculoskeletal injuries involving sprains, strains, and contusions result in roughly half of all lost workdays (Bureau of Labor Statistics 2005).

TYPES OF MUSCULOSKELETAL INJURIES

Sprains, Strains, and Contusions (Soft Tissue Injuries)

Sprains and strains are cases of torn tissue below the surface. A *sprain* is a stretching or tearing of a ligament. A *strain* is stretching or tearing of a muscle or tendon. Sprains and strains can be the result of a single event, like slipping on a wet floor, or from an accumulation of repetitive motions, such as lifting cases of product.

Contusions typically result from a collision or pinch between the body and an object, usually without breaking the skin. Blood then spills into tissues outside of blood vessels and results in a bruise. A bruise involving a large volume of blood can become swollen and tender to touch and is called a *hematoma*. Blood blisters and bone bruises are also types of contusions. The eye, mouth, and visceral organs like the spleen, kidneys, or lungs can also sustain contusions and may be life threatening and require surgical intervention.

In breweries, contusions are seen after falls, impacts from falling or moving objects, and by bumping a part of the body against brewing system equipment. Some examples include dropping a keg on a hand or foot, falling from an elevated workplace and hitting fixtures or the floor, and being hit by or thrown from a forklift. A special type of contusion where the skin is broken or cratered and openly bleeding can be difficult to manage since there is a lack of sound tissue to suture.

Unless there is open bleeding, all soft tissue injuries benefit from the same first aid and treatment program, known by the acronym RICE (Mayo Clinic 2022a):

R – **Rest.** Allow the injured area time to heal (e.g., temporary job reassignment).

I – **Ice.** Ice the injury immediately to reduce inflammation.

C – **Compression.** Wrap the area with an elastic bandage to control swelling.

E – **Elevation.** Keep the injured area elevated to reduce inflammation.

There are many opportunities for sprains and strains in brewery production, distribution, and front of house activities. Any activity that involves lifting, bending, pushing or pulling, climbing, twisting, or reaching overhead can tear tissues.

There is a saying among safety folk: slips and trips lead to falls. We know this reflexively, which is why a common cause of strains and sprains is trying to catch ourselves from falling after a slip or trip. Slipping on wet or icy surfaces can cause hyperextension of the knee or the leg resulting in torn connective tissues. If we fall,

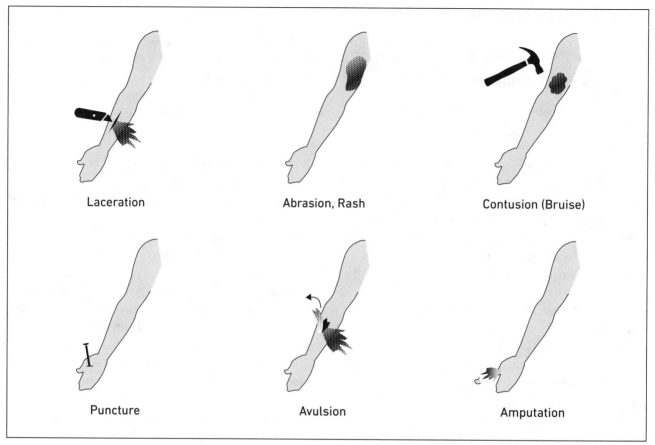

Figure 10.2. Soft tissue injuries.

we may end up adding contusion or other injury to the list of outcomes. Although injuries caused by slips, trips, and falls are usually considered separately from musculoskeletal disorders (Centers for Disease Control and Prevention 2020a), it is worth discussing them in the context of sprains, strains, and contusions.

Good housekeeping is the primary hazard control. Keeping hoses, cords, tools, and clutter out of the way is paramount. Maintain walking and working surfaces with good traction and free of puddles to the greatest extent possible. Keep stairs clear and provide railings where needed.

From a safe work practices standpoint, watch where you are going and be alert to changes in pitch, elevation, or floor friction. Generally, taking smaller strides with toes turned out slightly and keeping some bend in the knees will reduce the chance of slipping or hyperextension.

Wear appropriate, close-toed footwear in all job activities. Even for a casual bartender, flip-flops are never appropriate. Both housekeeping and PPE require a company-wide appreciation and effort; these subjects are discussed in chapters 8 and 9. Slips, trips, and falls are addressed in more detail in the next section on page 113.

Back Injury and Lifting

Brewing involves the lifting, carrying, and placement of heavy objects. Almost 40% of all days away from work due to musculoskeletal disorders in the US involve the back, significantly more than any other part of the body (Bureau of Labor Statistics 2018).

In breweries, workers routinely lift cartons of beer (30–36 lb.), hops (44 lb.), sacks of grain (55 lb.), and kegs (58–160 lb.). They bend over and drag heavy hoses—50 feet of 2-inch hose weighs 90 pounds when full. On a less routine basis, workers may move pallets (30–48 lb.), cylinders of CO_2 (190 lb.), or barrels of caustic (718 lb.). Is it any wonder that back injuries are commonplace among brewers?

Back injury can result from a one-time incident or from an accumulation of repeated small strains over time. (Repetitive motion disorders are discussed below.) Avoiding both acute and chronic back injury is done with a combination of safe lifting practices and the use of lifting aids, but please remember that back belts promote improper lifting. Safe lifting begins with a conscious plan of positioning, lifting, and navigating while carrying the

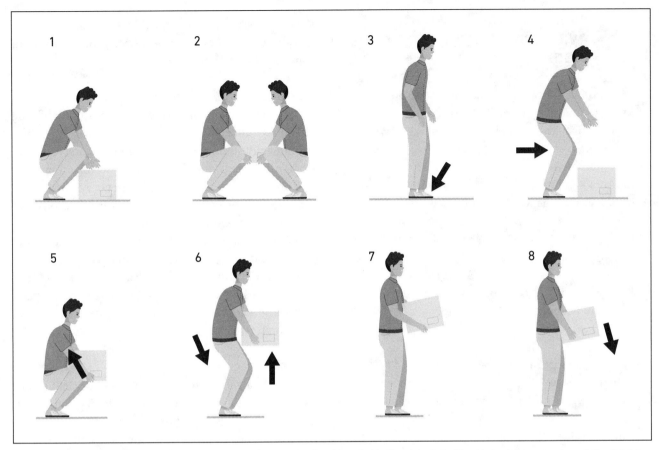

Figure 10.3. Proper lifting procedure: *1*, plan the lift and test the load and handholds; *2*, ask for help if load is too heavy or awkward; *3*, establish firm footing; *4*, bend at the knees keeping back erect; *5*, tighten abdominal muscles and pull load towards torso; *6*, lift with the legs and push hips forward to beneath the load; *7*, establish stable load, close to torso, with upright posture and steady gait; *8*, keep back upright at all times, turn with the feet, lower load by again bending at the knees.

load—refer to figure 10.3. If it is an uncommon load and its weight and center of gravity are unknown, conduct a test lift with planned positioning and then a small lift. If the load is too heavy, then lift with a colleague or use a lifting aid: hand trucks, keg carts, pallet jacks, and forklifts are all there to help you; for odd loads, consider a wheelbarrow, cart, platform truck, tilt truck, or lift table.

Repetitive Motion

Repetitive motion disorder is a type of musculoskeletal disorder, the outcome of a high frequency of a particular, repeated motion resulting in soft tissue injury over time. In beer production, repetitive motion tasks that can result in back strain are manual handling of materials like brewing ingredients, packaging supplies, and product in kegs and cases. Repetitive motion disorders are painful and chronic, often lasting a lifetime, yet they are avoidable with a catalog of well-known hazard controls. Repetitive motion disorders often affect the back, hands, wrists, elbows, shoulders, and knees. Individual anatomy varies, and what might not

bother one person becomes a chronic pain for another.

Repetitive motion disorders commonly affect the elbow-forearm-wrist-hand system, the neck-shoulders-back system, and the hips-legs-knees-ankles system. The hands and arms are most often affected by repeated grasping, twisting, lifting, overexertion, poor posture, or fatigue (National Institute of Neurological Disorders and Stroke n.d.). These motions are consistent with many manual or semi-manual packaging activities, such as depalletizing bottles and cans, packing off filled containers, hand labeling, and boxing variety packs. Tasks that involve sustained vibration in the hand, wrist, or forearm can hasten the onset of repetitive motion disorder symptoms.

Office tasks that can result in repetitive motion disorders include typing, numeric keypad or ten-key use, manually processing mailings, and other repetitive fine motor activities. Kitchen staff are also subject to repetitive motion disorders, particularly those who repeatedly prepare foods with knives, slicers, mandolins, and mortar and pestle.

REPETITIVE MOTION DISORDERS

Tendonitis – Inflammation of the tendon, the fibers that attach muscle to bone; occurs anywhere, but commonly in shoulders, elbows, wrists, knees, and heels

Bursitis – Inflammation of small, fluid-filled sacs (bursae) responsible for cushioning bones, tendons and muscles near joints; most common in the shoulders, elbows, and hips

Carpal tunnel syndrome (CTS) – Pressure on the median nerve in the palm of the hand results in weakness, numbness, tingling; affects the hands and arms

Epicondylitis ("tennis elbow") – Strain and inflammation of tendons in the elbow caused by repetitive motions of the wrist and arm

Tenosynovitis – Inflammation of the synovial membrane, the protective sheath that surrounds a tendon

Stenosing tenosynovitis ("trigger finger") – Difficulty in extending the finger caused by excessive gripping or rheumatoid arthritis; originates in the knuckle of the palm and affects the associated finger

De Quervain tenosynovitis – Inflammation surrounding tendons on the thumb side of the wrist, exacerbated by repeated hand or wrist movement; painful to grasp, make a fist, or turn the wrist

Ganglion cyst – Jelly-like lump forming along joints or tendons of the hand or wrist; noncancerous; may be surgically removed if painful

It is important to mechanize these operations when feasible. If the tasks are to remain manual, switch out operators frequently to allow stressed muscles and tendons to rest. Ensure that the worker is positioned at the best elevation, with a straight back, facing the work, and able to move objects or operate keys or controls with some bend in the elbow.

Posture

Simply repeatedly being in the wrong posture can bring about repetitive motion disorders and other strains. Additionally, jolting or vibrating movements, repetitive motion, and improper lifting can exacerbate improper posture and bring about quicker onset of symptoms.

Prevention of posture deficiencies prior to onset of symptoms is preferred. For seated tasks, make sure the operator can sit squarely with an erect back. Computer screens and process monitors should be 18–24 inches from the eyes, and slightly lower than eye level. For standing tasks, provide a cushioned surface to stand on to relieve lower back strain and rotate task type and duration frequently.

Arthritis

Arthritis is a general term for dozens of distinct conditions affecting joints, tissues around joints, and related connective tissues. Arthritis is more prevalent among older people, but persons of all ages can be affected. Joint pain and stiffness are common symptoms.

The most common form of arthritis is *osteoarthritis*, which is the "wear and tear" of joints and is commonest in knees, hips, and hands. The disease is progressive and may depend on genetics; also, the likelihood of it occurring increases with increasing age and body mass. Women, especially those 50 years and over, are more likely than men to develop osteoarthritis (Centers for Disease Control and Prevention 2020b). As cartilage within a joint breaks down, the bone undergoes degenerative changes. Osteoarthritis is characterized by pain, swelling, joint stiffness, loss of function, and disability.

Repetitive motions may hasten cartilage wear and bring on osteoarthritis earlier in life, so this ailment is a good reason to avoid repetitive motions. Osteoarthritis is often related to overuse or injury of a joint. This can happen when repeatedly flexing the knees, such as when frequently lifting bags of ingredients, manually palletizing orders, or lifting kegs. Proper lifting technique, job rotation, and use of mechanical lifting aids can help reduce the chances of developing osteoarthritis.

Rheumatoid arthritis is an inflammatory and autoimmune system disease where the body's immune system attacks the joint tissues. It is not so much caused by repetitive motions as other predispositions like age, gender, obesity, smoking, birthing and breastfeeding history, and inherited traits.

Gout is an inflammatory arthritis brought about by a condition known as hyperuricemia, a condition in which excess uric acid, a metabolic by-product, builds up in the body. Gout occurs when the excess uric acid precipitates as needle-like uric acid crystals in the joint voids, fluids, and tissues, resulting in excruciating pain. Gout particularly affects joints with voids in them, often the big toe or ankle (Centers for Disease Control and Prevention 2022).

Gout is well-known to affect those in the brewing trade. Men and post-menopausal women are most prone to gout attacks. Dietary factors known to increase the chances of gout are red meats, preserved meats, oily fishes (specifically anchovies, sardines, trout, tuna, and shellfish), fructose-containing drinks, beer, and hard alcohol. Obesity, physical inactivity, high blood pressure, diabetes, family history, and some medications are also predisposing factors.

Hernia

A *hernia* is when an organ or other body part pushes through the tissue or muscle that normally keeps it contained. There are over a dozen or so types of hernia, by far the most common being inguinal hernia. This occurs when a part of the intestine or some fatty tissue in the abdomen protrudes through muscle tissue into the crease of the thigh near the groin. Inguinal hernias are more common in men and cause pain and sometimes a visible bulge in the upper thigh, groin area, or scrotum (Hernia Clinic n.d., Cleveland Clinic 2023). Inguinal hernias are more likely to happen to older, less fit, or overweight persons. Frequent coughing or frequent straining during bowel movements are also associated with this type of hernia.

From an occupational injury perspective, inguinal hernias can be brought about by repeatedly straining the abdominal muscles due to improper lifting technique or repeated motions. The presence of an inguinal hernia may cause a dull ache, pain while lifting objects, or feelings of full bowels. Some individuals may be born with weaker than average abdominal musculature, but all inguinal hernias can be prevented with weight control, abdominal muscle fitness, and proper lifting technique.

APPLICABLE STANDARDS

At time of writing, OSHA has not been able to settle on an ergonomics standard. Despite this, the agency has named ergonomic hazards in regional and national emphasis programs as an area of concern.

From a compliance point of view, the best thing the employer can do is to identify and resolve ergonomic stressors in brewery operations. This begins with performing hazard assessments for tasks. Be sure to look beyond the brewing and packaging workforce to include service and business administration staff. Kitchen and bar staff are often on their feet on hard surfaces for their entire shift and have to lift heavy objects. Office workers may sit at poorly designed workstations and be subject to repetitive motions of the fingers, hands, and forearms. Addressing musculoskeletal disorders early protects workers' long-term health and avoids expensive workers' compensation claims.

Slips, Trips, and Falls from Ground Level

	SLIPS, TRIPS, FALLS – Linear Motion, Gravity
	Occurs: Plant-wide on any walking or working surface
	Outcomes: Sprains, strains, contusions, concussion; tissue and bone trauma; death. Equipment breakage; product loss.

This subsection deals with the causes and controls of slips, trips, and/or falls (STFs) from ground level, also called "on the same level." These incidents can result in a number of different injury outcomes, including sprains, strains, bruises, bone fractures, concussion, lacerations and punctures, or even death. Equipment or product may be broken or lost during STF incidents. The specific types of injuries have been discussed earlier in this chapter.

An STF happens either from the condition of the walking surface, obstacles that cause tripping, the way in which we walk, or a combination of these factors. The easiest way to define a walking-working surface (WWS) is any surface where you put your feet while working. This includes floors, ramps, stairs, ladders, and elevated working platforms. It also includes any "horizontal or vertical surface on or through which an employee walks, works, or gains access to a work area or workplace location" (29 C.F.R. § 1910.21(b)). So, even rappelling down the side of a 200-barrel fermentor counts as being on a WWS.

SURFACE TEXTURE AND ADHESION

Floor texture in a brewery setting can include anything imaginable. Production area floors need to consist of a waterproof, load-bearing material. Concrete, epoxy-coated concrete, or vitrified ceramic tile over concrete are the most common options. Concrete can be granular or polished smooth. Epoxies can have a polished-looking finish or have grit added for extra grip. Industrial tiles generally have a fine, granular surface that is less slippery than polished concrete or unamended epoxy. Epoxies and tiles can be made to varying degrees of slip resistance.

The typical brewery will not be using a tribometer to measure floor slickness, although it does occasionally happen during insurance claim investigations. Manufacturers of flooring materials may refer to the static coefficient of friction (SCOF) or the dynamic coefficient of friction (DCOF) measurements. The SCOF models the shoe-floor friction of a stationary worker, whereas the DCOF models that of a walking worker. The higher the resulting value from the tribometer, the higher the friction and lower the slickness.

Smooth or polished brewery floor surfaces get more slippery when wet, leading one to favor coarsely troweled concrete, amended epoxy, or vitrified tile. However, coarse floor surfaces are hard to sweep and toilsome to dry with a squeegee. Tiling with a moderate texture fired into the surface may be best, but it is the most expensive option. Inasmuch as the surface grip can be a hazard control to reduce STFs, there is always a cleaning or cost trade-off that has to be managed with other controls.

Brewery floors should be designed to pitch toward floor drains to keep floors dry and puddle-free. Because the weight of full cellar vessels can cause concrete floors to bow away from the slope, structural reinforcement of the concrete and additional pitch are both recommended. A structural engineer with an understanding of brewery design should be involved in floor design.

Corrosive cleaning chemicals, detergents, glycol refrigerant, and hydraulic oils are common in breweries and any of these will make a floor surface extremely slick. Often, slips on these wetted surfaces cannot be recovered, or caught, by the employee, and the resulting fall will add injury to the incident. Keep chemicals in secondary containment and clean up spills right away. A common PPE control is the wearing of chemically-protective rubber boots. Even when they come with lugged soles and are advertised as slip-resistant, these "chem boots" are no match for a slippery chemical on the floor.

Floor surfaces in bar, restaurant, and kitchen areas will likely have hard flooring made of concrete, tile, linoleum, or wood. Where these floors are worn, irregular, or unlevel, tripping will be more of a concern. The best remedy is to replace poor surfaces. Less effective controls can include refinishing the surfaces, or placing signs or reflective tape to alert persons passing through. When non-employees are injured due to a slip or fall, their medical and financial remedy will come from the business's commercial liability insurance, not workers' compensation. Customer injuries do not have to be recorded for OSHA, but a contractor's injury does need to be recorded on the OSHA 300 logs (p. 17).

Front of house and kitchen areas may use rubberized floor mats. These serve to reduce slipping in wet, greasy, or soapy areas, and they provide comfort for long hours of standing. These mats can pose an added chance for trips, if they are positioned to catch the worker's toes or the corners are worn and turned up. Worn mats should be replaced when they no longer lay flat.

Some retail areas and administrative offices may have carpeting. Carpeting is a common source of trips and should be kept flat and secured and be replaced when worn. If remodeling or painting is occurring, dropped cloths pose another tripping hazard. If employees cannot be routed around trip hazards, they will benefit from a toolbox talk about careful footing. In general, a good way to avoid STFs is to "walk like a duck." That is, bend slightly at the knees, and squat a little bit to lower the center of gravity, turn the toes outward, take smaller steps and plant the foot straight down.

Special events, outdoor dining areas, and inclement weather can introduce STF hazards. Grass, especially when worn down and damp, can be very slick. Dirt areas can be unlevel and have larger tripping hazards embedded in them. Sidewalks can become slippery with snow, ice, hail, rain, or leaves. Be aware of any perennially wet areas that have grown algae, as this is also slippery. Controls include good housekeeping, the use of salt or grit as appropriate, sandwich signs with warnings, and bollards and rope for directing foot traffic.

As a general summary of controls for avoiding STF incidents due to surface conditions, begin with engineering the best floor surface that is affordable, with proper pitch to trench drains. Keep drain grates in place. Railings are another significant engineering control. These should be used wherever code requires (stairs, fixed vertical ladders, and working spaces greater than 4 ft., or 1.2 m, off the floor), as well as anywhere else where an extra handhold will improve stability.

It is useful to build safe work practices into habitual behaviors. Watch where you are going and avoid distractions that cause you to look elsewhere. Walk with a lower center of gravity and a shorter gait. Choose the correct pathway, avoiding shortcuts that may lack railings or sound surfaces. Do not run in the brewery but walk deliberately. Identify hazardous surfaces and mitigate them as soon as possible.

Finally, wear footwear that is task appropriate. Entirely avoid open-toed shoes, sandals and flip-flops in a work setting. Kitchen staff may wish to wear dishwasher-safe silicone footwear. Footwear selection should take into account the compatibility of the upper with worker activities, adequate ankle protection (if needed), and a solid sole with ample breadth, surface contact, and lugging (ridges) to improve traction.

CHANGE IN PITCH OR GATE

Two variables important for inclined WWS are surface pitch and a person's gait. That is, what is the pitch or slope of the surface and what is the appropriate length of a person's stride for that incline?

A sloping pitch causes the force of a walking worker to increase on the lower elevation side of the step. Walking up a steep slope, the foot tends to slip backward (downhill) and causes the person to lose balance. Conversely, walking down an incline, the foot may want to slip ahead of the person (downhill), also resulting in a loss of balance. The steeper the incline, the greater the force pulling the body downward. Think about how you lean forward when going up steep steps, a ladder, or a hill.

Irregular surfaces and stairs or ladders that do not have evenly spaced rungs or steps are frequent causes of trips. This is because your brain looks at where you are about to walk and makes assumptions based on visual input and past experience. We expect stairs to be evenly spaced; when they are not, the foot may not fall directly on the tread. This does not happen when we hike up a steep trail with irregularly spaced rocks for steps, because we already expect the rise and run of the climb to be unique and we pay close attention. In the absence of engineering controls, like an elevator or textured, evenly spaced stairs, the best control is to reduce the gait and take premeditated steps.

RAISED OR DEPRESSED OBSTACLES AND CLUTTER

It is estimated that the average walking gait over level ground results in parts of the foot lifting no more than a quarter inch above the surface. Considering this is less than the height of a brewery transfer hose or 220-volt power cord, opportunities for tripping in the brewery are significant. In addition to hoses and power cords, brewers can trip on curbs, tools, supplies, dropped items, or refuse.

Similarly, depressed areas and foot-sized holes cause injury by allowing the foot to unexpectedly drop below the ground level. The main hazard on a production floor is trench drains with the top grate removed. Careless brewers often leave the grates out for long periods (usually for a small time-saving advantage). Placing a brewery discharge hose beneath one end of a grate reduces spillage onto the floor but creates a new trip hazard. A better solution is to cut a hole in the grate only large enough to admit the hose end, not a foot.

Table 10.4 **Occupational Safety and Health Standards (29 C.F.R. Part 1910) pertaining to slips, trips, and falls**

Short title of standard	Title 29 citation	Key requirements
Sanitation	§ 1910.141(a)(3)	(iii) Every floor, working place, and passageway kept free from protruding nails, splinters, loose boards, and unnecessary holes and openings
Walking-working surfaces	§ 1910.22	General requirements
Stairways	§ 1910.25	Stairways, railings, allowable dimensions
Dockboards	§ 1910.26	Dockboards, wheel chocks
Safety color code	§ 1910.144	Safety color code for marking physical hazards
Signs and tags	§ 1910.145	Specifications for accident prevention signs and tags
Exit routes	§ 1910.37	Maintenance, safeguards, and operational features for exit routes
Housekeeping	§ 1910.176	Handling materials – general

Depending on the property layout and the climate, employees, customers, or contractors are at risk from pets, children, objects dropped on the floor, gravel, snow, and ice. As mentioned before, loose carpets and irregular stair spacing can also result in trips.

Controls that can help reduce brewery trips begin with good housekeeping. In the case of a brewery cellar, this might be called good hosekeeping. Identify and correct bad behaviors, including leaving tools on the floor or on stairs, not removing clutter or refuse as soon as possible, and leaving floor drain grates off. Develop practices of stowing hoses, tubing, and cords between tasks and keeping them orderly and out of the lanes of foot traffic during use. Hoses can be hung on designated hooks along a wall or row of tanks. If a hose is loaded with sanitizer, for example, and coiled with the ends coupled, keep coils out of traffic lanes. Walk deliberately in the brewery and keep nonessential persons out of production areas. Where forklifts are in use, keep forklift traffic lanes clear of obstacles to reduce the chance of the operator veering into an occupied work area.

Floor surfaces should be kept free of chemical residues, puddles of water, and spilled product. Slip-resistant footwear helps but is not a singular control against slipping and tripping. Personal protective equipment will benefit from dedicated storage areas. Reusable PPE can be hung on hooks to dry after washing, while new PPE supplies can be shelved.

It is vitally important to create and reinforce a culture of organization. Providing panels for hanging fittings, hand tools, and lockout/tagout devices reduces the chance of those things being left on the floor, makes them easier to find, and creates a professional, orderly workspace. Provide shelving for corded tools, process measuring devices, and small containers of brewing additives and chemicals. Flat spaces like workbenches and stair treads are easy targets for setting down and forgetting an item. Reinforce good housekeeping within teams and leave time at the end of an activity for reorganizing the tools that were used.

APPLICABLE STANDARDS

OSHA does not have a standard openly devoted to slips, trips, and falls. Rather, it has numerous mentions of appropriate housekeeping and signage spread across a range of standards. The most relevant rules for breweries are summarized in table 10.4.

Falls from Height and Falling Objects

	FALLS AND FALLING OBJECTS – Linear Motion, Gravity
	Occurs: Plant-wide on any elevated surface; when using or moving articles overhead
	Outcomes: Sprains, strains, contusions, concussion; tissue and bone trauma; death. Equipment breakage.

Injuries from falls occur in breweries with some regularity. When these are not initiated by slips or trips on the same level (see previous section), they typically result from stepping off of, or losing balance from, a working surface above the level where the fallen worker ends up. According to the injuries, illnesses, and fatalities database maintained by the US Bureau of Labor Statistics (https://www.bls.gov/iif), fatal work injuries resulting from falls on the same level or from above are the leading cause of occupational fatalities after those related to transportation.

In this section, we discuss how to prevent falling from a stair, ladder, platform, brew deck, loading dock, the top of a vessel, a pallet rack, a roof, or any other elevated working surface. We also discuss being struck by an object falling from above.

WORKING AT HEIGHTS

Falls are serious and yet they are one type of accident that is almost entirely foreseeable and completely preventable. The perceived risk of a falling incident varies widely amongst individuals, due to whether a person has a fear of heights, how accustomed they are to working at heights, and the level of trust they have in whatever hazard controls may be in place. Part of the dilemma with low perceived risk of falling has to do with the worker failing to appreciate the role of gravitational acceleration and the force of impact.

It takes only about one-half of a second for a person to fall from a working surface height of four feet (approx. 1.2 m). The time it takes for a person to realize they are falling is most of that time, around one-third of a second. So, for a four-foot fall, there is insufficient time for the average person's reflexes to prepare for impact after they realize they are falling. The greater the mass of the person or falling object, the greater the force and the greater the energy transferred during impact. Also, the velocity increases as the distance increases.

Let's say we have a worker weighing 165 lb. (74.8 kg) falling off of a brew deck that is 6 ft. (1.83 m) off the floor. After falling one foot they are moving at about eight feet per second (8 ft./s, or about 2.4 meters per second) with a free fall energy of 175 foot-pounds (ft.-lb., or about 240 joules). By the time the worker hits the ground they are moving at almost 23 ft./s and carry an energy of 1,400 ft.-lb. (7 m/s and 1,900 J).

To put these results into perspective, 23 ft./s is about 16 mph (28 kmph), or the speed of a swift cyclist. Although data are scarce or subject to variable testing constraints, it is thought that energy of around 100 ft.-lb. (~136 J) is more than sufficient to fracture a human skull. If the impact is confined to a small surface area, such as occurs when falling onto an upward-facing rod, peg, post, or when being struck by a bullet, the energy required for a fatal injury can be much less. Indeed, many occupational fall fatalities are from heights less than the one used in this example.

Stairways

Stairways include conventional stairs, spiral staircases, and alternating step stairs (29 C.F.R. § 1910.25). Stair regulations are concerned with four things: railings, adequate width and head clearance for people, regularity of

STARING AT STAIRS

Alternating step staircases are quite uncommon, but there is a brewery in Portland, Maine using them to get from behind the bar to their business office above. You can tell a safety nerd from everyone else in a brewery taproom because he is asking permission to photograph a staff member coming down them.

Some safety instructors met at a rented house to rehearse their training at a national brewing conference. The stone steps to the front door were irregular, with one step being about one-third of an inch higher than the rest. Sure enough, even safety-minded people tripped on the unexpectedly high rise of the step.

the rise and run of the treads, and load-bearing capacity. Commercial architects are familiar with the requirements of this section and fire inspectors will doubtless be pulling out a tape measure during occupancy permitting. However, in older structures, stairways may not be compliant with OSHA requirements and may need to be refitted. Stairway rules are not particularly onerous, but they are very specific and they will afford safer transit for employees and customers when you are in compliance.

Ladders

OSHA's ladder requirements (29 C.F.R. § 1910.23) deal with portable ladders, fixed ladders, and mobile ladder stands. What makes a ladder different from a staircase is the steep angle at which they operate, the necessity of using the hands in ascending or descending, and the narrower rung that replaces the stair tread. Advantages of ladders include portability, low cost, and taking up less space. Disadvantages are many, and some obvious ones are the steep angle, the lack of railings, the minimal footholds, the fact that they are often electrically conductive, and that they are hazardous when not used specifically as designed.

The OSHA standard lists recommendations for general safety and specific ones for individual ladder types. Together with common sense safety practices, ladders come with quite a few instructions for safe operation. We have summarized these here and depicted key tips in figures 10.5 and 10.6.

Design Criteria for Ladders
- Rungs are parallel, level, and uniformly spaced
- Rungs are 10–14 inches apart (25–36 cm)
- Steps on stepstools are 8–12 inches apart (20–30 cm)
- Rungs are at least 11.5 inches (29 cm) wide on portable ladders and 16 inches (41 cm) wide on fixed ladders
- Ladders are free of puncture, cut, or snag hazards
- Metal and wood ladders meet specific criteria

Safe Work Practices When Using Ladders
- Ladders are only used in the manner for which they were designed
- Ladders should be inspected frequently, at least at the start of every shift where used
- Damaged or weathered ladders are immediately tagged and removed from use

- Employee must face the ladder when ascending or descending
- Use at least one hand to grasp the ladder when ascending or descending
- Never carry an object on a ladder that could cause loss of balance
- Select the best ladder for the job, generally in this preferred order (first to last): rolling platform stairs or ladder, straight or extension ladder, and stepladder
- Make sure the ladder is designed for the load; fiberglass stepladders are often color coded, but always read the label information
- If electrical hazards exist, do not use a conductive ladder
- Use a ladder with sufficient length
- Use a colleague to help stabilize the ladder and keep other people from dislodging the ladder
- Block doorways where a ladder could be disrupted by someone coming through

Fall Protection and Fall PPE

Fall protection cuts two ways: either an installed system that keeps one from being exposed to falling from height (i.e., an engineering control, like guardrails, railings, cages, etc.) or a PPE control that amounts to apparatus worn by the worker that somehow anchors them or limits the extent to which they can fall.

The engineering controls are contained in 29 C.F.R. § 1910 Subpart D, "Walking-Working Surfaces." Guardrails will have a top rail at 42 inches (107 cm) above the WWS and a second rail at half that height. Some minor variation in heights is allowed. Guardrails have to meet several prescriptive strength and dimensional criteria.

The OSHA standard allows for roping off an area that has a fall hazard into what is called a designated area. Holes must be covered and the cover cannot be easily dislodged. In addition to these controls, Subpart D also covers normal fixed railings for stairs and landings.

When the fall control is determined to be a personal fall protection system that is worn by the worker, it falls under 29 C.F.R. § 1910.140. With personal fall protection systems, OSHA provides prescriptive wording on the types and strengths of fasteners, anchorages, ropes, and so forth. With one's life literally "on the line," personal fall protection systems are not the place to shop for the cheapest equipment. Be sure that components are clearly labeled with test ratings to be sure they meet the standard.

A helpful trick to remember personal fall protection system components is A-B-C: anchorage, body harness, and connector. The anchorage is a secure point of attachment for the connector. Examples might be an eyebolt

Figure 10.5. Correct straight or extension ladder use.
Source: Brewers Association.

Figure 10.6. Correct stepladder use.
Source: Brewers Association.

Table 10.5 **Occupational Safety and Health Standards (29 C.F.R. § 1910) pertaining to working at heights**

Short title of standard	Title 29 citation	Key requirements
General requirements	§ 1910.22	Surface conditions, loads, repair
Ladders	§ 1910.23	General requirements, fixed ladder, mobile ladder stands
Stairways	§ 1910.25	General requirements, standard stairs, spiral stairs
Dockboards	§ 1910.26	Dockboards: loading, edges, wheel chocks
Duty to have fall protection and falling object protection	§ 1910.28	Employer has duty to protect employee from falling and from being hit by falling objects
Fall protection systems and falling object protection - criteria and practices	§ 1910.29	Guardrails, handrails, cages, designated areas, protection from falling objects
Fall protection training	§ 1910.30	Training requirements for fall hazards, fall protection equipment, retraining
Personal fall protection systems	§ 1910.140	PPE: harnesses, anchors, other requirements for personal fall protection, personal fall arrest, and work positioning systems
Powered platforms for building maintenance	§ 1910.66	Powered platforms for building maintenance.
Vehicle-mounted elevating and rotating work platforms	§ 1910.67	Vehicle-mounted elevating and rotating work platforms
Manlifts	§ 1910.68	Manlifts: general requirements, guards, mechanical specifications
Head protection	§ 1910.135	When required (ANSI Z89-1)
Safety color code for marking physical hazards	§ 1910.144	Yellow for marking physical hazards, including falls
Specifications for accident prevention signs and tags	§ 1910.145	Danger and warning definitions
Maintenance, safeguards, and operational features for exit routes	§ 1910.37	Stairs or ramps for non-level exit routes
Handling materials— general	§ 1910.176(g)	Guarding required against fall hazards

mounted into a structural component of a building, a hoisting eye welded into the top of a tank, or an exposed structural member that can be looped around. The connector is a load-rated device such as a rope or web lanyard, a lifeline, a deceleration device, or a combination of these, such as a lanyard connected to a deceleration device.

The body harness may be a safety belt, a harness that wraps the beltline and thighs, or a full-body harness that includes a waist belt and torso harness reaching over the shoulders. The selection of harness type has to do with the activity, the maximum falling distance, and

worker retrieval. The standard describes what harnesses are allowable under what circumstances, but a full body harness with a connector attachment between the shoulder blades is considered the protective minimum. With any personal fall protection system, the goals are to avoid allowing the worker to fall more than 3.5 feet (1.1 m) or hit a surface below the working surface.

Falling Objects

A falling worker is one kind of hazard, but so is an object or storage system falling onto a worker. The best approach to prevent falling object injuries is safe and secure inventory storage on shelving designed for typical loads. Stored objects need to be secured, such as banding or shrink-wrapping kegs, cartons, or supplies.

When workers are working above other workers, maintaining good housekeeping of tools and supplies is the best place to start. After that, having adequate toe boards can keep objects from being accidentally displaced or kicked off the elevated work area. Finally, PPE can include hard hats and sturdy work clothes and footwear.

APPLICABLE STANDARDS

Most of OSHA's requirements for fall protection are found in Subpart D, "Walking-Working Surfaces" (29 C.F.R. §§ 1910.22–.30). These rules are more prescriptive in nature than many other subparts and require close attention to specified dimensions. Fall protection for mechanical lifting devices is located in Subpart N, "Materials Handling and Storage." Numerous other mentions relating to fall protection, identifying fall hazards, and PPE for falling objects are scattered throughout Title 29 and summarized in table 10.5.

11

MATERIALS HANDLING AND MOTION HAZARDS

Manual Materials Handling

MANUAL MATERIALS HANDLING – Gravity, linear, rotational	
Occurs: Plant-wide	**Outcomes:** Sprains, strains, contusions, acute and chronic musculoskeletal disorders; slips, trips, falls; traumatic injury, death; damaged equipment or product

onsidering how often we manually handle materials and suffer injuries from doing so, OSHA regulations on manual materials handling are quite general and brief. Efforts by the agency to create ergonomic injury prevention regulations peaked in the 1990s, resulting in a published standard in 2000, which was repealed the following year.

At this time, unaddressed ergonomic hazards would be cited under the General Duty Clause, codified under 29 U.S.C. § 654. As mentioned in chapter 2 (p. 13), General Duty Clause violations can be difficult for OSHA to qualify. Injury avoidance really comes down to the employer recognizing that the cost of materials handling injuries includes low morale and high absenteeism, loss of productivity, increased insurance costs, and injured staff.

Since manual materials handling is often associated with surficial tissue and musculoskeletal damage, the reader should review the more detailed discussion of those injuries in chapter 10 (pp. xxx–xxx). These kinds of incident represent the largest segment of lost time due to injuries in the US workforce.

CLEARANCE

Adequate safe clearances must be maintained for aisles, docks, doorways, and passages where mechanical handling equipment is used. This equipment could be as simple as a hand cart or pallet jack, or as sophisticated as a forklift. Aisles and passageways shall be kept clear and in good repair, with no obstruction across or in aisles that could create a hazard.

Permanent aisles and passageways require appropriate markings. In some cases, there are specific dimensions or setbacks required, such as with stairways and electrical panels. Where mechanical handling and foot traffic coexist, pedestrians need to have adequate space to stay out of the way. Oftentimes, moving material results in reduced visibility for the person moving and directing the material.

Clearance limits for materials handling need to be indicated with signs. Other markings could include high-visibility floor markings, folding signs, stanchions with retractable belts, sawhorse barricades, A-frame signs, traffic cones, or steel fence panels.

HOUSEKEEPING

Avoid accumulating or storing materials in a way that constitutes a hazard from tripping, fire, explosion, or pest harborage. Keep stairways clear of materials at all times. Vegetation control, snow removal, and ice management are important considerations in outdoor areas. Housekeeping is dealt with in greater detail in chapter 8.

STORAGE

The manner in which materials are stored should not create a hazard in itself. Raw ingredients, intermediate materials, and packaged product must be stored securely on a sound floor or supportive structure. Bags, boxes, cartons, and kegs should be stored in tiers that are stacked, interlocked, and limited in height so that they are stable.

With kegs in particular, it is important to band them together on pallets with either cordage, plastic wrap, or other restraining system to prevent them from falling from above. Avoid stacking materials on top of walk-in coolers or other structures not designed for supporting loads.

GUARDING

Provide covers or guardrails to protect personnel from the hazards of process vessels, open pits, ditches, and the like. More on railings is provided in falls from height in chapter 10. On the remote chance that a brewery has a rail spur, a derail or bumper block is required to prevent a rolling car from contacting other cars being worked on or from entering a building or work area.

APPLICABLE STANDARDS

OSHA describes manual materials handling requirements in 29 C.F.R. § 1910.176. OSHA does not have an ergonomic standard at this time. Employers should be mindful to conduct work task hazard assessments and determine where manual materials handling could result in accident, injury, or illness and take appropriate measures to avoid such outcomes.

Engulfment and Grain Handling

GRAIN HANDLING – Simple asphyxiation, gravity, rotatory
Occurs: Bulk grain silos, spent grain silos, mills, grain conveyance
Outcomes: Pinch, tear, amputation injuries; falls from height; asphyxiation, coma, death; strains, sprains, back injury

Of the four main ingredients in beer, hops and yeast represent relatively small quantities to be handled. Water is conveniently managed in piping and tanks; gravity and pumps do the rest. This leaves grain in a class by itself.

Grain represents the second largest component of beer by volume, after water. It comes in hundreds of varieties of malt, as raw grain kernels, and as pregelatinized flakes. Packaging, storing, and dispensing grain gives rise to many common materials handling hazards. When grain is stored in bulk, engulfment, asphyxiation, and mechanical injury from pressure also come into play.

GRAIN STORAGE AND HANDLING

Most small breweries purchase, store, and handle grains and other fermentables in bags weighing 25, 50, or 55 pounds (roughly 11, 23, or 25 kg). Bagged ingredients may be moved by manual lifting, or through the use of non-powered equipment such as a pallet jack or motorized equipment such as a forklift. A typical pallet will hold 40 bags of malt for a gross weight of a little over one ton (around 900 kg). Manual materials handling is discussed in the previous section; proper lifting procedures are addressed in more detail in chapter 10 (p. 111). Opening bagged grains can be done by pulling a thread with a perforated seam, or by using a utility knife. Reviewing the section on lacerations is recommended if knives are used (p. 106).

Silos

Once production volume allows a return on investment for the capital expenditure, a brewery may add silos or grain tote dispensers. These hold bulk base malt that is required in large quantities. The production volume at which a base malt silo makes financial sense is about 2,000 to 3,000 barrels per year, depending on grain supplier load size and variables like grain cost and silo

WAIT! WEIGHT!

From an individual kernel to a cargo ship container, grain used in brewing comes in a range of sizes spanning 12 orders of magnitude. You have to choose the right way to handle it according to its weight.

Weight of grain in a ...	Transfer grain with ...
single kernel of barley: 0.00014 lb.	two fingers
sack: 50 lb.	two arms, two legs
pallet: 2,000 lb.	forklift
super-sack: 900 to 2,700 lb.	forklift, chain hoist, gravity and auger
silo: 10,000 to 50,000 lb.	air pressure, gravity and auger
bulk grain truck: 30,000 to 130,000 lb.	air pressure, gravity and auger
railroad hopper car: 220,000 lb.	gravity
cargo ship: over 100,000,000 lb.	bucket elevator

(nb. A short ton is 2,000 lb., or 907 kg. One pound is 0.454 kg, so a barley kernel is 64 milligrams; a cargo ship might hold over 45,359,237 kg, or 45,359 metric tons.)

capital costs. Bulk grain delivery comes in trucks ranging from 15 to 65 tons (14–60 metric tons). Brewery silo capacity varies widely, but capacities between 10 and 50 tons are common (between 9 and 45 metric tons).

While silos in the broader agricultural/animal context may take the form of towers, bunkers, ship's holds, silo bags, or piles, breweries use the tower type almost exclusively. A base malt silo looks much like a cylindroconical fermentor, but it has different fixtures for adding and removing grain. Two styles are the single-piece welded steel silo or the corrugated steel panels that bolt together. The smooth-sided single-piece silos are preferred because grain is less likely to adhere to the interior surfaces and there are no bolts to come loose due to pneumatic filling and seasonal temperature changes.

The fixtures on a typical silo include a vertical filling pipe with a cam-lock fitting on the bottom and a discharge point into the top of the silo, and a vent pipe that runs from the dome of the silo to the dispensing truck or to the open air. Grain is dispensed with positive displacement air pressure, sending grain up the filling pipe while displaced air comes out the exhaust pipe. Grain is pushed into the silo with a pressure of 3–3.5 psig (121.9–125.4 kPa). Higher pressures damage malt, creating dust and stressing the silo integrity.

At the bottom of the silo is the discharge gate, which may be a manual or a sliding gate valve that uses a programmable logic controller (PLC). When open, grain falls into a transfer boot, where the grain is picked up by the return air pipe pick-up boot on the bottom of the silo for a tubular conveyor. With spent grain silos, the discharge is usually a motorized screw auger with or without a gate valve.

The silo itself may be on load cells, or the grain may be conveyed to a weighing hopper or other device. The silo will have a manway on the side or the dome, or both. There may be a fixed ladder mounted to the silo or access from the roof or aerial catwalk.

There are hazards associated with silos both outside and inside the unit. Outside hazards include a chance of falling from height during inspection or repairs. Workers more than four feet (1.2 m) from the ground will require fall protection, including if climbing up a fixed ladder. Ladders should be kept behind a secure fence or wall or be equipped with a lockable closure to prohibit climbing by unauthorized persons. The grain discharge assembly is usually enclosed, but if it requires repairs, the system must be locked out to prohibit hand injury in the gate valve or auger. If work is being done on a spent grain silo dispense, the discharge screw should be locked out.

Inside hazards include falls and suffocation. A common reason for a worker to decide to climb into a silo is because there is a problem with discharging grain. This can occur because of moisture in the grain, which causes a plug to occur near the discharge point. Another cause is called bridging, where grain forms a bridge in the column of the silo with a hollow space underneath it. A worker standing on bridged grain can punch through and be buried almost instantly. Asphyxiation caused by the inability to breathe while buried is quickly fatal. Resolving bridged grain should be done remotely with a long pole, or by a trained permit-required confined space expert using a personal fall protection system (29 C.F.R. § 1910.140 and § 1910.146).

Other hazard controls that control admittance to silos are helpful. These can include limiting the keys to gain access to silo systems; signage warning of engulfment, asphyxiation, and falling hazards; and conducting regular training to alert workers.

Bulk Bags and Unloaders

Bulk grain sacks or bulk bags, technically called flexible intermediate bulk containers (FIBCs), may hold 550–3,000 lb. of grain (249–1,361 kg), with 2,000 and 2,200-pound bags being common (about 0.91 to 1.0 metric tons). These sacks are roughly cubical in shape and a single bag occupies a standard pallet. They are made of woven polypropylene plastic. On the top of the bag are four reinforced loops used for picking with a forklift or a hoist. A steel armature is typically used to hang the bag from the forklift or hoist in order to suspend the bag uniformly and keep it from lengthening while it is being emptied (see fig. 11.1). This framework is also called a bag lifting adapter, or spreader bar.

A common dispenser design involves hanging the sack above a hopper-shaped pan leading to a cylindrical chute. The sack has a fabric discharge spout at least 18-inches long sewn into its bottom. This spout is pulled down into the chute, where it can be opened by loosening a cord. Grain then falls downward into a boot containing the grain conveyor pickup. Grain can be weighed by installing load cells beneath the feet of the dispenser, or further along in the brewing process as long as the weight of grain inside the conveyor is counted. Occasionally, a brewery will purchase premixed grains according to its own recipes, and if the brew length is appropriate, say 30-40 bbl. (35-47 hL), a single bag will make a batch of beer.

The hazards associated with the use of FIBCs include forklift incidents, falling objects, and pinches and caught-between injuries. Placement of the sack by forklift requires a high lift, which can bring the load close to ceiling fixtures and even knock down lighting or bump sprinkler systems. More on forklifts is provided in the next part of this chapter, "Portable Powered Lifts and Trucks." Constructing homemade bag unloaders out of dimensional lumber or shelving is very unsafe and can lead to an unpredictable failure and falling materials weighing in excess of one ton.

As an alternative to using a forklift to position an FBIC, brewers can use an electric or manual chain hoist suspended from a steel I-beam in front of the unloader. If the I-beam is mounted to the grain unloading stand it must be firmly bolted into the floor because lifting the bag will cause substantial tipping forces on the system.

Figure 11.1. Bulk bags or FBICs. Bag mounted in unloader (left). Required FIBC warning label showing both unsafe and correct handling. Source: FormPak Inc. Reproduced by permission.

When hanging a full bag, hand pinching and crushing injuries are also possible. Be sure that the ground person and the forklift operator use established verbal or hand signals to communicate during this activity.

Some unloaders will have a pneumatic knife valve or gate (sliding plate) valve. Be certain the power to such valves is locked out during bag installation or removal. The grain conveyor pickup is another source of possible injury, as there may be another knife valve, a screw auger, or other moving parts. Again, be sure to apply lockout/tagout if putting the hands into any potentially dangerous component.

GRAIN CONVEYANCE IN PRODUCTION

Most commonly in small to medium-sized breweries, bulk grain and milled grain are moved using tubular conveyors. Tubing may be food grade PVC or ABS plastic, although polycarbonate or Teflon® may be used in special applications. Within the tubing is a movable conveyor that may be of drag chain/cable or steel screw auger design. Drag chain designs will use plastic disks mounted at intervals, or occasionally, cup-shaped bucket auger elements.

The hazards associated with augers in tubular conveyance systems include electrical, pneumatic, and rotational and linear mechanical energy. These systems are typically fully enclosed during operation, so the primary concern is that repair and maintenance work is done under lockout/tagout procedures. Conveyors may become blocked due to any number of reasons. Cleaning out the system usually involves gaining access at both ends and either toggling the auger backward and forward or blowing compressed air into the tubing. The latter can produce prodigious

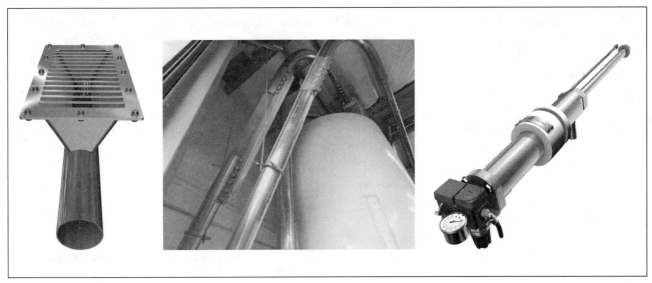

Figure 11.2. Components of a tubular grain conveyor (left to right): grain pickup typically mounted to hopper, sack unloader, silo, or grist case; tubular conveyor system; pneumatic cable tensioning device. *Source: Cabelvey. Reproduced by permission.*

amounts of irritating grain dust, requiring local ventilation or respiratory protection.

Other recommended controls include using only properly rated electrical controls and an adequate conveyor drive motor. Consult a conveyor expert when designing a new system, as there are limits to the steepness of lift, angles of bend, and grain volume throughput that particular systems can achieve. As with other grain-related activities, regular housekeeping to control dust accumulation is important.

Pneumatic grain handling is often the preferred method for truck to silo transfers and, less commonly, within the brewery too. Grain is moved by air powered by a blower. In some cases, such as when dispensing from a silo, a compressor may be installed to add a slight positive pressure to the vessel. This frees kernels that might otherwise stick to the sidewalls using the process of fluidization.

MILLING

Milling is the process of cracking grain kernels open to expose their endosperm to the brewing process. Most mills are dry roller mills, which include the hazards of pinch points, drive belts and pulleys, and electrical hazards. Some mills are more leaky than others and can liberate fines through their crevices. Another mill type is the wet roller, where the grain is hydrated prior to being milled. Wet rollers eliminate the problem of dust but require washdown procedures. Wet roller mills are more efficient than dry roller mills. A third type of mill is a hammer mill, usually used for special applications like spice milling.

The point of milling grain for brewing is not to make flour, but to yield large granules a bit smaller than the whole kernel. It is best to refer to cracking or breaking of grains to avoid a war of semantics with fire safety officials. Milling a grain into flour creates explosive dust, but milling brewing grains into coarse particles does not produce sufficient combustible dust. One can even show the marshal what the milled grain looks like. Nevertheless, some fire marshals will assert that milling is creating sufficient dust to risk a dust explosion and will want to apply OSHA's grain handling facilities standard. In light of this, the standard is listed in table 11.1. It should be mentioned that high-efficiency brewing systems (HEBS) and some distilling processes use a much finer grind than in conventional brewing, so accumulation of explosive dust could occur in those settings.

The more plausible hazard with grain dust is inhalation. Breathing in grain dust can be irritating, resulting in coughing and sneezing, as well as eye discomfort. It is well-known to brewers that some individuals are more prone to being irritated by grain dust than others. Exposure may trigger asthma, bronchitis, or some other respiratory sensitivity. For these individuals in particular, providing voluntary disposable dust masks can be helpful, as can any improvements in air handling and fugitive dust control.

Grain dust is not toxic or carcinogenic and is often referred to as a nuisance dust. OSHA mandates an 8-hour TWA PEL of 10 mg/m^3 (20 C.F.R. § 1910.1000 Table Z-1). The National Institute for Occupational

Safety and Health (NIOSH) and American Conference of Governmental Industrial Hygienists (ACGIH®) have a more conservative TWA exposure limit of 4 mg/m³ (US Department of Labor 2021).

To assess grain dust exposure, an industrial hygienist will collect a known volume of air from the worker's breathing zone over a period of three to six hours using OSHA method PV2121 or an equivalent method. Dust exposure is determined by weighing dust captured on a 5 µm pore size, 37 mm diameter PVC filter, and expressed in units of mg of total dust per cubic meter of air (US Department of Labor 2021).

If the PEL is exceeded, the employer needs to use engineering controls to get exposures below the PEL or institute a full respiratory protection program (discussed in chapter 9). If the PEL is not exceeded, the employer can permit the use of disposable dust masks without a respiratory protection program, so long as their use remains voluntary.

Controls that limit dust generation and accumulation will mitigate both explosion and inhalation hazards. Mills are best set up in their own room. Electric supply, switches, and lighting in the mill area should be rated for class II/division 2. This typically means that the mill motor, power disconnect, switches, and lighting will meet National Electrical Code (NEC) "dust ignition-proof" enclosures. The use of LED lamps is preferred because they are solid state and are not subject to breaking like fluorescent, incandescent, or halide bulbs.

A good housekeeping regimen is the best prevention of the accumulation of fine matter. Address crevices in any machinery that releases fines and seal them with an appropriate gasket, caulking, or tape. Sweep up and dispose of dust after each milling, or whenever there is a noticeable accumulation of fines. Dry sweeping, while discouraged, should be done gently. A shop vacuum can be used, but the plastic nozzle and hose tend to build up a static charge, which can create a static spark ignition source.

Never let dust accumulate greater than ⅛-inch thickness. Dust accumulations also attract vermin and can contaminate product with lactobacilli. Document the routine dust management practices in an SOP. This can act as the written housekeeping plan required under OSHA grain handling standard (29 C.F.R. § 1910.272(j)).

If your brewery has been unable to make a convincing case to the fire marshal that hazardous dust conditions will not be created, your options include installing a positive displacement ventilation system, installing a dust collection system and baghouse, or hiring a testing lab to conduct on-site testing to see how much dust is present outside the mill and conduct explosivity tests on the grain fines.

Periodically de-energize the mill and inspect the rollers for wear, foreign matter, and bearing integrity. Keep all housings and machine guarding installed to prevent crushing or amputation. If there is a manual dump hopper above the mill, ensure that the bottom of the hopper has a fixed screen or other guard above the rollers. This will keep hands and tools from coming into contact with the rollers. If utility knives are used in bag opening, tether the knife so it cannot reach the bottom of the hopper.

Milled grain drops towards the bottom of the mill where it is picked up by a tubular conveyor and sent to either a grist case or directly through a hydrator assembly into the mash tun. Control dust from conveyance downstream of the mill as you would at the mill.

Some breweries, especially larger ones, may be using a wet mill. Wet mills crush brewing grain under controlled hydration conditions. This results in no dust production and offers a higher brewing efficiency than mashing in with dry malt. Wet mills usually have built-in washdown cycles to reduce the presence of lactic acid spoilage organisms.

SPENT GRAIN HANDLING

Spent grain is wet, heavy, and more difficult to convey than dry grain. The classic method for spent grain management in small breweries is to rake spent grains out of the lauter tun into 55-gallon barrels, hoppers, bins, dumpsters, or similar. These heavy containers must then be moved to a position where they can be transported out for off-site management as animal fodder, composting feedstock, or municipal solid waste.

When spent grains are removed from the brewhouse with automated systems, it is usually by automated discharge from the lauter tun to a receiving hopper. At the bottom of the hopper is the pickup for a powerful positive displacement pump, which pushes the wet grain to either a waste bin or a spent grain silo through high strength piping.

The other possible source of wet spent grain is if a mash filter press is used in place of a lauter tun. These systems, common only in larger breweries, squeeze wort out of converted mash grain using a plate and frame filter press. The resulting spent grain is drier than that from a lauter tun and, consequently, this method improves ingredient usage efficiency. The spent grain usually falls beneath the filter press onto a belt conveyor, which carries the spent grain to a waste receiver like a bin or a truck.

Mash filter presses have the possibility for crushing or amputating fingers and for lifting injuries due to the weight of the plates. Mash filters should be treated like brew decks in terms of situational awareness, allowing the presence of authorized individuals only, and having written procedures for operation and when conducting maintenance (e.g., lockout/tagout procedures).

If left to sit around, particularly in warmer weather, spent grain will putrefy into an offensive-smelling waste that can attract vermin, stinging insects, birds, and flies. A spent grain silo confines some of the stink and controls pests to a good degree, but spent grain should always be removed from the brewery premises as quickly as feasible.

APPLICABLE STANDARDS

The primary standard that may apply to grain handling and milling in the brewery is OSHA's grain handling facilities standard. This is included along with other applicable standards in table 11.1. Some argue that malted cereal grains do not produce combustible grain dusts, and others cite the coarse grind required in brewing produces very little fine, combustible dust. While these suppositions are backed up by no anecdotal or reported dust explosions in brewing, there is always an outside chance of obtaining the right combination of accumulated dust, fine particle size, a means to get the dust airborne, and a source of ignition resulting in a dust explosion. The employer may consult with their OSHA authorities on applicability, or they may take the preventive approach of doing as much as is required by the standard and preempting any concerns a fire marshal might have.

GENERAL

This section pertains to portable, powered devices used to move a wide variety of materials, equipment, or personnel. Examples include forklifts, powered pallet jacks, scissor lifts, and boom cranes. Fixed handling devices, like conveyors, process piping, and vacuum and pneumatic systems, are discussed in the final section of this chapter (p. 139). Pressurized systems hazards are discussed in chapter 12.

Portable Powered Lifts and Trucks

	POWERED LIFTS AND TRUCKS – Linear, gravitational, tension, pressure, rolling, electrical, chemical
	Occurs: Plant-wide
	Outcomes: physical trauma due to blunt force, falling objects, caught between, asphyxiation; equipment and packaging damage

The main distinction between mechanized handling machines is whether they are designed primarily to move freight or workers. For the most part, a powered industrial truck (PIT) moves freight, whereas a mobile elevated work platform (MEWP) moves workers.

Cranes are separate from PITs and MEWPs, and are generally found at breweries only during construction or remodeling. There are several classes of cranes, each with their own standard: overhead and gantry cranes (29 C.F.R. § 1910.179 and 29 C.F.R. § 1926.1438), tower cranes (29 C.F.R. § 1926.1435), cranes and derricks in construction (29 C.F.R. § 1926.1400 et seq.), sideboom cranes (29 C.F.R. § 1926.1440), and other specialized crane rules for marine cargo, rail freight, and helicopters. It is recommended that the brewery ensures any crane contractor hired be duly licensed and compliant with applicable OSHA standards and state or local requirements. Crane use, rigging, and hoisting should not be performed by those without specialized training. Crane use is not discussed further in this book.

POWERED INDUSTRIAL TRUCKS

Materials handling equipment designed primarily for moving freight falls into the category of powered industrial trucks (PITs). These may be ride-on units like forklifts, or walk-behind devices like powered pallet jacks. All these tools translate freight across floors or terrain and also lift it to some degree, whether that be inches or many feet. Non-motorized pallet jacks, hand carts, dollies, etc., are considered manual materials handling tools and are discussed in the first section of this chapter.

Table 11.1 **OSHA and other consensus standards mentioning grain handling**

Short title of standard	Title 29 citation	Key requirements
Grain handling facilities	1910.272(c)	The employer shall develop and implement a written housekeeping program that establishes the frequency and method(s) determined best to reduce accumulations of fugitive grain dust on ledges, floors, equipment, and other exposed surfaces. (See national consensus standards NFPA 91 and 654.)
Exit routes, emergency action plans, and fire prevention plans	1910 Subpart E, Appendix (4)	Assure that hazardous accumulations of combustible waste materials are controlled so that a fast-developing fire, rapid spread of toxic smoke, or an explosion will not occur.
Duty to have fall protection and falling object protection	1910.28	Employer has duty to provide fall protection and protection from falling objects; head protection required for falling objects hazard.
Fixed ladders	1910.28(b)(9)	Requirements for fixed ladders extending more than 24 feet (7.3 m) above a lower level. Requirements for fall protection vary depending on whether ladder existed in place before and after November 19, 2018.
Fall protection systems	1910.29(b)	Guardrails
Housekeeping	1910.140(a)(3)	(iii) To facilitate cleaning, every floor, working place, and passageway shall be kept free from protruding nails, splinters, loose boards, and unnecessary holes and openings.
Handling materials	1910.176(c)	Storage areas shall be kept free from accumulation of materials that constitute hazards from tripping, fire, explosion, or pest harborage. Vegetation control will be exercised when necessary.
Permit-required confined spaces	1910.146	Entire standard applies due to hazards of engulfment, asphyxiation, and mechanical hazards.
	1910.146(b)	Definitions of engulfment and permit-required confined space.

National consensus standards from other professional organizations	
Organization	*Relevant code or standard*
National Fire Protection Association[a]	NFPA 91: *Standard for Exhaust Systems for Air Conveying of Vapors, Gases, Mists, and Particulate Solids*. 2020 ed.
	NFPA 654[b]: *Standard for the Prevention of Fire and Dust Explosions from the Manufacturing, Processing, and Handling of Combustible Particulate Solids*. 2020 ed.

[a] "Codes & Standards," National Fire Protection Association, accessed February 16, 2023, https://www.nfpa.org/Codes-and-Standards.
[b] This is due to be consolidated into PFPA 660 in 2024.

Fuel Types and Classes

PITs may be powered by batteries or petroleum combustion. Historically, battery-operated trucks used lead-acid batteries, but lithium-ion batteries are becoming more common. Lithium-ion battery banks cost more initially, but charge faster, do not require time to cool off before use, are lower maintenance, and have lower toxicity than lead-acid systems. Lead-acid batteries also have the downside of emitting flammable hydrogen gas during charging. Future power sources may include hydrogen fuel cells or fuel cell-battery hybrids.

Fuels used in internal combustion PITs include propane (as liquified petroleum gas), gasoline, and diesel. A major concern with internal combustion PITs is the production of carbon monoxide (CO) in the exhaust. If PITs are to be used in enclosed areas, like refrigerated storage or tractor trailers, efforts to reduce, monitor, and ventilate CO should be foremost. Carbon monoxide is a chemical asphyxiant that starves blood cells of oxygen. It is colorless, odorless, tasteless, can sometimes mimic the flu, and can take overexposed individuals by surprise.

The OSHA regulations for PITs used in manufacturing are found in 29 C.F.R. § 1910.178. OSHA classifies industrial trucks or tractors into 11 categories based on fuel type and the conditions of use related to the nature of fire hazards in the workplace. These are shown in table 11.2.

All new and used PITs need to meet design and construction standards of ANSI B56.1-1969 (29 C.F.R. § 1910.178(a)(2)). PITs should not be modified unless by a qualified mechanic. In paragraphs (b) and (c) of the standard, OSHA delineates a wide variety of regulated equipment. OSHA also provides an eTool that has substantial additional content regarding subtypes of equipment, operations, situational safety, and training (US Department of Labor n.d.ᶜ).

Many breweries are likely to not have the kind of fire hazards requiring specialized PIT designations. If this is the case, type E, G, D, or LP lifts will be adequate (see table 11.2). If a brewery also has distilling operations or is handling bulk spirits for producing hard seltzers or ready-to-drink mixtures, it may be considered a Class I/Group D location. If a risk of grain dust explosion exists it may be considered a Class II/Group G location. Consult OSHA tables in 29 C.F.R. § 1910.178(c) to aid in equipment selection.

It is likely that EE, EX, or DY designations will be required in these settings.

Seated Operation

PITs where the operator is seated on the device include all except Class III. Most common in a brewery are Classes I and IV with solid or cushioned tires (table 11.3). Cushioned tires are best on smooth floor surfaces and have easier maintenance than pneumatic tires.

While in use, operators should be certain to always use the seat belt and keep their hands inside the operator's cage. Many forklifts have a seat occupancy interlock that keeps the forklift from being able to move when there is no operator in the seat.

Standing Operation

Classes II and III (table 11.3) may be designed for the operator to stand on the device behind the controls or walk behind or beside the device during operation. Equipment in these classes include electrically powered narrow aisle trucks ("order picker"), electric motor hand trucks or hand/rider trucks, standing operator powered pallet jacks ("rider pallet truck"), and powered low-lift jacks that incorporate a scissor mechanism. Hand-operated pallet jacks that are raised by manually pumping a hydraulic

Table 11.2 Industrial truck and tractor fuel types and fire safety designations

Designation	Fuel	Description
E	Battery	Electric; has minimum acceptable safeguards against inherent fire hazards
ES	Battery	Electric; as type E, but with additional safeguards to prevent emission of hazardous sparks and limit surface temperatures
EE	Battery	Electric; as with type ES, but has the electric motors and all other electrical equipment completely enclosed
EX	Battery	Electric; units differ from E, ES, or EE units in that the electrical fittings and equipment are so designed, constructed, and assembled that units may be used in certain atmospheres with flammable vapors or dusts
G	Gasoline	Gasoline; has minimum acceptable safeguards against inherent fire hazards
GS	Gasoline	Gasoline; has additional safeguards to the exhaust, fuel, and electrical systems
D	Diesel	Diesel; has minimum acceptable safeguards against inherent fire hazards
DS	Diesel	Diesel; has additional safeguards to the exhaust, fuel, and electrical systems
DY	Diesel	Diesel; as with type DS, but does not have any electrical equipment including the ignition and are equipped with temperature limitation features
LP	Liquified petroleum gas	LPG; similar to G but uses LPG as fuel; has minimum acceptable safeguards against inherent fire hazards
LPS	Liquified petroleum gas	LPG; has additional safeguards to the exhaust, fuel, and electrical systems

cylinder are not considered motorized and are not covered by the PIT training requirements.

OPERATOR CERTIFICATION

Powered industrial trucks should only be driven by certified operators, just as a license is required to operate an automobile, truck, or motorcycle over the road. Only persons aged 18 years and older may operate a forklift in a workplace setting. Every single person who operates a forklift must be up to date on operator certification training. This course includes classroom instruction and a field practicum including the type of equipment and workplace setting that the operator will be working in. The certificate is good for three years and can be given by a qualified individual employed at the brewery, at a trade school, or other off-site trainer. OSHA requires all operator training and evaluation be conducted by "persons who have the knowledge, training, and experience to train powered industrial truck operators and evaluate their competence" (29 C.F.R. § 1910.178(l)(2)(iii)).

Operator certification includes an understanding of the specific equipment being used, the setting in which it is used, how to conduct inspections, operating and maneuvering equipment, assessing loads, and safe pickup and offloading. Specifically, a PIT training program should cover the following:

- Truck-related training on equipment equivalent to workplace PITs
  - Warnings, precautions, operating instructions
  - Difference between PIT and automobile
  - Truck instrumentation and controls
  - Starting, operating, and controlling motor
  - Limitations of the equipment
  - Capacity and stability, including data plate and stability triangle
  - PIT inspections, maintenance, and recordkeeping
  - Refueling (if combustion) or recharging (if electric)
  - Steering, maneuvering, picking and dropping loads
  - Visibility and limits to visibility
  - Forks and ancillary equipment, such as tilt control, back plate, mast types, and any specific attachments, like drum grappler or super-sack spreader bars
- Workplace-related training
  - Surface conditions
  - Load composition and types
  - Load handling: manipulation, picking, stacking, unstacking, and unloading
  - Pedestrian traffic areas and established rights-of-way
  - Hazardous locations
  - Inclines, ramps, and docks
  - Narrow aisles, congested or restricted areas
  - Enclosed areas where the breathable atmosphere could be compromised
  - Any other unique or hazardous areas
- Refresher training
  - Required every three years
  - Required if the operator is observed to be operating unsafely
  - Required if the operator has been in an accident or near miss incident
  - Whenever the operator is assigned to operate a different lift device
  - Whenever a condition in the workplace changes in a manner that could affect safe operation of the truck

Table 11.3 **Common powered industrial truck classes**

Class	Fuel	Description
I	Battery	Electric motor rider trucks for smooth surfaces
II	Battery	Electric motor narrow aisle trucks
III	Battery	Electric motor hand trucks or hand/rider trucks
IV	Internal combustion	Internal combustion engine trucks (solid/cushion tires) for smooth surfaces
V	Internal combustion	Internal combustion engine trucks (pneumatic tires)
VI	Battery or combustion	Electric and internal combustion engine tractors (e.g., aircraft tugs)
VII	Battery or combustion	Rough terrain forklift trucks

SAFE USE OF POWERED LIFTS AND TRUCKS

Safe operation of powered materials handling systems relies on all types of hazard controls depicted in the hierarchy of controls. The best place to start is to understand the types of PIT and MEWP accidents and injuries that are most common (table 11.4). By understanding how injury results from various actions or inactions, we can choose the best available control options.

Pre-Operation

A pre-operation inspection is required at the beginning of each shift where the equipment is being used. This is best accomplished with a checklist developed for a specific piece of equipment. First, ensure that the standard safety features are in place and operational. These are the brake, horn, mirrors, lamps, and seat belt. Some machines will have additional safety features, like a mast-tilt limiter, backup alarm or light, seat interlock, or a mounted fire extinguisher.

Next, examine the tires for wear or chips. The mast and fork assembly should be complete, with fork blades properly engaged in their holding notches, a load backrest, and a mast that is properly lubricated and not showing any signs of hydraulic oil leaks from the lift or tilt cylinders and associated hoses and fittings. There should be no parts missing or deformed. Any modifications that have been made should be with purpose-built aftermarket components installed by a qualified mechanic.

It is important to keep forklifts clean and operational. Cleanliness includes keeping the mast and related components free of excess lubricants and the data plate legible. Remove from service and repair any forklift found to be damaged or lacking necessary safety systems. Conduct repair and maintenance according to the manufacturer's recommendations.

Driving and Maneuvering

Only trained and certified workers may operate a PIT. Ensure that the training includes the specific types of trucks being used. Due to the inherent hazards of PITs, operators need to use the equipment responsibly. This means there should be no horseplay, practical jokes, or allowing riders.

Table 11.4 **Common accidents with powered industrial trucks and mobile elevated work platforms**

Event	Hazard controls
Tips over on a person	Maintain pedestrian corridors; assign rights-of-way; travel with loads low to ground; control speed on turns; face inclines with load on uphill side; wear seat belt; keep travel surfaces are free of oils, spills, and debris
Lift strikes a person	Maintain pedestrian corridors; assign rights-of-way; use horn and backup alarm; never approach a fixed wall or structure with a pedestrian in between
Load or stack falls on a person	Secure loads with straps or stretch wrap; stack sensibly on secure shelving; keep pedestrians away during stacking; never walk below a lifted load or a lift engaged in stacking; travel with loads low to ground; control speed on turns; face inclines with load on uphill side; ensure lift has overhead guard; wear seat belt
Collision injures operator	Wear seat belt; be aware of surroundings, control speed, and avoid distractions
Operator injures exposed body part	Wear seat belt; keep body parts inside operator lift cage; be aware of surroundings; control speed
Passenger falling from forks or lift	Do not carry passengers; do not carry persons on platform or fork blades or without purpose-made lift platform; use fall protection system on lift platform; substitute other system like ladder or rolling platform, or scissor lift for forklift
Injury getting in or out of vehicle	Be aware of surroundings; avoid rushing; use handholds; wear appropriate work clothes
Body part caught in vehicle mechanism	Avoid loose clothing and PPE; operate the lift from the correct operator's position
Platform occupant pushed into ceiling	Be familiar with controls; use a lift with a cage structure or equipment with railings for lifting personnel; establish verbal communication plan; allow lifted persons to control lift, if possible

If unsafe operation is observed, if there is a near-miss incident, or an accident or injury, the involved operator should receive refresher training and recertification.

Every brewery should establish house rules for their PITs to include rights-of-way, traffic lanes, speed limits, and any other policies. Right-of-way rules describe whether the forklift or pedestrian has the right of way and should be consistent throughout the entire operation. Rights-of-way can be maintained with signage, floor markings, designated pedestrian walkways, horn use, and eye contact between the PIT operator and pedestrians. Mirrored safety eyewear is ill-advised because it prevents direct eye contact. Right-of-way training should be given to all persons in the production and warehousing areas, not just equipment operators.

Use proper footing and the handhold, if available, when entering or exiting the lift. Wear a seat belt always, even if only using the equipment momentarily. Make seat belt use habitual and mandatory.

Read and know how to utilize information on the data plate (fig. 11.4). Never exceed the rated load. Learn what it takes to ensure loads are stable and balanced. Understand load stability and the stability triangle (*see* p. 135). Do not raise or lower the load while traveling and always travel with the load in a low position. Keep safely away from dock edges, inclined surfaces, and overhead hazards like lighting and sprinkler systems. When loading into trailers from a dock, require the use of vehicle chocking, parking breaks, and dock plates to secure the trailer in place. Be aware of other PITs, vehicles, and pedestrians in the work area.

Strive to maintain clear visibility of the work area and make sure there is adequate clearance when raising, loading, and operating a forklift. Never give rides or use the forks to lift people. Personnel should only ever be lifted with a suitable work platform and while using fall protection apparatus.

Whenever possible, it is recommended to mark the floor with high-visibility lines to define corridors for PITs and for pedestrians. Folding signs, barrier tape, or stanchions can be used to temporarily indicate pedestrian pathways. Learn where bollards, corners, and other protrusions are and mark them with high visibility coloration.

Handling Loads

In handling loads, both the capabilities of the PIT and the characteristics of the load should be known before picking and moving the load. The lifting capability of the equipment is conveyed on the data plate, most often found on the dashboard of the PIT (*see* fig. 11.4).

Understand the load composition, that is, its weight, center of gravity, and how best to position the load for picking up. Do not exceed the safe load capacity of the PIT. To ensure this, consider the load composition,

Figure 11.3. The major components of a forklift are standard. Forklift training is necessary and important! © *Getty Images/Nerthuz (forklift)*

forklift mast tilt, surface slope, and maximum required lift height. When load composition is asymmetrical, pick it up only slightly and observe the behavior of the PIT. Does the PIT lean forward, requiring the mast to be tilted back? Does it lean to one side, meaning the load should be side shifted?

Equipment Limitations

Avoid using a PIT for tasks outside of the capability of the device or the operator. When erecting brewery vessels and other large components, first assess whether the PIT is appropriate and adequate for the job. The process of lifting a vessel from lying on its side to a

vertical position can often exceed the stability of the PIT. A crane or other hoisting device may be a better solution. If an attempt is made for an unusual lift and the PIT becomes unstable, stop work and find a safer means of performing the work.

Stability Triangle

Operators should learn about the forklift stability triangle and apply its principals in everyday equipment use. The stability triangle is a triangle in two dimensions and forms a tetrahedron in three dimensions. From the uppermost vertex of this shape hangs an imaginary plumb line with a weight (the plumb bob), which is free

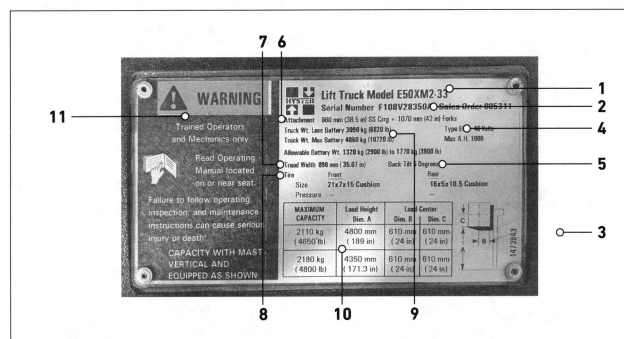

1. Model number
2. Serial Number
3. Mast Type: may be 2-stage, 3-stage, or 4-stage depending on lift height capability; not always shown for single stage masts, as in this example
4. Fuel Type: standardized code, see table 11.2
5. Back Tilt: the degrees back the mast can tilt to help keep loads secure on the forks
6. Attachments: what attachments have been added to the forklift, examples include side shifter, layer picker, drum grappler
7. Tread Width: the operational width of the forklift

8. Tire Size: usually shown as width then inside diameter (rim diameter); types include: solid, pneumatic, and hybrid solid with cushioning
9. Truck Weight: overall weight of the truck
10. Forklift Diagram: offers one or more data points important for understanding the balance and lift of the forklift
    A. Maximum fork height at different loads
    B. Front-back load center
    C. Vertical load center
       If equipped with left-right shift, other stability data can be shown
11. Only trained operators who have read and understood the operator's manual should operate forklifts

Figure 11.4. Use the information from the forklift data plate to know and understand what the limitations of your forklift are.
Source: Brewers Association.

to swing in any direction. The plumb bob represents the center of gravity of the system. In a standard forklift at rest, the bottom plane of the tetrahedron is between the hubs of the two front wheels and a center point beneath the counterweight in the rear, such that the center of gravity exists beneath the operator, low and centered. See figure 11.5, where the center of gravity is represented by the ◕ symbol (the Secchi disk).

As the mast is lifted, loads are added, or slopes or movement are experienced, the center of gravity of the entire system will move upward, sideways, and front to back of this initial point (figs. 11.6–11.10). The system could be an empty forklift, a loaded lift, or a lift in motion. If in motion, it could be undergoing instances of momentum and centrifugal force due to accelerating, decelerating, or turning.

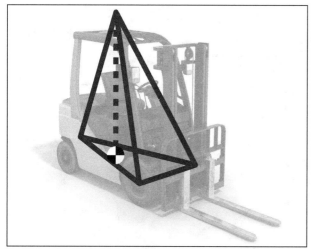

Figure 11.5. Stability triangle of unloaded forklift at rest.

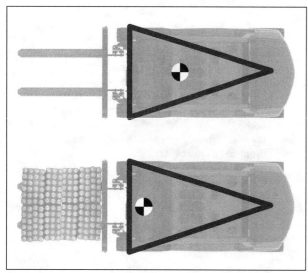

Figure 11.6. Stability triangle of unloaded and loaded forklift at rest as seen from above.

Figure 11.7. Stability triangle of loaded forklift at rest on level surface as seen from the front, side, and top.

Figure 11.8. Stability triangle of loaded forklift at height on level surface, still within the stable bounds. Front, side, and top views.

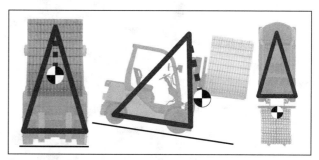

Figure 11.9. Stability triangle of high loaded forklift on inclined surface. Forklift will tip over forward. Ensuring the load always faces upslope will maintain the center of gravity in the stability triangle. Stopping quickly on a flat surface can cause the same kind of instability due to forward momentum. Front, side, and top views are shown. Notice not all views must be outside of the stability triangle for the system to tip over.

Figure 11.10. Stability triangle of loaded forklift at height on side slope. Forklift will tip over sideways. Shifting the load to the upslope side of the forks can help to achieve stability. Front, side, and top views are shown. Notice not all views must be outside of the stability triangle for the system to tip over.

The value of the stability triangle is in recognizing under what circumstances the center of gravity will move outside the walls of this imaginary volume, even for a brief moment. Imbalance can cause the truck to tip forward, drop its load, and tilt rapidly backward in response. In some cases, forward tilt can throw the operator out of the seat into the mast. If carrying a high load, a maximum safe load, or a load on an incline, the stability triangle gets very small—another way to think of it is that the plumb bob is hanging very high on a short line and doesn't have to move much to hang outside of the triangle. Under these circumstances, the forklift is prone to tip over sideways or forward. If the operator is not wearing their seat belt they can be easily thrown out of the cab and end up underneath the tipped equipped. This type of accident is the single most frequent cause of forklift fatalities.

MOBILE ELEVATED WORK PLATFORMS

Mobile elevated work platforms, or MEWPs, are platforms used mainly for lifting workers into position. The term platform is meant to convey any box, bucket, or railing-enclosed platform designed to hold one or more workers. MEWP is a general term that encompasses scissor lifts, aerial lifts, powered platforms, and vehicle-mounted platforms.

MEWPs come in a variety of classes and can be powered electrically, with internal combustion, or, in some cases, with manual cranks. All new and used MEWPs need to meet design and construction standards of ANSI A92.2-1969 (29 C.F.R. § 1910.67(b)(1)). For frequent use in a brewery setting, the most likely devices will be the scissor lift and the boom lift (fig. 11.11).

Scissor lifts fall into Group A. They may be controlled solely from the chassis (Type 2) or from both the chassis and the platform (Type 3). A scissor lift with dual controls will therefore be called a 3A lift. Type 2 and 3 lifts can be driven while extended. With a scissor lift, the platform does not move horizontally outside of the tipping axis.

Group B includes boom lifts with the same control type options as scissor lifts. A boom lift with dual controls would be classified as a 3B lift and can be driven while extended. Importantly, a boom lift work platform can be moved laterally beyond the tipping axis, making these fundamentally more challenging to operate than scissor lifts. Fall restraint systems are required for boom lifts.

MEWPs require the use of fall restraint or fall arrest systems. A fall restraint, such as a lanyard, prevents a worker

Figure 11.11. Common mobile elevated work platforms: scissor lift, dual controls (*left*) and boom lift, dual controls (*right*).
© GettyImages/anmbph

from being exposed to a fall. Fall arrest systems limit the employee's fall or prevent them from coming into contact with a lower surface. These often involve a full body harness and self-retracting lifeline. A belt and lanyard can be used to limit the fall to two feet, but a full body harness is required for arrest up to six feet. Rules laid out in 29 C.F.R. § 1910 Subparts D, F, and I may be applicable; see §§ 1910.28, .29, .67, and .140 in particular. Further discussion on fall protection is included in chapter 10 (see p. 118).

In general, the rules for MEWPs in OSHA's general industry standards are more scant than those for PITs, but training and daily inspection prior to use are still required. In developing procedures for the use of a MEWP in the brewery, it is recommended that the employer create a program consistent with the PIT requirements.

APPLICABLE STANDARDS

The American National Standards Institute's ANSI B56.1 standard series originated in 1969 and formed the basis for much of OSHA's powered lift standards. These are listed in table 11.5. There are two essential OSHA standards for powered lifts in manufacturing: one for PITs (29 C.F.R. § 1910.178) and the other for MEWPs (29 C.F.R. § 1910.67). Breweries may need to comply with one or both, depending on their equipment.

OSHA also publishes numerous additional guidelines. Two important ones are the compliance assistance guide and the age prohibition guidance. Also shown in table 11.5 are three consensus standards relevant to forklift design and safety from ANSI, NFPA, and ISO.

Where machine guarding exists as a perforated metal screen, wire barrier, or sheet of expanded metal, a machine guarding gauge is a helpful tool. One such gauge—the "gotcha tool" in common parlance—simulates the size and distance of different parts of the human arm and measures whether a body part could reach the point of the mechanical hazard. If the hazard is close to the guard, a finger must not be able to reach it. If the hazard is farther behind the guard, then you measure if a hand, forearm, or upper arm could reach through. Figure 11.13 shows this tool.

Table 11.5 **OSHA and related consensus standards pertaining to powered lifts**

Short title of standard	Title 29 citation	Key requirements
Powered Industrial Trucks	1910.178	Applicable to any electric or internal combustion powered lift truck, fork truck, motorized hand truck or motorized pallet jack; includes training, inspections, and safe operations.
Mobile Work Platforms	1910.67	Applicable to mobile elevated work platforms (e.g., scissor lifts, boom lifts, bucket lifts, etc.) where a worker is positioned for work atop a vehicular device. Requires training and personal fall arrest system.
Fall Protection	1910.28, .29, .140	Fall restraint, fall arrest, and fall protection systems for use with mobile elevated work platforms.

OSHA Compliance Assistance and National Consensus Standards	
Compliance Assistance for the Powered Industrial Truck Operator Training Standards	OSHA's compliance assistance directive CPL 02-01-028 draws from 5th edition of ANSI B56.1 *Safety Standard for Powered Industrial Trucks* to modernize OSHA PIT training requirements and clarify training requirements.[a]
Prohibition Against Young Workers	Safety and health information bulletin "Protecting Young Workers: Prohibition Against Young Workers Operating Forklifts" prohibits employees under 18 years of age from operating forklifts.[b]
Safety Standard, Improved Surfaces	ANSI/ITSDF B56.1: *Safety Standard for Low Lift and High Lift Trucks.* Effective October 7, 2010; defines safety requirements relating to the elements of design, operation, and maintenance of low lift and high lift PITs controlled by a riding or walking operator, and intended for use on compacted, improved surfaces.
Fire Safety Standard for Powered Industrial Trucks	NFPA 505: *Fire Safety Standard for Powered Industrial Trucks Including Type Designations, Areas of Use, Conversions, Maintenance, and Operations.* Identifies industrial truck types for use in locations classified as hazardous, truck conversions, and maintenance/operation of electric and internal combustion PITs.
International Consensus Standard for Industrial Trucks	ISO 3691-1:2011: *Industrial trucks — Safety requirements and verification — Part 1: Self-propelled industrial trucks, other than driverless trucks, variable-reach trucks and burden-carrier trucks*

ANSI, American National Standards Institute; ISO, International Organization for Standardization; ITSDF, Industrial Truck Standards Development Foundation; NFPA, National Fire Protection Association.
[a] US Department of Labor (2000).
[b] US Department of Labor (2003).

Motion Hazards: Machinery, Vehicles, and Flying Objects

MOTION – Linear, rotatory, vibratory, rolling, reciprocal, projectile	
Occurs: Materials handling, forklifts; milling and mixing; pulleys, gears, hydraulic systems, fans; palletizing machines, packaging line components; fleet vehicles	**Outcomes:** Contusions, crushing, lacerations, punctures, fractures, eye injury; hair pulling, amputation; spinal injury, paralysis, death.

Figure 11.12. Examples of older machines or those missing guards: *Left*, older model rotary filler with pinch points between filler nozzles and bottles and entanglement hazard of carousel (shown not in operation); *Middle*, conveyor gear drive pinch point; *Right*, gear drive with warning sign, but easily reached drive gear and chain pinch point.

MACHINERY DESIGN AND INSTALLATION

Machines can often be designed in a way to eliminate mechanical hazards or reduce them. During packaging, older conveyors often used drive motors with chains and gears or belts and pulleys, both systems requiring the need for hazard enclosure. Newer designs employ enclosed, direct drive motors that move the conveyors without chains or belts, leaving only a small risk of pinching beneath the conveyor belt or plates, which can be mitigated by lesser guarding.

In some case, robotics have taken the worker entirely out of the area where they could be injured either by mechanical hazards or repetitive motion. Such robotic devices include autonomous depalletizers and palletizers, order loaders, keg handlers, and variety pack builders, among others. Keep in

Figure 11.13. A machine guarding gauge is used to make sure body parts cannot reach moving parts through machine guards.

mind that controls around the robot need to be in place to prevent a person from getting injured. Those controls may range from traditional metal mesh guards to light curtains or microwave (radar) proximity detectors.

When purchasing machinery, brewers are advised to be critical of equipment design and look for built-in controls. Simply labeling a pinch hazard with a pictogram may at first seem adequate, but there may be a better, more systemic, solution that can be used.

Machines often vibrate, even invisibly, and this can cause them to "walk" across the floor and prematurely fail due to loosening of fasteners. Equipment prone to vibration should be fixed to a floor or wall surface with suitable expansion bolts or mounted on rubber dampeners, or both.

SAFE WORK PRACTICES

Machines should be inspected before every use. Considerations are rigidity and support; the condition of bearings, drives, and other moving parts; proper fitting of attachments; and the presence of appropriate guards or other protective systems. Do not hybridize or modify equipment or use substandard parts.

OSHA generally allows the employer to put into place whatever hazard controls are effective for the situation. Brewers are also, generally speaking, a creative lot. Together, this makes for a wide-open field when it comes to controlling mechanical hazards. This is predicated on understanding where stored (potential) mechanical energy is, and how it can be released as kinetic energy. Once we understand how to avoid the unintended or unexpected release of mechanical energy, developing controls is mostly straightforward.

One creative solution is using a simple wood dowel as a poker to dislodge packaging errors like a mis-laid lid or an improperly seated bottle on a filler. Granted, the stick itself could be described as an engineering control, but the invention of it, and the determination to use it instead of one's hand, is behavioral and can be considered a safe work practice.

Another safe work practice is to identify and restore any degraded mechanical hazard guarding. Workers will defeat guarding if they are frequently having to get past it. Consider why this is the case. Is a repair or an adjustment needed?

ADMINISTRATIVE CONTROLS

Practice and workplace reinforcement of methods will help get the most out of mechanical hazard control.

Training on each piece of equipment should be mandated. Do not expect a person to know how to use a device just because its operation seems obvious. Also remember to document that training.

Conduct hazard assessments of activities involving these hazards. Develop SOPs that will outline the equipment's use and provide safety advice. Lockout/tagout (LOTO) should be a part of every workplace that has mechanical hazards and applied when machines are being serviced or repaired. See chapter 19 for more on this requirement. When a LOTO procedure is required, mention it in the SOP so it does not get overlooked.

Signage is important with these hazards. Use simple pictograms to convey the nature of the hazard (crushing, pinching, flying object, etc.). Signage is reinforcement, but not an adequate control by itself.

PERSONAL PROTECTION

Just as with selecting other controls, understanding the mechanism of a mechanical hazard directs one toward the best PPE. Most commonly, eye, hand, and overall body protection is appropriate. Flying objects occur at such speeds and with such unpredictability that safety glasses should be worn whenever there is such a hazard. Reflexes may occasionally save the day, but most often it is a pair of safety glasses. Their use should become habitual.

If the worker is reaching into moving parts or working around flying debris, good hand protection may be required. Leather, canvas, or synthetic work gloves will provide more physical protection than disposable nitrile gloves.

Finally, wear tough work clothes that will deflect, or at least slow, impacts from machines or flying materials. Most of the time this will mean a work outfit consisting of long pants with the pant legs over suitable work boots, and a short- or long-sleeved work shirt. The best materials for these are synthetic, 100% cotton, or cotton-polyester blends in poplin, canvas, or duck-type weaves. If warmer layers are needed, they can be cotton-flannel, wool-synthetic blend, or have polyester fill. Clothing material recommendations for greatest physical impact protection may not be as thin or as cool as those recommended for hot environments.

APPLICABLE STANDARDS

OSHA's standards in 29 C.F.R. § 1910 Subpart O do not specifically describe machine guarding for random fixtures, the agency choosing instead to categorize rules by specific classes of machines. Table 11.6 lists some OSHA standards most relevant to brewery operations. If one encounters mechanical hazards that could cause nipping, pinching, entanglement, or amputation in any brewery equipment, it is prudent to fall back on general guarding principles like those established in §§ 1910.212 and .219.

Systems not specifically called out in the rules but where guarding might be warranted include grain mill hoppers, packaging line devices (rinsers, fillers, crowners, lidders, laser or X-ray inspection stations, and labelers), packaging conveyors, palletizers and depalletizers, and robotic pallet loaders and the like. A hazard assessment will help identify hazards associated with these machines.

Table 11.6 Occupational Safety and Health Standards (29 C.F.R. Part 1910) pertaining to motion hazards

Short title of standard	Title 29 citation	Key requirements
Machine guarding	1910.212	General requirements
Woodworking machinery requirements	1910.213	Rules for table saws, circular saws
Abrasive wheel machinery	1910.215	Detailed prescriptive rules for bench grinders, portable grinders, including guard design and tool rests
Safety color code for marking physical hazards	1910.144	Safety color code for marking physical hazards
Specifications for accident prevention signs and tags	1910.145	Specifications for accident prevention signs and tags
Lockout/tagout	1910.147	LOTO to be used when servicing or maintaining machines and equipment in which unexpected energization or start-up could occur
Powered industrial trucks	1910.178	Applicable to any electric or internal combustion powered lift truck, forklift, motorized hand truck or motorized pallet jack; includes training, inspections, and safe operation
Powered platforms	1910.66	Powered platforms for building maintenance
Vehicle elevating and rotating work platforms	1910.67	Vehicle-mounted ladders and baskets
Manlifts	1910.68	Manlifts: general requirements, guards, mechanical specifications
Cooperage machinery	1910.214 [Reserved]	Regulations not specified, but a barrel rolling machine should have an interlocked cage or similar control

12

PRESSURE HAZARDS: COMPRESSED GASES, PNEUMATICS, AND HYDRAULICS

Pressure Hazards: Compressed Gases, Pneumatics, and Hydraulics

	**PRESSURE – Mechanical potential energy, projectile, flow**	
	Occurs: In cylinders, compressors, vessels and hoses; plumbing of all types	**Outcomes:** Mechanical and line-of-fire injury due to fast moving solids, liquids, and gases; skin penetration, eye injury, blunt force trauma, burns and frostbite; simple or chemical asphyxiation, engulfment; damage to equipment

PRESSURIZED FLUIDS AND GASES

In any brewery there are compressed gas supplies and machinery that can pressurize fluids or gases, even solids. Pressure hazards consist of potential energy that is ready to become kinetic energy. When pressure is released, the conversion of potential energy to kinetic energy can do tremendous damage to equipment and persons. Hence, pressure hazards are very dangerous for workers. All it can take is for a valve to be opened at the wrong moment or for a fitting to fail.

COMPRESSED GASES AND AIR

In a brewery, gases under pressure can be found in compressed gases that are dispensed from cylinders, gases generated on-site and stored using compressors and pressure tanks, and gases produced in vessels under pressure during brewing processes. Sources of gas that are purchased, like compressed gas cylinders, will have an accompanying safety data sheet (SDS), but those produced at the brewery will not. Brewers can keep a fact sheet on each gas produced on-site as an ersatz SDS. It may be helpful for the reader to review the "Hazard Communication" section of chapter 16.

Compressed Gas Cylinders

Carbon dioxide (CO_2), nitrogen (N_2), and oxygen (O_2) are the commonest cylinder gases purchased by breweries. Gas blends are common if producing or serving nitrogenated beers, or if running a long-draw draught system. Premixed CO_2/N_2 blends (called beer gas) can be bought. A common blend is 75% N_2/25% CO_2, but other proportions may be found. A brewery may form its own gas mix by passing CO_2 and N_2 through a gas blender.

Depending on the laboratory facilities, a brewery may also have hydrogen (H_2) or helium (He) gas cylinders. A shop outfitted for gas welding could have O_2, acetylene, and argon (Ar) or other inert gas. Propane may be used to power forklifts, pilot brewing systems, and warehouse heaters; in more remote brewery locations, propane may fire the boiler and/or a direct-fire kettle. Shop areas may have small quantities of pressurized, flammable gases such as propane or MAPP gas for brazing, or aerosol cans of paint, cleaners, or pest control products.

Compressed gas cylinders, tanks, and cans are at risk of exploding when exposed to a building fire. Besides the mechanical hazard that rapid pressure release poses, compressed gases often have additional hazardous properties. Of those listed here, H_2 and acetylene react explosively.

An O_2 leak can increase the flammability of things like hair and clothing. Carbon dioxide, N_2, and Ar can all displace O_2 in the air and cause asphyxiation. Gases used in breweries are colorless and most are odorless. They may readily mix with the air in a room and cause hazardous conditions that are not noticeable at first.

OSHA requires employers to visually inspect gas cylinders to make sure they are in "a safe condition." This may be interpreted as being labeled, identified as full or empty, stored in a secure manner so as not be knocked over, kept away from hazards such as flammable materials or corrosives, and fitted with a safety relief device.

Cryogenic Bulk Tanks

A cost-effective way to store and dispense gases that are consumed in larger quantities is with a cryogenic bulk tank, often just called a bulk tank or bulk cylinder. Cryogenic liquids are those with boiling points below −130°F (−90°C). Liquid N_2, O_2, and Ar all have boiling points around −300°F (−184.4°C). Carbon dioxide turns to a gas at −109.3°F (−78.5°C), which means it is easier to store than the other gases and is not always defined as a cryogenic liquid.

Cryogenic bulk tanks hold lower pressures than standard gas cylinders, typically 250 psi or less (≤1,724 kPa), but they hold a larger quantity because the gas is liquified and kept at a very cold temperature. The construction of a cryogenic tank includes a stainless steel inner pressure tank, an outer steel tank, and a space between them that contains insulation and is kept under a vacuum.

Breweries may keep bulk CO_2 or N_2 in cryogenic bulk tanks. Bulk O_2 storage involves additional OSHA requirements. A major difference in the dispensing systems for bulk gases is that an expansion tank is needed to allow the liquid gas to vaporize before it can be used as a production gas. Sending liquified gas to a piece of equipment or process can cause it to rupture. The rapid expansion of liquified gas also makes fittings extremely cold—direct contact with valves or piping can cause frostbite.

Controls for bulk liquified gases are the same as for any gas under pressure, with additional protections needed against rapid expansion, freezing, and runaway leaks. An essential control is to make sure each component of the system can be quickly isolated by valves and that those valves during rapid freezing can still be operational and handled safely. Bulk gas release episodes can also change the makeup of air in a workspace, potentially causing asphyxiation.

On-Site Gas Generation and Capture

As an alternative to purchasing compressed gases for all production and dispensing needs, a brewery may capture N_2 from the atmosphere, or CO_2 from fermentations. Nitrogen generators separate N_2 from air by removing O_2 with either molecular sieve adsorption or membrane filtration. Both types operate under pressure and discharge O_2 to the atmosphere. If the N_2 is meant to be stored, a pressurized receiver can be filled either by system pressures or with an assistive compressor.

Carbon dioxide recapture systems were once only available to large breweries, but smaller systems are now being engineered. These work by condensing liquified CO_2 from fermentation exhaust gas using glycol refrigeration. A foam trap pre-cleans the gas stream, then ethanol vapors, water vapor, and hydrogen sulfide (H_2S) are removed through various methods. The resulting CO_2 is then stored in bulk at pressures around 200 psi (1,379 kPa).

On-site gas generation and capture systems introduce more pressure hazards into the brewery, but they are of the same sort as already exist. Controls will be recommended by the equipment manufacturer.

DID YOU KNOW?

The volume of a cylinder is not the volume of the cylinder. It is expressed either as the weight of compressed or liquified gas it holds, or as the volume of gas it holds when the gas is outside the cylinder at ambient conditions.

A 20-pound CO_2 tank holds 20 lb. of CO_2 when full. One pound of CO_2 occupies 8.741 cubic feet at one atmosphere and 70°F, so that cylinder could also be said to hold 175 cubic feet.

Compressed Air and Pneumatic Control Systems

Brewers produce "house air" on-site using widely available air compressors. A trunk line of air is often run through the brewery carrying the largest practical flow rate of air at about 100 psi (689 kPa). This line then tees off to different production areas and devices. Many brewery systems use compressed air, not the least of which are air-actuated valves, keg cleaning machines, and many controllers and actuators of automated brewhouses and packaging lines. There are many applications for compressed air in a brewery:

- Aerating wort prior to pitching (an alternative to using compressed pure O_2)
- Pushing beer through indirect contact
  - Powering product transfer diaphragm pumps
  - Powering beer pumps in long-draw draught systems
- Rotational and linear motion controllers
  - Grain silo or bulk grain sack knife valves
  - Glycol ball valves
  - Automated steam valves
  - Product and gas controllers in brewhouse, cellar, and packaging
- Process equipment
  - Keg cleaning machines
  - Pneumatic keg vacuum lifters
  - Label dryers, air knives
  - CO_2 recovery systems and N_2 generator gas receivers
  - Bulk grain delivery systems
  - PET bottle formers

- Shop areas
  - Pneumatic shop tools
  - Parts cleaning
  - Tire inflation

Positive displacement compressors work by intaking ambient air and then packing it into a reduced volume with rotary vanes, screws, or a piston. Air is then pushed into a rigid metal tank that can withstand high pressures. When the tank fills to a pressure set point, the vacuum pump turns off. This filling until shutting off is called the duty cycle.

Compressors come on without warning and are loud during the duty cycle. It is advantageous to locate air compressors in areas where sound is shielded. These areas include basements or storage areas, the boiler room, outside the building, or anywhere personnel do not regularly occupy. Compressors are prone to "walking" or creeping across the floor when in operation. They should be securely bolted into concrete or an appropriate floor substructure.

Commercial compressors seen in breweries typically have a maximum operating pressure of 125 to 175 psi (862–1,207 kPa). A gauge on the tank reflects the reservoir pressure. If there is a lot of demand for air because of multiple devices, or if a device requires a large volume for its operation (keg cleaners are like this), then not only is pressure important, but also how much air can be supplied. Size, or capacity, is listed in cubic feet per minute (cfm). Operating too small a compressor will cause the compressor to cycle continuously and fail to deliver expected pressures or volumes. This can also lead to overheating of the machine, machine failure, and a risk of fire.

Even with a regulator to reduce tank pressure, the lines going out to brewery equipment will be filled to that same pressure. Excessive pressure can result in the rupture of tubing or failure of fittings or pneumatic devices. This in turn could result in flying object injury and damage to eyes. Components are installed between the tank and the airline to condition the air and to regulate pressure. Compressed air contains condensed water vapor, oil and wear metals from the compressor, and whatever contaminants came in the original air. An assembly to filter out contaminants and a valve to downregulate the air pressure in the main line should be installed and inspected and emptied periodically. Pressure tanks have a condensate valve on the bottom, which also needs draining.

Portable tools using compressed air commonly require 80–100 psi at 3–10 cfm (552–689 kPa at 0.09–0.28 m³/min.). Even pneumatic actuators and controllers used in automation are usually designed to this pressure range even if they only require a small volume displacement of air.

Beer Vessels Under Pressure

Stainless steel kegs are given a maximum allowable working pressure of 60 psi during filling or cleaning (Brewers Association n.d.). The best control for limiting overpressurization of devices is to place a secondary regulator on the local air supply, near to where it connects to devices. In this way, one leg of an air distribution system serving a canning line, for instance, could step down from 175 to 100 psi. Another leg going to a keg cleaning station could step down to 40 or 50 psi, according to the operator's margin of safety. While a keg may be able to withstand 300 psi (2,068 kPa) before rupturing, we are also protecting the entire air distribution system by running the lowest pressure that will power devices adequately.

Fermentors, bright beer tanks, and serving vessels in a brewery operate at pressures of between 10 and 40 psig (68.9–275.8 kPa gauge). The Brewers Association recommends that any tank operated at greater than 15 psig (103.4 kPa gauge) should be manufactured and certified according to American Society of Mechanical Engineers (ASME) Section VIII Division 1 Boiler and Pressure Vessel Code (Brewers Association 2021a). Tanks manufactured and rated to ASME code are made with heavier steel, and have rigorous manufacturing requirements for shape and weld integrity. These tanks cost more in materials and manufacturing labor. Unrated tanks cannot be retroactively certified.

It must be noted that most beer tanks used by smaller breweries are not ASME-rated, even though almost every state requires ASME compliance for tanks over 15 psig. To use a non-rated tank safely, never pressurize the vessel above 15 psig, always use a pressure relief valve or pressure-vacuum relief valve (PRV and PVRV, respectively), and conduct frequent inspections. When inspecting tanks, observe the condition of welds and fittings, look for deformations of the manway or other fixture points. Always have a pressure gauge easily visible on every vessel.

IS IT PSI OR PSIG?

When talking about the pressure in brewery tanks it is easy to be confused by *psi* and *psig*. What is that gauge on the tank really telling us?

Let's start with psi, pounds per square inch. This is a measure of *absolute pressure*. Absolute pressure is the total amount of pressure that the vessel contents are under: this is the sum of the force of the Earth's atmosphere and any additional applied pressure, such as from carbonation and/or dispensing gas. At sea level and a temperature of 59°F, the atmospheric pressure—the force of the Earth's atmosphere—is one atmosphere (1 atm), which is equal to 14.7 psi (101.3 kPa). So, at sea level, a beer that is under 12 psi of applied pressure in a tank would be under an absolute pressure of 26.7 psi (12 + 14.7).

Now let's look at psig, pounds per square inch gauge. This is a measure of the applied pressure beyond atmospheric pressure. Almost all gauges on beer tanks and gas dispense systems use psig. Gauges like these have been factory-adjusted, or are adjustable by the user, to remove the effects of the atmosphere at a specific elevation. Just like you would tare a scale to ignore the weight of your bucket when measuring grain, the psig gauge ignores the atmosphere and measures only the pressure of the beer applied in the vessel.

Do not forget that atmospheric pressure decreases as elevation increases, and it also changes with the weather. Let's say you have brewery in Durango, Colorado at an elevation of just about 7,000 feet (2,134 m) above sea level. Atmospheric pressure up there exerts only about 11.3 psi. Our earlier example of beer under 12 psi applied pressure in a tank (i.e., 12 psig) would only be under 23.3 psi, more than 3 psi less than at sea level. This will have a marked effect on the carbonation level of the beer when dispensed.

When talking about larger gross pressures, like those in gas cylinders and air compressors, psig isn't so important. But in the context of beer carbonation or nitrogenation pressures, knowing the pressure of the beverage is critical.

Table 12.1 **Typical brewery pressures**

Gas container/system	Typical max pressure	Key controls
Cylinders of compressed gas	2,200 psi (15,168 kPa)	Cylinder security; dual gauge regulator; dedicated storage and signage
On-site generated CO_2	220 psi (1,517 kPa)	Regular system maintenance; cleaning of foam trap; managing process plumbing to and from unit
Air compressor, typical[a]	175 psi (1,207 kPa)	Mounted to floor; secondary regulators downstream of supply; maintaining air dryer and filters
Incidental gas releases	<20 psi (<138 kPa)	Understand where and how produced; allow for dissipation
Beer cellar vessels, CO_2	22 psig (152 kPa gauge)	Pressure gauge; maintained PRV or PVRV; secondary regulator on gas supply; ASME-rated tank
Beer cellar vessels, N_2	40 psig (276 kPa gauge)	Pressure gauge; maintained PRV or PVRV; secondary regulator on gas supply; ASME-rated tank

ASME, American Society of Mechanical Engineers; kPa, kilopascals; psi, pounds per square inch; psig, psi gauge; PRV, pressure relief valve; PVRV, pressure-vacuum relief valve
[a] High-pressure, multistage compressors up to 5,000 psi (34.5 MPa) exist, but are uncommon in breweries.

CONTROLS FOR PRESSURE HAZARDS

Even though pressurized systems in a brewery range widely in terms of pressure, volume, and the substance being pressurized, the hazard controls used are nearly universal. All brewery workers should learn the basics about how to recognize pressure in use, how pressure converts to kinetic energy and can move things quickly and dangerously, and how to operate valves as pressure controls.

Many pressure hazard controls are similar for gases, compressed air, hydraulics, and pneumatics. We will take a more detailed look at fittings, valves, and labeling. A quick rundown is provided in the sidebar on p. 147.

Proper Fittings and Tubing

Fittings connect the ends of pipes or tubing to the next component of a system. Fittings need to make a leak-proof connection with threaded, barbed, welded, glued, or plug-in connections. They must be compatible with the pipe's contents and withstand the rigors of how the piping is used in the brewery. A pressurized system may have as few as two to many dozens of fixtures and they must all remain free of leaks.

A longstanding challenge in fluid and gas plumbing systems is the variety of conventions in threading, diameters, and gauges of material. Safety with pressurized systems requires the right connections, made with compatible unions affixed with the correct tools.

Where barbed connections to hose are made, the choice of clamps is significant because hoses may carry significant pressure. In general, all hose clamps should be of a continuous band of stainless steel type, which are normally affixed with a purpose-designed tool.

Small tubing, like braided hose or vinyl beer jumpers in a draught system, will take an ear clamp, such as those made by Oetiker. They may be referred to as continuous band, stepless, a crimp clamp, an ear clamp, a two-ear clamp, or a self-tensioning clamp. These types of clamp possess a small operating diameter range and most require special pliers to open and remove them. Breweries should be stocked with supplies of all necessary diameters, types, and tools.

Increasingly, plastic tubing is used to replace copper or galvanized steel water piping. PEX tubing is made of crosslinked high-density polyethylene, hence the name. PEX tubing can be attached with stepless clamps, push-fit connections, expansion connections (special tool required), or traditional copper or brass crimp or compression fittings. This tubing is designed for hot and cold water, but it is successfully used by breweries in glycol refrigeration loops.

Worm drive clamps are all-too-common tubing clamps that open and close with a screwdriver or hex socket tool. The bolt threads run through slots cut into the steel band in a rack and pinion fashion. The band is weaker than an ear clamp and the bolt threads dig into the tubing material. While worm drive clamps are accepted in automotive use where one isn't often removing the hose, in a brewery they should not be used on any fitting where a hazardous failure could occur. They are appropriate where there is little to no pressure and

INCIDENTAL GAS RELEASE

Sure, fermenting beer produces CO_2, but breweries can make other gases that are not so highly valued. Sometimes, they can even be annoying or dangerous.

- Forklift charging stations release highly flammable H_2 gas. Keep these areas well ventilated. Gaseous H_2 rises and usually dissipates, but it is explosively flammable.
- Propane or diesel forklifts, boiler furnaces, and direct fire kettles can all produce carbon monoxide (CO) in their exhaust. Invest in monitors to sound an alarm for dangerous CO levels.
- Chlorofluorocarbons and hydrofluorocarbons (CFCs and HCFCs) are refrigerants used in older equipment. When released they can damage the ozone layer. Older refrigeration systems should be serviced by trained technicians so that CFCs do not escape. Newer equipment uses environmentally safer substitutes like cyclopentane, although cyclopentane is flammable.
- Some fermentations, especially traditional lagers, produce enough hydrogen sulfide (H_2S) and dimethyl sulfide (DMS) to stink up the place. These gases are smellable in the cellar, but are highly unlikely to approach the PEL.
- A blocked sewer vent line or an overfilled grease trap can allow stinky sewer gases to arise from floor drains. Keep drain traps clear and hydrated.
- Rotten spent grain—whew, stinky stuff— can give off methane, H_2S, and butanal. Manage spent grains quickly.

damaged tubing can be easily replaced. Two examples are the waste line coming from a tap tower drain and the foam on beer (FOB) waste line connecting to a waste bottle or drain. The only advantages to worm drive clamps is diameter adjustability and requiring common tools. They are better off being rarely used in a brewery.

Especially when fitting the air supply to pneumatic controllers, brewers are increasingly using pneumatic push-in fittings, which are also called push-to-connect fittings. Though the tubing is more rigid than other gas hoses, installation of push-in fittings is quick and reliable.

When liquids are being transferred in the brewery (makeup water, wort, beer, clean-in-place cleaners, etc.) they are often sent through brewery hose, which consists of rubber-impregnated woven fiber, often with specific layers depending on application. These hoses will have large stainless steel barbs attached with a stainless steel band. The barbed end fits into the hose and the other end will be a sanitary fitting that couples to other hoses and fittings on pumps, tanks, keg fillers, and everywhere these fluids come from and go to.

Valves

Simply put, a valve opens and closes to regulate in varying degrees the flow of something moving within a pipe, tube, tank, canal, or other material conveyance. While we think of them as everyday components, valves are essential emergency devices and should not be taken for granted.

A valve will fail if it is not properly mounted into a pressurized system or if its components are incompatible with the contents or pressure requirements. Correct fitting and compatibility are discussed above. Valves also fail as a result of wear on moving or static parts and due to misoperation by a person or automated controller. Valve seats are inner parts of valves that are frequently subject to wear and, oftentimes, corrosion as well. When selecting or repairing valve seats, consult manufacturer or supplier resources to determine the best valve seat material for your application. Rebuilding isolation butterfly valves with an improved seat, such as one made of hydrogenated nitrile butadiene rubber (HNBR), will extend preventive maintenance cycles and reduce the chance of failure during use.

The manner in which a valve is operated relates to the system longevity and also the safety of the operator and those around them. As a matter of general practice, opening or closing valves too quickly can induce a shock wave throughout the plumbing, sometimes referred to as "water hammer." If a process does not require instantaneous valve operation, operate it more slowly. If it does need to operate quickly, be sure that plumbing is secured to rigid surfaces—rubber dampeners can be used.

When operating a valve, if the mechanism seems difficult to turn or moves erratically, it may be in need of repair or replacement, or it may be that backpressure

UNDER PRESSURE, UNDER CONTROL

Pressure hazard controls have a lot to do with engineering and administrative controls. Here are some of the most important ones:

Respect the Container
- Never exceed the designed pressure rating of components
- Use ASME-rated equipment when possible
- Protect containers with pressure relief devices
- Avoid incompatible materials, for example, ammonia refrigerant attacks brass, copper, zinc, galvanized steel, and aluminum

Be Careful with Cylinders
- Secure all gas cylinders from falling over using a strap or chain attached to a suitable structure
- Keep cylinders away from combustibles and sources of ignition
- Store cylinders where they cannot be bumped or have objects fall on them from above
- Move cylinders with a cart made for the purpose
- Never kick a cylinder off of a truck or ramp
- Do not roll cylinders on their side

Use Properly Fitted Fittings
- Attach hoses to barbs using continuous band stainless steel clamps
- Know when to use coatings or Teflon® tape on threaded fittings
- Use the correct wrench to install regulators to avoid rounding over
- Check fittings often—temperature changes can loosen fittings
- Quick release fittings have specific applications—consult with manufacturer

Overregulate—It's OK
- Install an on-off valve as close as possible to the supply
- Install valves to allow isolation of pressurized components
- A primary regulator indicates supply pressure
- A secondary regulator allows lowering to desired use pressure
- Check regulator settings frequently; somehow, they just . . . change

Label Whenever You Are Able
- Label storage areas with the specific gases kept there
- Maintain SDSs for purchased gases, hydraulic fluid, and refrigerants
- Keep cylinder labels visible
- Label process lines with contents

in the system is hindering valve operation. When valves are acting up, safely replace them or lockout the system while the valve is being serviced.

Certain valves can be damaged over time by overtightening. This can be true of needle valves, which are used to control fine flows of gases and liquids. The needle is a tapered, pin-like component that screws into a tubular seat in the valve body. Closing such valves with excessive pressure can wear a depression into the needle, which prevents complete closure over time.

Table 12.2 **Sanitary coupling gaskets: appearance, compatibility and applications**

Material	Color	Dot Identification	Applications
Nitrile rubber (Buna-N, NBR)	Black or white	One red dot	All brewing uses except boiling liquids, acetone, and nitric acid; resistant to oxidizers and bases; inexpensive, but wears out. Temp range: −30 to 200°F (−34.4–93.3°C)
EPDM rubber (peroxide cured)	Black, white, or gray	Three green dots	All brewing uses except hydrogen peroxide; better resistance to acids, bases, and alcohols than nitrile rubber; resists wear better than nitrile rubber. Temp range: −30 to 300°F (−34.4–148.9°C)
Fluorine rubber (Viton®, a.k.a. FKM)	Black or white	One white, one yellow	All brewing uses except acetone; excellent chemical resistance; expensive; wears out; poor for steam in place. Temp range: −30 to 400°F (−34.4–204.4°C)
Silicone	Milky white or clear	One pink dot	All brewing uses except strong acids; resistant to bases; takes wide temperature range. Temp range: −40 to 450°F (−40.0–232.2°C)
PTFE (Teflon®)	Blue or white	No dot	Cryogenics; avoid wide temperature swings. Temp range: −100 to 500°F (−73.3–260.0°C)
PTFE (Gylon®)	Blue or white	No dot	Cryogenics; avoid wide temperature swings. Temp range: −346 to 500°F (−210–260°C)
Tuf-Steel® (PTFE and 316 stainless)	Gray	No dot	High-temperature uses: steam, fryer oil. Temp range: −350 to 550°F (−212.2–287.8°C)
PTFE (EPDM core)	Black and white, or black and blue	Three green dots	Combines chemical and thermal properties of PTFE with malleability of EPDM; very expensive. Temp range: −30 to 300°F (−34.4–148.9°C)
PTFE (FKM core)	Black and white	One white, one yellow	All brewing uses; good chemical resistance; takes wide temperature range. Temp range: −30 to 400°F (−34.4–204.4°C)

Sources: Zabkowicz (2020), Austenitex (n.d.), D&D Engineered Products (n.d.)
EPDM, ethylene propylene diene monomer; FKM, Fluorine Kautschuk Material; NBR, nitrile butadiene rubber; PTFE, polytetrafluoroethylene

The order in which valves are opened or closed may affect the process being conducted as well as operator safety. Opening the waste drain valve to a kettle full of hot wort will be catastrophic, for example. Another dramatic valve mishap would be to move the contents out of a fermentor without first opening another valve to admit CO_2 to replace the displaced beer. The vacuum that results when the hydrostatic pressure exerted by the contents suddenly drops as the beer exits the tank will almost certainly crush the vessel like a beer can. Both manual and automated valve operations for complex tasks should be "choreographed" to avoid hammering, vacuum or pressure mistakes, or outright failure of piping or tubing systems.

Valves may have an all-open/all-closed design, but most allow a range of flows depending on the valve position. Many valve designs exist, but they fall into just a few categories: multiturn, quarter-turn, and self-actuated, the last of which includes relief valves and check valves. Multiturn and quarter-turn valves can be manual or automated (figs. 12.1–12.4).

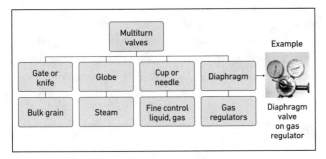

Figure 12.1. Valves as controls—multiturn types.

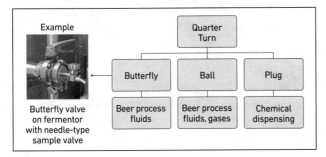

Figure 12.2. Valves as controls—quarter-turn types.

Automated Valves

All valves should have one of two override capabilities: they can be manually closed even if under automation control, or, if a pressurized system faults out of normal operation (e.g., power failure, emergency shutdown, or pressure exceedance), the valve will default into the safest possible position. This safe position could be open or closed depending on the system.

Among their many possible roles, programmable logic controllers (PLCs) can be used to open and close valves. A brewery using PLC automation may install a controller to operate a valve in a specified manner during a fault condition. This type of controller is sometimes called a safety controller. A separate PLC panel specifically containing safety controllers, a "safety PLC panel," is a good idea for larger breweries with multiple automated systems. Keeping the operational PLC panel separate from the safety PLC panel allows workers to interact with and fine tune operational input and output devices while making it less likely a safety block will be altered by accident.

Self-Actuated Pressure Relief Valves

Relief valves and the check valves are self-actuated flow controllers. Usually, a pressure relief valve (PRV) or pressure-vacuum relief valve (PVRV) is designed to open when a set pressure or vacuum is exceeded. These are safety valves that keep tanks from exploding or imploding. Relief valves may be factory set to a specific pressure or

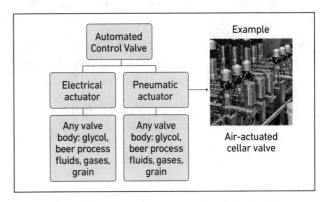

Figure 12.3. Valves as controls—automated control valves.

they can be user adjustable. For larger cellar vessels, relief valves may be counterbalanced to allow easier triggering. If PRVs and PVRVs become sullied with beer proteins they may not function correctly in an emergency. Routine scheduling of removal, inspection, and hand detailing for relief valves should be a part of every brewery's workflow.

In addition to PRVs and PVRVs, brewery tanks, kegs, and some cylinders can be protected with a frangible disk or fusible plug. A frangible disk is a piece of engineered sheet metal, usually scored and marked with its failure rating, that is fitted onto a tank flange. If the pressure or vacuum exceeds tolerances, the disk ruptures and the pressure can quickly equalize. Rupture disks don't foul easily, but they are disposable, rather costly, and, when they blow, usually cause the beer in a tank to be ruined.

A fusible plug is a threaded plug made of soft metal like bronze or tin. It is installed into compressed gas cylinders and laboratory autoclaves. If the vessel is overheated, as happens when leaving it too long on a hot plate or when there is a building fire, the plug melts and the vessel pressure is relieved prior to the vessel exploding.

A check valve is a one-way valve that allows flow in one direction only. When used for safety, for example, a check valve might allow wastewater to discharge in the direction of the sewer but not allow sewer water to flow back into the brewery.

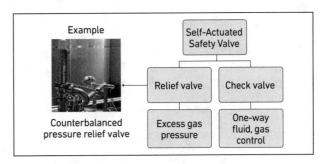

Figure 12.4. Valves as controls – self-actuated safety valves

Process Labeling

In addition to the engineering controls—proper fittings, valves, and pressure relief—a useful administrative control is the labeling of process lines. Labeling can help in identifying leaks, ordering replacement parts, and improving situational awareness in the brewery. The color conventions used are established in the ANSI/ASME A13.1 consensus standard (*see also* table 12.3).

Table 12.3 **Pipe marking color standards**

Pipe contents	Text color	Field color	Brewery application examples
Fire-quenching fluids[a]	White	Red	Sprinklers, sprinkler standpipes; kitchen extinguishing systems
Toxic/corrosive fluids	Black	Orange	Clean-in-place (CIP) hard plumbing, CIP cart; chemical treatment process lines; ammonia refrigerant
Flammable fluids	Black	Yellow	Natural gas, propane, oxidizers (e.g., O_2)
Combustible fluids	White	Brown	Heating oil plumbing between delivery point and storage tank and between storage tank and burner; hydraulics
Potable water, process and boiler water, steam, other water	White	Green	Drinking water, hose bibs/plumbing; boiler feed water, steam supply, steam condensate return; emergency shower and eyewash supplies; non-potable wash water
Compressed air	White	Blue	Compressed air; compressed non-toxic, non-corrosive, non-flammable gases (e.g., CO_2, N_2, Ar)
User-defined	White Black White White	Purple White Gray Black	Beer; glycol refrigerant; miscellaneous

Source: *Scheme for the Identification of Piping Systems,* ANSI/ASME A13.1-2020 (New York: American Society of Mechanical Engineers, 7 October, 2020).
[a] "Fluids" can apply to liquids, gases, and gas mixtures.

APPLICABLE STANDARDS

There are hundreds of references to pressure in 29 C.F.R. Part 1910, but the vast majority pertain to specialty industries like commercial diving or hazardous materials used in large quantities. Some examples are listed in table 12.4. Furthermore, many pressure hazards are in fact mechanical motion hazards caused by pressure-related system failures or rapid releases of gases or liquids. There is no single convenient regulation for pressure hazards as there is with, say, occupational noise. It will be up to the employer and employees to know how things work, where they can go wrong, and put into action sound controls throughout pressurized systems.

Table 12.4 **Occupational Safety and Health standards pertaining to pressure hazards**

Short title of standard	Title 29 citation	Key requirements
Compressed gases (general requirements)	1910.101	Employer shall inspect and keep safe; follow Title 49 rules and Compressed Gas Association (CGA) consensus standards[a]
Flammable liquids	1910.106	May be applicable to breweries producing, storing, mixing, or packaging bulk ethanol
Compressed gas and compressed air equipment	1910.166–.169	OSHA's compressed air standard has little to no applicability for brewery applications of compressed air
Oxygen-fuel gas welding and cutting	1910.253	Managing cylinders, protective equipment, hoses, and regulators

[a] 49 CFR Parts 171–179; CGA Pamphlets C-6-1968, C-8-1962, P-1-1965, S-1.1-1963, 1965 addenda and S-1.2-1963.

WARNING

POTENTIAL COMBUSTIBLE DUST HAZARD

ENSURE PROPER VENTILATION

13
FLAMMABLE AND EXPLOSIVE HAZARDS

Flammability and Explosivity

	**FLAMMABILITY – Chemical, thermal**	
	Occurs: Building structure; petroleum fuels, laboratory solvents, lubricants, paints, combustible dust	**Outcomes:** Burns, physical trauma, asphyxiation, death; building damage

THINGS THAT CAN CATCH FIRE

Only two things can burn at the average brewery: the building and its contents. Many components of a building's structure and fabric are combustible, and a brewery's contents will include many combustible items like packaging materials, not to mention the chemical substances in the building like fuels, lubricants, solvents, bulk spirits, paint supplies, and cleaners. Less common, though still a possibility, is that the brewery environs can burn, as we see with wildfires, lightning strikes, or industrial accidents. More on emergency planning relating to fire response is provided in chapter 8.

Burning, or combustion, describes a type of fast chemical reaction where oxygen and a fuel combine to release heat, light, and, usually, gaseous by-products. In chemical terms, these reactions are exothermic redox reactions, meaning they generate heat and are driven by the transfer of electrons from a reductant (fuel) to an oxidant (usually oxygen). Combustion reactions require an initiation energy to get started. Most often this comes

as a flame or spark, but it can be in the form of radiant heat, a static discharge, or a chemical initiator.

When the reaction uses up all the available fuel or oxygen in a very fast reaction, we call that an explosion. This occurs in the cylinder of an internal combustion engine, the cartridge of a bullet, when a gas leak in a building explodes, or with high-explosive charges.

Fires, on the other hand, proceed at a slower rate and usually have ample fuel supply and enough heat energy to continue to propagate. The relationship of fire ingredients is often shown as a triangle or tetrahedron (figs. 13.1–13.3).

Building Materials and Supplies

Incidents of breweries burning down have many causes, but they are made more likely by many breweries being sited in warehouse districts or other older, repurposed structures. These buildings may not be up to current fire safety code because they are grandfathered in, they have been modified without permit oversight, or the brewery's electrical load exceeds the building's utility service. Often, the modified wiring in these older structures has never been recorded

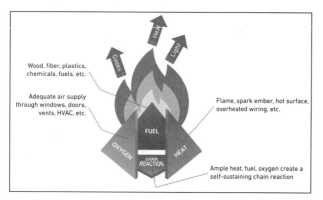

Figure 13.1. In a building fire, the building constituents and oxygen burn until the fuel is exhausted or the fire is extinguished by limiting the access of oxygen.

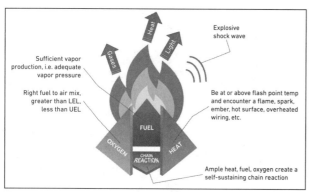

Figure 13.2. In a flammable liquid fire, the liquid evaporates enough to form a combustible mixture with air. If there are sufficient vapors and oxygen, and the minimum "flashpoint" temperature is achieved, the mixture will explode. If more liquid fuel is still present, the explosion makes it hot enough to burn, continuing the emergency. LEL, lower explosive limit; UEL, upper explosive limit.

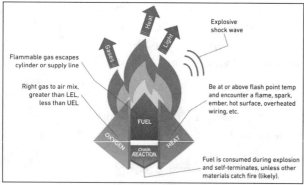

Figure 13.3. In a gas explosion, the gas and air mixture explode catastrophically. The explosion may self-terminate if all the fuel is used or if the shock wave removes enough oxygen so that combustion cannot continue. Otherwise, the explosion may ignite combustible materials in the vicinity. LEL, lower explosive limit; UEL, upper explosive limit.

in the final as-built drawings. It is all too common to see breaker circuits with no labeling. Hiring a qualified commercial electrician to map and label the building's wiring and circuit loads is a prudent investment that can save your brewery and the lives of those working in it.

Older buildings, those of certain construction types, or buildings in certain occupancy classes may not be required to have a sprinkler system. Sprinkler systems are proven to help control fires quickly and their presence often determines whether the structure is salvageable after a fire. Your local fire marshal, municipal planning department, or architect can assess the sprinkler requirements for a brewery installation.

Even without the liabilities of an older structure, a brewery may have poor housekeeping practices. Operators may have allowed cardboard, wood pallets, and plastic packaging waste to accumulate. Sloppy storage of fuels, flammable chemicals, and maintenance supplies will contribute to the risk of fire. The grounds outside the building may be overgrown with dried weeds or littered with combustible waste. The kitchen operation, if there is one, might not have an adequate fire suppression system; or perhaps the crew likes to barbecue in the side lot with a charcoal grill.

Every business owner is advised to periodically walk the entire space, deliberately looking for overaccumulated combustible materials, improperly stored flammables, and uncontrolled sources of ignition. During this time, make sure all exits are clear, have signs and emergency lighting, and that all electrical panels have 3 ft. (0.9 m) of clearance in front of them. Remember that breaker panels and electrical disconnects are emergency systems and should have ready access. The reader is encouraged to also review sections in this book on housekeeping (p. 75), emergency planning and signage (p. 79), and electrical hazards (p. 179).

Petroleum Fuels and Solvents

It would be novel to directly power a brewery with wood fires, windmills, hydropower, or waste cogeneration. But the reality for most breweries is that boilers, furnaces, and forklifts will be fired with a petroleum energy source, including natural gas, propane, fuel oil, or diesel.

The tendency for a liquid or gaseous fuel to sustain combustion requires exceeding a minimum vapor temperature (*flash point*), an adequate accumulation of vapors (dependent upon the *vapor pressure* and *lower explosive limit*), an ongoing fuel and oxygen supply, and a source of ignition capable of initiating the combustion reaction. Fuels that are gaseous in the first place—propane, butane, natural gas, and acetylene, for example—produce plenty of vapors and travel easily to sources of ignition. These fuels should be the first priority when controlling fire and explosion hazards.

Following gaseous fuels, liquid fuels with high volatility—gasoline, alcohols, paint thinners, liquified petroleum gas, and aerosol cans, for example—should be the next priority. If a volatile liquid fuel escapes its container it will produce sufficient vapors to potentially travel and ignite. The least volatile petroleum products—namely, diesel, fuel oil (bunker oil), enamel paints, and concentrated pesticides—can still catch fire and will complicate an existing fire resulting from another source.

Fuels listed for use in boilers, direct-fired kettles, forklifts, and kitchen appliances should be managed in enclosed piping systems and purpose-made containers. Fuels or solvents kept on hand in smaller containers should be stored in a location away from sources of ignition and away from oxidizers. Typically, this involves a secondary containment pallet, a bermed floor area, plastic totes, or a flammables storage cabinet. Providing absorbent spill cleanup materials and keeping flammables away from drains is also recommended.

Safety data sheets and labels should be maintained in sound condition near to the storage area of such materials. An SDS can provide valuable emergency guidance on things like fire control, first aid, PPE, and spill cleanup.

Flammable Liquids and Vapors

OSHA, the Department of Transportation (DOT), the Environmental Protection Agency (EPA), and the National Fire Protection Association (NFPA) all have differing definitions of flammable and combustible liquids. When a flammable product is shipped to the brewery it will be classified and labeled according to DOT rules (49 C.F.R. §§ 172.101, .417, and .419). When the product is in use at the brewery, it is defined by OSHA's four categories (29 C.F.R. § 1910.106(a)(19)). If and when it becomes a flammable waste product, it will most likely follow the EPA's ignitable waste rules (40 C.F.R. § 261.21, but possibly §§ 261.30–.35). The consensus standard NFPA 30: *Flammable and Combustible Liquids Code* breaks down these products into flammable and combustible classes, and further into subclasses, based on flash point and boiling point; these classes being labeled IA, IB, IC, II, IIIA, and IIIB.

INSURANCE CLASSES OF CONSTRUCTION

Frame
Burns easily. Roof, exterior walls, and floors are of combustible material such as wood. Masonry veneer (brick face) or metal cladding don't change the construction class.

Joisted Masonry
Hard to ignite, burns slower. Masonry or fire-resistive external walls rated for not less than one hour, with combustible roof and floors. Used for block constructed buildings, but sometimes includes heavy timber structures.

Non-Combustible
Does not easily burn. Can lose integrity during a fire. Roof, exterior walls, and floors are of non-combustible material such as steel. Used for warehouses and manufacturing plants.

Masonry Non-Combustible
Usually still standing after a fire. Masonry or masonry and steel construction. No wood trusses in roof structure. Walls rated for not less than one hour. Used for warehouses, offices, and shopping centers.

Modified Fire Resistive
Usually still standing after fire. Masonry or masonry and steel construction; minimum 4 inches thick (102 mm). Roof structure is pre-poured concrete or masonry over steel. Rated for more than one hour and up to two hours. Used for mid-rise offices and apartments.

Fire Resistive
Usually still standing after fire. Masonry or masonry and steel construction; minimum 4 inches thick if solid (102 mm), 8 inches if hollow (203 mm). Roof and floors concrete over steel. Roof, walls, and floors rated for more than two hours. Used for high-rise buildings and parking garages.

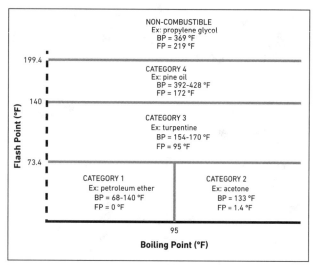

Figure 13.4. OSHA's categorization of flammable liquids (29 C.F.R. § 1910.106(a)(19)) based on flash point (FP) and boiling point (BP).

It is usually the vapor of a flammable liquid that catches fire before the liquid itself, since the vapor can mix with air to make a readily combustible mixture. Even though it takes several things for this happen, these conditions do align with some regularity.

First, a liquid must produce an adequate supply of vapors through evaporation. This is measured with a quantity called vapor pressure, which is discussed more in chapter 16 (see p. 197). Vapor pressure increases with ambient temperature and decreases with increasing molecular weight. So, acetone, with a molecular weight (MW) of 58.1 grams per mole (g/mol), will appreciably evaporate on a warm day, while kerosene (approximate MW of 170 g/mol) will not make enough vapors to turn over a jet engine on a cold day. Devices like preheaters, glow plugs, and carburetors are designed to make low-volatility substances more ignitable.

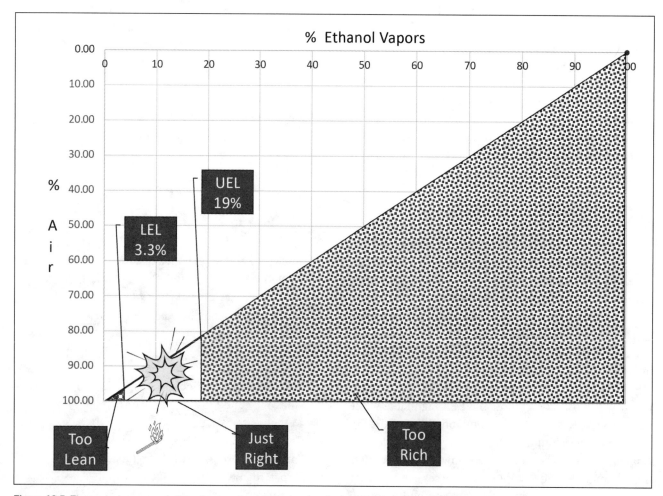

Figure 13.5. The explosive range of ethanol vapor-air mixture showing the lower explosive limit (LEL), upper explosive limit (UEL), and the explosive region between 3.3% and 19% ethanol vapors in air.

Once the substance has met the physical demands of volatilization, the mixture still needs to reach the right proportions of fuel and air, and be at the minimum temperature at which the vapors will ignite in the presence of a source of ignition. The minimum concentration of a vapor that will burn in air is termed the *lower explosive limit* (LEL), or the equivalent lower flammability limit (LFL). Below this, the fuel-air mixture is too lean to combust. There is also an upper bound to the ignitable mixture called the *upper explosive limit* (UEL): above this concentration, the mixture is too rich to burn.

Between the LEL and UEL is the very dangerous state of having an explosive atmosphere on your hands. Handheld monitors for hazardous gases and vapors are usually set to 10% of the LEL to give persons a chance to escape. It is more important to stay below the LEL than get to a place above the UEL for the simple reason that above the UEL there will probably be a coincident deficiency of oxygen, which is itself an asphyxiation hazard.

It should be noted that the explosive concentration range for some chemicals is very broad. The sterilizing agent ethylene oxide has an explosive range of 3% to 100%. Other the other hand, the range for ammonia gas is 18%–23%. Ammonia is listed by the DOT as a non-flammable gas. Ammonia is actually flammable, but with a relatively high LEL and a narrow band of explosivity—persons will succumb to ammonia asphyxiation long before they are caught in an explosion. (In the hazardous materials industry this is known as a "two-for-one special.")

The other physical property of flammable chemicals we need to know is the *flash point*. This is the minimum temperature at which a vapor and air mixture will ignite given a source of ignition. An example of this is the minimum temperature at which a car's engine will turn over.

To summarize, it takes an adequate vapor pressure for vapors to be produced from a flammable liquid and accumulate to form a minimum explosive mixture with air, following which the mixture must encounter a source of ignition while at or above the flash point. It sounds like a lot of conditions have to converge in one place to cause an explosion, and that is true, but such conditions do converge with regularity all around us. Accomplishing the following hazard controls for flammable liquids can prevent a catastrophe:

- Remote and local alarms should be installed.
- SOP and training to call 9-1-1 immediately and alert people in the vicinity when an alarm is triggered.

UNUSUAL OR UNEXPECTED FIRE HAZARDS

As the saying goes, the only certain thing is uncertainty. Or call it Murphy's law: if something can go wrong, it will go wrong. Here is a short list of unexpected ways a fire hazard can manifest (your combustion may vary):

- Electric forklift charging stations generate explosive hydrogen (H_2) gas— make sure there is adequate ventilation.
- Piles of oily rags can overheat and catch fire from the heat of evaporation—lay them flat to dry and store in a lidded metal can once dried.
- Lithium batteries used in hybrid and EV vehicles, electric bicycles, scooters, and (maybe one day) hoverboards react violently with fire suppression water— designate a separate parking area.
- Dust can accumulate on electrical contacts, causing overheating and ignition—inspect and clean contacts periodically.
- Vermin can chew through electrical insulation or build nests in warm machinery—maintain a good pest control program.
- Kitchen appliances left on overnight can start fires—the last staff member to leave should always make a walkaround inspection.
- Worn insulation from using extension cords for fixed systems—extension cords are for temporary use.
- The building or grounds can become a target of arson—keep the building secure and don't let flammables accumulate outside.
- A pigeon might drop a discarded lit cigarette down a chimney, starting a fire in a bird's nest—yes, really ("Bizarre causes of fires revealed," *Irish Independent*, December 27, 2013).

- Maintain container integrity, labeling, and secondary containment.
- Close containers between uses.
- Maintain a spill response kit and be trained in its use.
- During a spill, act quickly to shut off the release, then reduce surface area of spill with absorbents, if you have the training and the supplies.
- Deploy compatible fire extinguishing foam to create a barrier between oxygen and the fuel.

IGNITION SOURCES

Preventing fires through good housekeeping, product substitution, and safe working habits is preferred over all other fire and explosion controls. Identifying and eliminating ignition sources and fuel accumulations are paramount. When the fuel, oxidizer, and ignition source can't get together, the fire can't occur.

Electrical Systems

Electrical systems can be the source of ignition because of their potential to create a spark or because they can overheat and catch fire. Sparks are created when there is a sufficient voltage (electric potential) between two points. Normally this is avoided with proper grounding, which routes any current that short circuits into the Earth with a continuous, low resistance conductor. Humans are less conductive than grounding wire, and since electricity takes the path of least resistance, stray charge flows to ground rather than jumping across air between conductors or to a human body.

Systems can overheat and catch fire due to the decrepitude of wiring or appliances, or to overtaxing circuits with low current supplies. Dust from grain milling and grime from other activities can accumulate on conductors in breaker panels, disconnects, switches, and thermostats, causing them to overheat and burn up. Insulation can be lost over time due to drying out, physical wear, or rodent damage. Older wiring may not have integral grounding and so failed wires can catch structural components on fire.

The incorrect use of extension cords and power strips is a leading source of building fires. Specific causes include damaged cords and overloaded circuits. Unfortunately, the convenience of extension cords and power strips leads to many unsafe practices, like using them as permanent wiring and overloading branch circuits with too many devices. Conduct frequent inspections to make sure temporary cords and power

IGNITION COGNITION

How do fires get started? Let us count the ways.

1. Open flames
- Burners, ignitors, pilot lights
- Welding flames, sparks
- Kitchen stoves, grills, pizza ovens
- Portable torches, candles, luminaria
- Heaters fueled with propane or kerosene

2. Heat
- Transformers, chillers, boilers
- Coiled and energized extension cords
- Overloaded circuits
- Worn motor bearings
- Friction
- Hot surfaces

3. Spark
- Thermostat, electric switch
- Electrical fault, short circuit
- Dust
- Static discharge

4. Spontaneous combustion
- Oily rags
- Petroleum and oxidizer reaction
- Decaying biomass (e.g., compost)

5. Hanging around the shop
- Angle grinder, grinding wheel sparks
- Welding slag
- Smoking, toking, ear candling

strips are used according to the following safe practices and other hazard controls:

- Extension cord safety
  - Use only as temporary wiring for tasks like remodeling, maintenance, or repair.
  - Use only when necessary because of lack of access to a fixed branch circuit.
  - Do not use as a substitute for permanent wiring; establish a criteria, for example, no more than 60 days.

- Select a cord wire gauge appropriate for the load.
- Use UL-certified cords only.
- Use cords with a grounding conductor.
- Never run power through a rolled or coiled cord (inductive heating is a fire hazard).
- Never run a cord under a carpet, through a door jam, or other source of wear against the insulation.
- Never run a temporary cord through a wall.
- Avoid stringing multiple cords in series ("daisy chaining").
- Remove worn, cracked, or otherwise damaged cords from service immediately upon identifying damage.
- Power strip fire avoidance
  - Only plug power strips into hard-wired outlets, never extension cords or other power strips.
  - Do not use as an alternative to fixed wiring.
  - Do not permanently mount power strips, even if provided with mounting holes.
  - Remove damaged power strips and those with dirt accumulations in the receptacles from service immediately upon identifying damage.

One additional broad general safety recommendation to avoid fire or electric shock is to never connect or disconnect a temporary supply with load on the circuit. That is, make sure the switch to whatever tool or appliance is being connected or removed is in the power off position. Connecting or disconnecting a live circuit will cause a spark between conductors and can expose the worker to electric shock.

Heat Producing Appliances

Portable heaters have been the source of many fires; only use them according to manufacturer recommendations. Do not leave a heater unattended or, if you insist on doing so, make sure that the heater is stable, that nothing can fall on it, and that it is far enough away from combustible materials that it will not overheat anything. *Never* leave a propane or diesel space heater (salamander) unattended. Carbon monoxide accumulation is an additional hazard with these units.

Shop heaters, building HVAC systems, furnace flues, kitchen ductwork, and fire suppression systems should be inspected and cleaned on a regular schedule, typically at least twice per year. Air conditioning and glycol refrigeration units generate significant heat in the process of cooling. The units must be kept clear of combustible materials that impede air circulation. Type ABC (multipurpose) fire extinguishers should be professionally inspected and on hand wherever a fire may get started.

Open Flames

Welding releases hot sparks and slag that can ignite a fire readily or after a period of smoldering. Only weld in areas previously cleared of combustible materials. Use extreme caution if using a cutting torch or brazing torch in demolition or plumbing tasks.

Columnar propane heaters or tiki torches may be used in outdoor service areas. These should be kept away from all combustible materials like canopies, tablecloths, and wooden structural elements. Indoor burning of wax candles and real Christmas trees are prohibited by many jurisdictions. Fire extinguishers should be readily available near any novelty open flames.

FUEL FOR THOUGHT

What is actually burning? Let us count the ways.

1. Building fabric
- Structure, frame, cladding, roof
- Wall, ceiling, and floor treatments
- Electrical insulation
- Foam sealants

2. Building contents
- Packaging: fiber, plastic, foam, wood
- Furniture, textiles, window coverings
- Holiday decorations
- Waste, refuse, recycling

3. Hazardous materials
- Laboratory chemicals, solvents
- Glues, inks, cleaners
- Bulk spirits
- Lubricants, petroleum fuels
- Compressed oxygen, propane, or acetylene
- Pest control, janitorial, and kitchen chemicals
- Paints, stains, thinners

Other Ignition Sources

Numerous incidental ignition sources can exist in shop areas. Sparks from angle grinders, welding torches or similar tools can catch shop debris on fire. Even in the absence of sparks, oily rags can heat up and spontaneously catch fire. Products that are known to start fires in this way include linseed oil; oil-based paints, stains, and thinners; and petroleum fuels like gasoline and kerosene. Best management practices for oily rags include laying them out in a single layer to dry then keeping them in a lidded metal can. When the can is full, ship them to a commercial laundry that will accept them, or cover the rags with water and take them to a municipal hazardous waste collection site.

Another type of spontaneous combustion is when an oxidizer and organic material combine. Disposing of pool chlorine tablets and hydraulic fluid together has caught more than one garbage truck on fire. Nitric acid, when combined with wood, alcohols, or glycols, can make an explosive mixture that may catch fire with or without another ignition source.

In certain parts of the country, severe lightning is relatively common. Buildings have caught fire from being struck by lightning. In these areas, or as a matter of safe facility design anywhere, a lightning arrest system can be installed. This is essentially a system for routing the electrical energy of a strike into Earth ground. The NFPA 780 *Standard for the Installation of Lightning Protection Systems* provides guidance.

⚠ FUEL FOR THOUGHT

Where does a fire get its oxygen?

1. Atmospheric oxygen
- Through windows, doors, and holes in building structure
- HVAC systems, fans, and vents
- Leaking compressed oxygen cylinders

2. Oxidizers—they're O_2 too
- Oxidizers: peracetic acid (PAA), hydrogen peroxide (H_2O_2), nitric acid (HNO_3), sodium percarbonate
- Epoxy hardener

Grounding and Bonding

Static discharge of electricity can make a spark that will ignite available vapors or combustible dust. A spark is caused by an electric potential between two points. This can happen, for example, when a rubber belt is used in a conveyor system or when transferring flammable liquids through a rubber hose. Averting static charge buildup is done with a procedure using grounding wires. In the case of flammable liquid transfer it is called grounding and bonding (fig. 13.6).

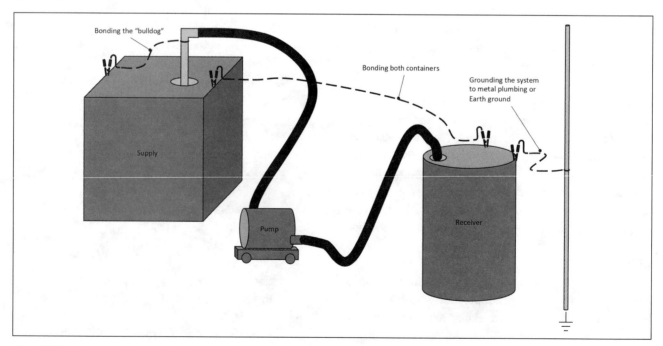

Figure 13.6. Transferring grain alcohol from a steel tote to a steel barrel using grounding and bonding technique.

APPLICABLE STANDARDS

While breweries do not commonly have a lot of flammable materials on hand, keeping the premises from burning requires sound building construction, electrical code compliance, good housekeeping, and responsible handling of flammable and combustible materials (table 13.1).

Table 13.1 **OSHA and other consensus standards pertaining to flammability and explosivity**

Short title of standard	Title 29 citation	Key requirements
Hazardous materials	1910.106(e)	(2)(ii) Allowable containers. (5) Fire control. (6) Sources of ignition. (9) Housekeeping. Control leakage, prevent escape of flammable liquids, clean up spills promptly. Maintain aisles for fire protection equipment. Combustible waste stored in closed metal waste cans and disposed of daily. Keep grounds free of tall grass, weeds, trash, or other combustible materials.
Flammable liquids	1910.106(a)(11)	May be applicable to breweries producing, storing, mixing, or packaging bulk ethanol.
Exit Routes, Emergency Action Plans, and Fire Prevention Plans	1910 Subpart E, Appendix (4)	Assure that hazardous accumulations of combustible waste materials will not occur.
Materials handling	1910.176(c)	Storage areas shall be kept free from accumulation of materials that constitute hazards from tripping, fire, explosion, or pest harborage. Vegetation control will be exercised when necessary.
Laboratory standard	1910.1450 (Appendix A)	Proper housekeeping includes appropriate labeling and storage of chemicals, safe and regular cleaning of the facility, and proper arrangement of laboratory equipment.
Grain handling	1910.272(c)	The employer shall develop and implement a written housekeeping program that establishes the frequency and method(s) determined best to reduce accumulations of fugitive grain dust on ledges, floors, equipment, and other exposed surfaces. (See national consensus standards NFPA 91 and 654.)

National consensus standards from other professional organizations	
Organization	*Relevant code or standard*
National Fire Protection Association[a]	NFPA 30: *Flammable and Combustible Liquids Code.* 2021 ed.
	NFPA 77: *Recommended Practice on Static Electricity.* 2019 ed. This standard covers safe work practices to avoid static charge ignition of flammable and combustible materials. OSHA may use NFPA 77 to determine if employer took adequate precautions.
	NFPA 91: *Standard for Exhaust Systems for Air Conveying of Vapors, Gases, Mists, and Particulate Solids.* 2020 ed.
	NFPA 329: *Recommended Practice for Handling Releases of Flammable and Combustible Liquids and Gases.* 2020 ed.
	NFPA 654[b]: *Standard for the Prevention of Fire and Dust Explosions from the Manufacturing, Processing, and Handling of Combustible Particulate Solids.* 2020 ed.
	NFPA 780: *Standard for the Installation of Lightning Protection Systems.* 2023 ed.

[a] "Codes & Standards," National Fire Protection Association, accessed February 16, 2023, https://www.nfpa.org/Codes-and-Standards.
[b] This is due to be consolidated into NFPA 660 in 2024.

14

EXTREME TEMPERATURE HAZARDS

Thermal Burns and Heat Disorders

	**BURNS, HEAT DISORDERS – Thermal**	
	Occurs: Brewhouse, kitchen, boiler and steam for contact burns; plant-wide for heat disorders	**Outcomes:** Burns, scarring, organ failure, death, PTSD; heat disorders can cause confusion, fever, organ damage, coma, death

Since we are warm-blooded organisms with in-built thermoregulating mechanisms, humans function normally across a fairly wide range of temperature conditions. But when heat energy becomes extreme or long-lived, serious injury or death can result.

There are many types of burns and temperature exposures in a brewery. Wort burns from kettle boilovers represent one of the most serious and potentially fatal hazards facing brewers. Clean-in-place (CIP) processes involve temperatures around 170°F (77°C) and present burn hazards from both scalding temperatures and corrosive chemicals. Direct contact burns from hot vessels and steam are also too common. If there is an on-site kitchen, numerous additional heat hazards will be present.

This subsection deals with *thermal burns*, where the body is exposed to a hot substance or hot surface, and *heat disorders*, where the body absorbs or loses body heat due to environmental factors. Burns discussed elsewhere include electrical burns and radiation burns in chapter 15 and chemical burns in chapter 16.

THERMAL BURNS

Burns caused by contact with hot liquids or surfaces range from a short-lived stinging to an extraordinarily painful, slow-healing, disfiguring burn. Oftentimes, the only difference between these two outcomes is a fraction of a second in duration of exposure. In cases with large burn areas or extremely deep injuries, permanent organ failure can occur in days, weeks, or even months after the incident took place.

Besides the obvious pain and physical scarring, thermal burns often injure us psychologically, resulting in post-traumatic stress disorder, depression, and anxiety (see p. 226). Disfigured individuals may be treated differently by others, as they can be seen as unpleasant to look at, incompetent, or pitiable.

 ## AN OUNCE OF BURN PREVENTION IS PRICELESS

If ever the saying that prevention is better than protection were true, it is when talking about burns. There are many burn hazards present in a typical brewery, including boiling wort, hot CIP, steam and steam-heated surfaces, welding, and kitchen jobs involving open flames, hot surfaces, hot oil, and scalding liquids. If you avoid getting burned, you avoid many or all of the follow potentialities:

- Extreme pain
- Secondary injuries from falling, jumping, or bumping
- Long, painful recovery
- Skin grafts and cosmetic surgeries
- Permanent organ injury
- Physical disfigurement
- Long-term psychological trauma
- Negative consequences to the business

Burn Physiology

A thermal burn is an injury where tissue is damaged by rapid heating. The proportion of body surface area affected, the amount of heat energy transferred, and the duration of the exposure all increase burn severity. Tissue damage can include the skin, underlying muscles and connective tissues, nerves, fatty tissues below the dermis, and even internal organs.

Laypeople commonly refer to first-, second-, and third-degree burns. Medical personnel will likely use the terms superficial thickness, deep partial thickness, and full thickness (fig. 14.1). In a superficial thickness

Table 14.1 **Exposure times and temperatures in the brewery and kitchen**

Incident	Approx. temperature in °F (°C)	Duration required to produce 2nd or 3rd degree burn
Brewery		
Direct steam contact from brewhouse vessel, boiler blowdown, laboratory autoclave, or pasteurizer	212+ (100+)	instantaneous
Boiling wort splash	213 (100.5)	instantaneous
Caustic cleaning spray	170 (76.7)	0.2 sec
Hot water contact from HLT or heat exchanger	160 (71.1)	0.5 sec
Hot α-mash immersion	154 (67.8)	1.1 sec
Hot ß-mash immersion	146 (63.3)	2.8 sec
Surface contact with kettle or steam pipe	140–160 (60.0–71.1)	0.5–5 sec
Kitchen		
Hot fryer oil splash	350+ (177+)	instantaneous
Commercial dishwasher steam	212+ (100+)	instantaneous
Simmering water burn	185–200 (85.0–93.3)	instantaneous
Commercial dishwasher hot water and dishes	160–190 (71.1–87.8)	50 ms–0.5 sec
Hot beverage service	160–180 (71.1–82.2)	0.1–0.5 sec

burn, the skin is reddened (caused by excessive blood flow to the area, a condition known as hyperemia), but is generally unbroken. This type of burn can be caused by momentary contact with a hot liquid, hot surface, or exposure to radiant energy, such as with welding or a sunburn.

Burns that injure the dermis and perhaps the tissues below the skin usually involve blistering or charred flesh. These burns, whether considered second or third degree, will have hyperemia around more deeply damaged tissue (fig. 14.2). The destroyed tissue is called the zone of coagulation. This unpleasant term describes tissue in which the proteins in the tissue have been denatured, meaning they have irreversibly lost their folded shape and function due to heat energy.

Between superficial redness and the zone of coagulation is the zone of stasis. This is where our hope lies in the case of effective burn response and treatment. When the zone of coagulation can be stopped from spreading, the zone of stasis is the place where healing begins. If the injury is not treated effectively, the zone of stasis can be lost.

In cases of burns over a large part of the body, life-threatening outcomes become more likely. Some of this is due to the body's natural response to a serious burn. First, the body releases cytokines, which are a wide variety of messenger proteins that trigger responses in other cells. A *cytokine storm* occurs when the body produces a problematic amount of cytokines as a consequence of immune response. In burns, cytokines can increase capillary permeability to help in tissue repair, but this has the side effect of causing inflammation and

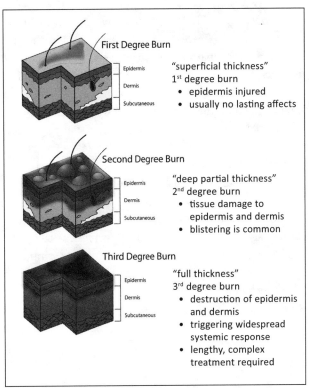

Figure 14.1. Burn severity. Degrees of burn indicating the relative depth of tissue damage. Source: K. Aainsqatsi (Wikimedia Commons; adapted for post), licensed under CC BY-SA 3.0.

fluid buildup (Arturson 1980). Cytokines can depress the immune system, making the wound vulnerable to infection (Ikeda and Kobayashi 1998).

Other side effects of severe burns include bronchial constriction, increased basal metabolic rate, reduced heart muscle contractability, and peripheral vasoconstriction. Some brewery workers who experienced significant

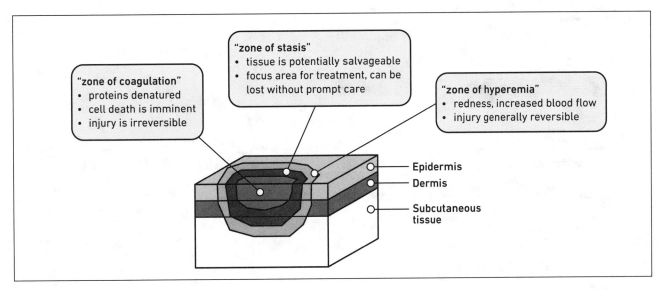

Figure 14.2. Burn injury zones. A partial thickness burn showing the necrotic zone of coagulation and outer zones.

wort burns from kettle boilovers have died from these secondary events or from organ failure caused by them.

In light of the pathophysiology of burns, first aid and emergency medical responses differ greatly depending on the type of burn suffered. For superficial burns with little to no blistering, run cool, clean water over the burn. Never apply ice. Further treatments to manage pain and avoid infection can be found with simple dressings and over-the-counter medicines. With second-degree burns, it is best to seek medical attention even if initially it doesn't seem too serious. For more serious burns with partial or full thickness injury, seek medical attention immediately. Keep the injured worker calm, elevating the injury, if possible. Do not put water or ointment on such a burn. Do not attempt to remove boots or close-fitting clothing, as doing so may deglove the area. Remove foreign matter like fabric or jewelry, if it can

be done safely. An individual with a serious burn will likely experience pain relief from endorphins initially, followed by extreme pain and symptoms of shock.

Typically, no scars will result from a superficial burn. Allow one to two weeks for the region to fully recover. Treatment for severe burns requires great attention to infection control and regular removal of dead tissue, an unpleasant experience for which pain management is difficult. Burn injuries with partial to full-thickness injury are very slow to heal, increasing the risk of infection and other complications; skin grafting may be required. Serious burns require long-term hospitalization. During this time, it is common for intravenous fluid replacement and antibiotics to be administered. Organ failure can precipitate from continued inflammation, so anti-inflammatory care may also be administered. The majority of severe burn

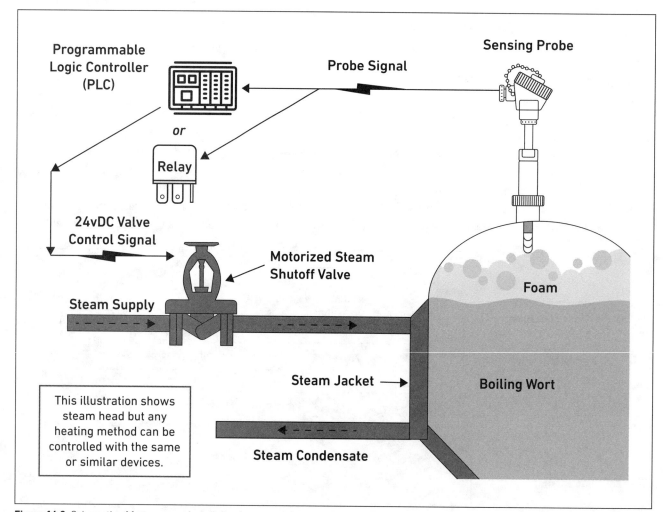

Figure 14.3. Schematic of foam sensor installation for new equipment or retrofit. Foam sensing probe signal activates PLC or relay to discontinue heat energy to kettle. When foam subsides, signal ceases and heating will resume. *Source: Brewers Association.*

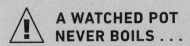

A WATCHED POT NEVER BOILS . . .

. . . And an unwatched kettle boils over! Avoid boilovers with the concurrent use of multiple hazard controls.

Safe Work Practices and Administrative Controls
- In a manually operated brewhouse, be present on the brew deck whenever steam is running.
- Do not stand in the "line of fire" when opening the kettle hatch or controlling steam.
- Keep extraneous people off the brew deck during boiling.
- Train those in the vicinity about boilover hazards and response actions.
- Have an emergency plan in case of boilover.
- Follow a preventive maintenance schedule for inspection and cleaning of foam sensors, steam valves, programmable controls, and lockout control points.
- Check all wort transfer plumbing and valves before knockout.
- Walk deliberately to avoid slips and trips and to maintain grip and balance.
- Avoid being rushed, fatigued, or distracted.
- Avoid horseplay at all times.
- Abstain from alcohol and drugs while brewing.

Process and Recipe Controls
- Do not exceed the kettle's designed capacity for starting wort volume.
- Boil some hops early, even with hazy IPAs, to achieve foam stabilization after the hot break.
- Higher protein grain bills will amplify foaming.
- Trim steam when the wort first comes to the boil to avoid overheating.

Engineering Controls
- Install and maintain a foam cutoff switch connected to a heat source cutoff switch or automated steam valve.
- Use a fixed digital thermometer to monitor wort temperature.
- Use a foam control additive.
- Keep a cold water service hose handy to knock down foam.
- Use a see-through hatch cover and interior lighting.
- Consider installing a hatch interlock that prohibits opening the hatch when kettle pressure exceeds ambient pressure.
- Consider a panic alarm with a strobe and claxon to alert those in the vicinity.
- Make sure an experienced supplier designed the brewing system and it was installed by licensed tradespersons.

PPE
- Use lined rubber gloves with the cuffs turned up.
- Wear thick PVC or butyl chemical boots with your pantlegs over the boots.
- Always wear eye protection to protect against splashes.
- Never wear shorts, porous footwear, or personal listening devices on the brew deck.

fatalities occur because of sepsis, multi-organ failure, or inhalation injury (Mehmet 2013).

Direct Contact with Hot Liquids

Burns resulting from direct contact with liquid can be larger and more severe than momentary contact with a hot surface. This is due to the way a hot liquid can spread over the body, can soak clothes, and, in some cases, is sticky. Contact duration of only one-fifth of a second with a 170°F (76.7°C) cleaning solution can result in partial-thickness to full-thickness burns.

Kettle Boilover

Boiling wort is a very concerning brewery hazard. Not only is the temperature high enough to cause instantaneous severe burns, the wort's sugar content makes it sticky on the skin. Boiling wort has a higher heat capacity than water at the same temperature. Heat capacity refers to the amount of heat energy it takes to elevate the temperature of a material. A higher heat capacity means it takes more energy to get to a given temperature. In other words, boiling wort has had more energy put into it than boiling water. Since injury severity correlates to the amount of energy being transferred, boiling wort transmits more hazardous energy than boiling water.

There are a great many controls that can prevent boilovers. It is best to employ all of of them if you can: see the "A Watched Pot Never Boils . . ." sidebar. Three controls that you should never be without are a foam sensor/cutoff switch, a food-grade foam control agent, and maintaining diligent attention during the boil.

A foam sensor detects the presence of foam or liquid at a fixed elevation in the brew kettle. The two most common technologies used are a conductivity probe and a liquid level probe. Both need to be kept clean and may require modifying CIP procedures. The height of the probe should be halfway between the surface of a preboil warm wort and the bottom of the manway opening.

Quite often, brew kettles do not come fitted with a foam sensor. Retrofitting is straightforward, however. Weld a 2-inch sanitary clamp fitting on the dome of the kettle, usually near the center of the vessel and install the probe through the fitting (Brewers Association 2021b). This facilitates easy inspection, cleaning, testing, and replacement. Next, connect the sensor to a process logic controller (PLC) or simple relay. When foam is detected, the PLC or relay will trip, switching off the heating supply either by closing a steam valve or switching off a furnace flame or electric element (fig. 14.3).

Foam control additives are food-grade mixtures designed to break surface tension in liquids. They are effective in combatting foaming in the brew kettle but some brewers have an aversion to using them, either because they are a foreign or unnatural substance or because of fears that the finished beer will have poor head retention. If using manufacturer-specified dosage rates, head retention will not be affected and hop utilization will increase.

Foam control additives come in two types: defoamers and antifoamers. Defoamers work against existing foam, whereas antifoamers work by preventing foam from forming. In practical terms, it does not matter which type of foam control additive a brewer uses.

Foam control additives need to have limited solubility in wort to work most effectively. Most commercial products are a formulation containing the principal antifoaming agent (also called the vehicle), an emulsifier, and a hydrophobic component, sometimes referred to as the activator. Most antifoamers used in brewing are based on silicone oil. Defoamers are typically based on a food oil or mineral oil. The food oil used can be olive oil, which is high in linoleic acid, which is a yeast nutrient that goes into yeast cell wall construction. The emulsifiers help the vehicle disperse into the wort. These may be silicone surfactants, fatty esters, larger alcohols, or ethoxylated surfactants like polysorbates or polyethoxylated tallow amine, an animal product. The hydrophobic component can be long-chain alkane or alkene carboxylic acids, alcohols, esters, or amides (Johnson 2009).

Having an operator present whenever wort is heating or boiling in the kettle is a good working practice to establish. This allows liquid level, foam production, kettle temperature, and heating to be constantly checked. The operator needs to be positioned out of the line of fire from the kettle opening. Other tasks requiring a brewer may be for hop additions, adjuncts and flavor additions, wort sampling, or boil-off volume measurements. Fully automated breweries may have no one near the brewhouse vessels. It is important for these brew kettles to be fitted with lockable or interlocked manway hatches, or at least a means to keep unauthorized people away from the kettle.

Hot Cleaning Solutions

Throughout the brewery, acid and alkaline cleaning solutions are commonly used at temperatures ranging from 140°F to 170°F (60–77°C). Commercial dishwashers can operate up to 190°F (88°C). Hot cleaning processes present a chemical hazard in addition to the thermal hazard.

Cleaning solutions are pumped through CIP systems, and through hoses, fittings, vessels, and pumps. Failures at any point can result in exposure. Among the common injuries in breweries are cases where hot cleaning solutions are sprayed onto a worker or into their eyes. Getting sprayed on can happen from being distracted, fooling around, or over-pressurization and subsequent failure of fittings. Serious burns from cleaning solutions occur instantaneously or within fractions of a second upon contact.

Engineering controls to reduce injuries from cleaning solutions include using chemically and thermally compatible pumps, hoses, tubing, and fittings. Numerous cases exist in breweries of hoses coming loose from fittings. Hot chemical solutions have a tendency to deform plastics, especially the low-density vinyl tubing used in keg and serving vessel jumpers, on keg washers, and throughout packaging systems.

Tubing should be fixed with correctly-sized stainless steel stepless clamps. Do not use worm drive hose clamps as they are subject to stripping and can dig into and weaken hose materials. Avoiding band style clamps cannot be stressed enough. Not only are they unable to properly handle high pressure events without failing, the area between the shank and the hose lining is also prone to harbor and proliferate microbiological contaminants. Tri-clamp fittings are only one of many sanitary brewery connection options. They should be professionally attached to brewery hoses with stainless banding or ferrule. For connecting brewery hoses to equipment, use sanitary fittings with a gasket. Gasket materials are discussed in chapter 12 (see table 12.2).

An important safe work practice that should become habitual is the double-checking of connections and valve positions prior to use. Starting at the source of the flow, let's say in this case a CIP tank that contains hot caustic, trace the flow through connections, hoses, pumps, and valves to the destination. Make sure that flow is going to go where it is supposed to and that pressure caused by the resultant flow is managed through either recirculation or providing a way out of the destination for hot liquids or heated air. Only after this quick visual recheck should the flow commence.

Kitchen Scalds

Some hazards in a commercial kitchen are more serious than even those in a brewery. Food fryer oil at over 350°F (177°C) and steam released from commercial dishwashers can instantly cause debilitating burns. Fryer oil scalds are especially serious due to oil's tendency to cling to the skin (Washington State Department of Labor and Industries 2009).

Cooks may be stirring, carrying, and pouring boiling liquids while working in cramped spaces on slippery floors. Slipping while carrying a hot liquid can result in both burns and contusions or lacerations. Splashes, splatters, and steam clouds can burn the eyes and cause permanent damage.

Kitchen staff should be trained to recognize and avoid hot liquid hazards through shift meetings, safety meetings (toolbox talks), and mentoring from senior staff. Personal protective equipment should include aprons, long-legged pants, and close-toed, non-slip footwear. Dishwashers require heavy duty, long gauntlet rubber gloves. The best gloves for this work have a layer of fabric thermal insulation embedded in the rubber. The best elastomers for hot kitchen cleaning are heavy nitrile, neoprene layered on latex, and heavy PVC. Hot mitts (oven mitts) and tongs should be provided and used to supplant direct hand contact with hot foods.

Kitchen staff should communicate continually in close quarters. This is mainly done with familiar verbal statements like "coming through," "behind," "hot behind," "incoming," "sharp," and "need a hand." A quiet kitchen may be an unnecessarily dangerous kitchen.

Tactile communication, such as by touching another's shoulder, can be helpful, but at no times should touching be inappropriate. Never touch the front torso, lower back, or buttocks, or brush pelvic regions together. Apologize promptly if such contact happens by accident. Report inappropriate touching; anxiety and distraction caused by harassment increases the likelihood of a workplace incident.

Exposure to Steam

Unpressurized steam and boiling water have the same temperature, but steam causes worse burns quicker because it holds more energy per gram and can transfer heat more directly to the skin. In a brewery, steam burns can occur during boiler blowdown or when placing one's head over a boiling kettle or whirlpool. Opening the hatch on a boiling kettle can rapidly release steam, especially if the kettle was under any amount of pressure.

Exposure to steam in the kitchen can occur when uncovering microwaved foods, releasing pressure cookers, reaching over boiling liquids on the stovetop, and from opening dishwashers. Use hot mitts at arm's length to open a container expected to release steam. Open the container pointing away from you and others.

The presence of steam is normally predictable. Steam burns are best avoided by positioning oneself outside the line of fire. In the case of a steam fitting failure resulting in a random release of steam, the best prevention is installation by a qualified steamfitter and regular inspections and preventive maintenance. It is not uncommon for a steam fitting to begin leaking and then staff wrap a rag around the leak. Such situations need to repaired, not bandaged.

Contact with Hot Surfaces

During wort production and in the cleaning steps that follow, surfaces that were in contact with steam or hot wort a few minutes prior can cause deep tissue burns after half a second to 5 seconds of direct contact. Don't presume you'll move your hand quickly enough if sensing a hot surface. There are ample scenarios where one is forced into extended contact by slipping and falling onto a hot surface, entering a vessel before it has cooled, or even the double whammy of jumping back during a wort scalding episode only to be pinned against a very hot surface.

To the extent possible, hot surfaces should only exist in out-of-reach places. Adequately constructed brewhouse vessels will have insulation between the steam jacket and the vessel exterior. Direct fired kettles are notorious for very hot surfaces, especially close to the firebox near the bottom of the vessel. Steam pipes need to be insulated, not only for worker protection, but to reduce heat losses that warm up the brewery and waste energy.

Administrative controls include applying signs to hot surfaces that cannot otherwise be protected, educating those in the area about hot surfaces, and limiting access to authorized persons. Signs on pipes containing steam supply, steam condensate return, and hot water are all colored with a green field and white lettering, regardless of temperature. Refer to table 12.3 for ANSI/ASME and globally harmonized pipe labelling color conventions (*see* p. 152).

Protective equipment, in addition to long pants and work shoes or boots, should include gloves sufficient for contact with hot surfaces. If no liquids will be contacted, gloves may be leather, knitted carbon fiber, or terrycloth. Heavy cowhide gloves are preferred for the hot sparks and surfaces of welding work. Where both heat and liquid contact are expected on the brew deck or around the boiler, terrycloth base gloves coated with heavy PVC or neoprene rubber are popular and affordable.

Hot surface protection in the kitchen includes terrycloth, quilted cotton, or silicone rubber hot mitts or potholders. Some cooks will use dish towels from their commercial laundry supplier, but these are less effective than purpose-designed holders. Wet woven fibers transmit heat readily, so frequently changing out kitchen holders is necessary. The surprise of receiving a burn in the kitchen can cause the worker to reflexively drop what they were handling, complicating the incident.

HEAT DISORDERS

Heat-related illness, or heat disorder, refers to several conditions related to the body's core temperature regulation. When core body temperature falls below 98.6°F (37°C), we are in the realm of *hypothermia*. When the body is overheating it is experiencing *heat stress*. Advanced heat stress, where the body's thermoregulatory system has largely shut down, is called *hyperthermia*.

Some sources define hyperthermia as advanced heat stress, called *heat stroke*, when the body exceeds 104°F or 105°F, equivalent to 40°C or 40.5°C (Wasserman, Creech,

⚠ UNDERLYING CONDITIONS OF HYPOTHERMIA AND HEAT STRESS

Both hypothermia and heat stress have three common defining features. The body is dehydrated, it is low on food energy, and there is some sort of temperature stressor. The temperature stressor could be a hot work environment; working in direct sunlight; heavy or tight clothing; cold, wet, and windy weather; direct physical contact with hot or cold objects or surfaces; or a physiological condition of the worker. Knowing the three features present in persons suffering this condition is critical to selecting preventive measures and treating affected individuals, but the real goal should always be prevention.

and Healy 2022; Korey Stringer Institute 2019). However, it can also mean any condition in which the body's core temperature is elevated and climbing. In reality, there is enough variation between individuals and their preparedness for heat stress, and the rate at which heat stroke progresses, that reversing the elevated temperature is probably more important than determining exactly what temperature the victim is experiencing (Trinity et al. 2010).

Heat Stress

Heat stress is a general term applied to when the body is struggling to remain at a baseline core temperature of 98.6°F (37°C). At the onset of heat stress, the body is maintaining temperature by evaporative cooling from the skin and clothes and by redirecting blood flow closer to the skin where excess heat can radiate from the body. If there is a high level of physical exertion involved, muscles are generating extra heat that the body also tries to eliminate.

This early phase where the body is overheating and struggling to maintain a stable core temperature is commonly called *heat exhaustion*. It is often accompanied by one or more notable and peculiar symptoms, discussed below.

When the body's ability to maintain core temperature fails, heat exhaustion rapidly switches to *heat stroke*. Some of the symptoms of heat stroke are different than heat exhaustion, some are the same. Always call 9-1-1 if you know someone is suffering from heat stroke. Do not delay calling for medical assistance.

Heat stress can also be caused by factors other than environmental heat. This is the case with fever, sepsis, the ingestion of a toxic substance, or withdrawal from an addictive substance. It is important to rule out these other situations, if possible, before markedly cooling the body (Wasserman, Creech, and Healy 2022).

Early Signs and Peculiar Symptoms

So many heat stress symptoms are common to a variety of ailments. These symptoms include headaches, dizziness, irritability, and tiredness. This is why it is important to recognize unique symptoms that are more likely to be associated with a heat disorder. The following are seven early signs to watch out for:

- **Sunburn.** Sunburn is painful in the short term and is accompanied by increased cancer risk in the long term. If you see it appearing on a coworker, say something and get them out of the sun.

- **Heat cramps.** These sharp and painful cramps usually occur in muscles doing a lot of work during a hot activity. They can occur in the extremities, but also affect the diaphragm, the neck and shoulders, and any other large muscle groups. Heat cramps are a sign of dehydration and overworked muscle groups.

- **Heat rash, or prickly heat.** When the body is sweaty and dirty and being rubbed the wrong way by clothing, pores in the skin can get fouled with dirt, clothing fibers, dead skin cells, and so on. Clogged pores become inflamed and perspiration cannot escape. This painful rash affects the groin, armpits, neckline, and any places where the skin is rubbed. It is best prevented with looser, clean clothing and regular cleaning of the skin.

- **Irregular heartbeat** can occur throughout the progression of heat disorder, but it can occur earlier than people realize. If a worker observes this unsettling feeling, they should be excused from the work activity until the heart returns to normal.

- **Thirst.** If a worker becomes extremely thirsty, their fluid intake is inadequate for heat stress prevention. Allow adequate rest cycles and encourage fluid intake. The old saying is, "If you're thirsty, it's too late." Perhaps a better saying is, "If you're thirsty, take a lot of fluids as soon as possible," because you need to make up for what was missing. Take in preventive fluids if hot work is to continue.

- **Dark urine and/or infrequent urination.** Urine from a hydrated worker will be clear to yellow. Workers should need to urinate several times per day. Dark yellow or yellow-orange urine is a sign of moderate dehydration. Dark orange or even brownish colored urine represents severe dehydration. With darker urine comes reduced urination frequency. Take in significant volumes of liquid, avoiding sweet sodas and alcohol.

- **Heat syncope or head rushes.** When the body is redirecting more blood to flow to the extremities, experiencing blood volume loss due to dehydration, or systolic blood pressure is dropping, syncope (fainting, or loss of consciousness) or head rushes become more likely. If a person stands up quickly they may "see stars" or almost blackout, as oxygen deficiency in the brain becomes momentarily noticeable. Persons experiencing heat syncope (pronounced *sing-kuh-pea*) can fall over and receive other injuries.

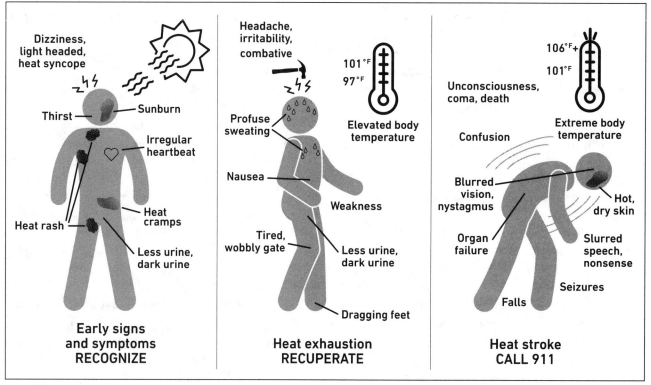

Figure 14.4. The three stages of heat stress: early warning signs with peculiar symptoms; heat exhaustion with profuse sweating and poor motor control; and heat stroke, where bodily cooling systems have failed and core temperature rapidly rises.

Heat Exhaustion

Heat exhaustion occurs when the body is still struggling to keep its core temperature regulated. It can be brought about by simply being in a hot environment or working the body hard in a warm or hot workplace, causing the body to heat up internally. Heat exhaustion describes a common symptom of the illness: the person becomes *exhausted* from the processes attempting to keep the body cool.

A worker can often recover from heat exhaustion if the condition is caught early and adequate restorative measures are taken. Bear in mind that some individuals are more prone to heat exhaustion than others. This includes anyone who is dehydrated, undernourished, or both, and so drawing extensively on their body's food and water reserves. Because the body's thermoregulatory mechanism relies on dumping excess heat from the body, those with excess fat layers or poor cardiovascular health are at a disadvantage.

During heat exhaustion, the body is attempting to regulate temperature by several mechanisms. The heart pumps more blood to surface of the skin to allow radiational cooling. The visible effect of this is redness in the face, ears, and the extremities. The body can differentially constrict certain blood vessels while dilating others. This results in lower blood pressure in the brain and internal organs, and higher blood pressure in regions where the blood is being redirected. When the brain and organs receive less blood flow and, therefore, less oxygen, symptoms such as nausea or vomiting, headache, dizziness, or fainting can occur. Because of the lowered blood flow to some parts, metabolites can accumulate and cause cramps. The heart is put under stress by needing to pump more blood when dehydrated. Systolic blood pressure can drop, the heartbeat can race, and the pulse can become weak.

Efforts to cool the body through evaporation leave the skin cool and damp, that is, clammy. Water loss through sweating causes the blood to lose volume, thus affecting blood pressure. Water is also required for metabolism, so a shortage results in poor energy production. The worker may be seen panting for breath. Salt stains may be observed, especially on dark clothing.

Common activities that increase the chance of heat stress occurring include wort production, heavy lifting tasks, kitchen work, and working outdoors in retail service, at festivals, or while performing grounds maintenance. Age, alcohol use, and diseases like hypertension, diabetes, and obesity all increase the likelihood of suffering heat stress.

The undisputed best controls for heat exhaustion are those that prevent it. Proper hydration and acclimation are most important. If a hot work environment is occurring day after day, the individual must work extra hard to hydrate. Since the worker will likely be at a hydration deficit at the end of the shift, they must take on adequate fluids to replace what was lost as well as fluids necessary for the next day's hot work. The best field metric for adequate hydration is passing pale colored urine. Thirst comes on after dehydration has already begun, so do not use thirst as primary indicator.

The CDC recommends gradual introduction in hot work activities over at least five days. Numerous heat stroke fatalities have been recorded on the first day of work in unacclimated but otherwise fit workers. Acclimation is lost at about twice the rate as it is accrued. For this reason, the first couple of days in the workweek pose a higher chance for heat illness (National Institute for Occupational Safety and Health 2017).

Build hot work activities into rotations through cool or shaded areas. Rotating out a worker into a cooler environment allows their body's thermoregulatory system to normalize before returning to the hot area. This, in effect, puts heat stroke farther from reach. To the extent possible, schedule the hottest activities during cooler parts of the day. For brewers, this can mean starting before dawn for wort production. For beer festival exposures, provide ample shade, drinking water, and rest periods. For all workers, reduce caffeine and sugary drinks consumption and avoid alcohol altogether.

Work clothing should be loose-fitting and made of quick-drying fabric. Choose synthetic blends over denim or canvas duck fabrics. Sweatbands, bandanas, and absorbent hats can reduce sweat in the eyes. If working in the outdoors, light-colored clothes and a wide hat are recommended. Kitchen staff should wear lightweight chef's garb, while still adequately covering the legs and feet against other hazards.

Beyond prevention, engineering controls to change the environmental conditions can be effective. These include air conditioning or evaporative cooling systems, shop fans, and improved piping insulation. With outdoors work, shielding from the sun with umbrellas or canopies is helpful.

Administrative controls include scheduled breaks in cool or shady areas and job rotation. Personal protective equipment can include cooling neck bands, sweat bands, or sun hats for outdoors. Outdoors workers can avoid sunburn by applying sunscreen lotion and by wearing loose, skin-covering clothing.

Heat Stroke

Heat stroke, sometimes called sunstroke, occurs when the body's thermoregulatory mechanisms fail entirely. When suffering from heat stroke, a person's core temperature may exceed 104°F (40°C). In such circumstances, the body must be cooled from the outside in. Heat stroke can result in permanent organ damage or death, as well as other workplace injury outcomes due to lapse of vision, motor skill deterioration, or collapse. If you suspect heat stroke, call 9-1-1 without delay.

Unlike heat exhaustion, heat stroke involves the cessation of perspiration, leaving the skin hot, dry, and red. Blood pools in the skin and extremities, resulting in dangerously low blood pressure. The brain and vital organs are starved of oxygen. Most heat stroke fatalities are due to the heart malfunctioning and being unable to pump blood effectively. Lack of blood oxygen to organs leads to nausea, vomiting, and neurological sequelae like confusion, delirium, or hallucinations. Suffering individuals may be non-communicative, mumble, or speak nonsense. Heat stroke victims may be conscious, experience heat syncope, become unconscious, or die in minutes. Heat stroke victims can also become agitated and combative, which can be stressful and unsettling for the rescuer.

Responding to Heat Stroke and Heat Exhaustion

Heat stroke victims need emergency hospital care. After calling for emergency medical services (EMS), get the patient to shade or a cooler environment. Sit them up if they are conscious. Make sure they are breathing adequately. Peel back wet clothes and cool with a fan, wet towels, or compresses. Important: do not give the patient fluids. If facilities are available for a bath, place the person into 90°F (32°C) water, if conscious. If you expect a delay before the ambulance arrives, the person can be laid in ice or shallow cold water, if conscious. This can be painful. Do not leave the patient unattended. Give encouraging support to the patient and maintain communication if they are awake. The attending EMS will administer intravenous fluids. A heat stroke victim will not return to work the same day, or perhaps not for several days. They may experience lingering neurological affects such as confusion. Return them to work under a doctor's recommendation.

If you suspect someone is suffering from heat exhaustion, pay attention to their signs. If they pass out, or if you notice they are dry and hot to touch, exhibiting adversarial behavior, speaking nonsense, have a zombie-like demeanor, or are vomiting, the person may actually be suffering from heat stroke—call 9-1-1 immediately. Remember that heat stroke victims should not take fluids or food.

When a heat exhausted individual has been identified, there are several helpful ways to restore their well-being. Because dehydration, low energy, and temperature stressors are the causes, treatment should address them directly. Ensure staff familiarize themselves with how to respond to a coworker suffering from heat exhaustion. The Brewers Association has resources, as does OSHA (Brewers Association 2022; US Department of Labor n.d.[d]).

Persons who have recovered from heat exhaustion may be able to return to work the same day if they have responded well to bodily cooling and rehydration. If they resume work, consider a less demanding activity and have a coworker check in on them regularly. Remind the affected individual to take extra care in rehydrating after work, as they will likely still be at a hydration deficit.

Hypothermia and Frostbite

Hypothermia is the condition where the body's core temperature drops below normal (typically below 95°F, or 35°C) as the body struggles to produce and maintain enough heat for thermoregulation. When extremities become cold enough due to poor circulation, permanent tissue damage called frostbite can occur. In some cases of frostbite, the damaged body parts require amputation; in other cases, lasting nerve damage and reduced mobility may result. There are a few possibilities in breweries for hypothermia and frostbite to occur.

Walk-in freezers and refrigerated storage are cold places where workers can be stationed for extended periods of time. In addition to being cold, these spaces usually have a lot of moving air, which tends to cool the body faster. Winter groundskeeping, snow shoveling, and winter festivals are other activities that can bring about hypothermia. In the event of a blizzard, a brewery may find itself sheltering members of the local community.

Hypothermia is remarkably similar in some respects to heat stress. The body is also more at risk for hypothermia when dehydrated, low on food energy, and there is a temperature stressor. Losing body heat because of inappropriate clothing or extended contact with a cold environment will increase the likelihood of hypothermia.

 HEAT STROKE TAKES NO PRISONERS, BUT DOES TAKE AN AMBULANCE

There are two sorts of heat stroke. There is the "classic" type brought about solely by a hot environment—this is termed non-exertional heat stroke (NEHS). The other type is brought about by strenuous activity in a hot environment, which is termed exertional heat stroke (EHS).

Preexisting conditions or circumstances that mean an individual is more likely to suffer NEHS include:

- Older age
- Obesity
- Chronic diseases, e.g., heart disease, hypertension, diabetes, Parkinson disease
- Those using diuretics, antihistamines, alcohol, or cocaine
- Anyone with a history of heat disorders
- Persons whose circumstances mean they may lack access to healthcare services and advice, have poor nutrition, have unhealthy living conditions or habits, or are chronically stressed

In contrast, EHS results from a combination of exertion and a hot environment and may follow a different pattern to NEHS:

- EHS is common in younger, healthier individuals
- Exertion heats muscles, increasing cooling needs that may then not be met
- EHS can come on very quickly
- Early signs of EHS can be easily missed

Remember: if you suspect heat stroke, call 9-1-1 immediately.

Likewise, symptoms of hypothermia may sound familiar if we replace sweating with shivering. The person may become confused, have a headache, and feel extremely exhausted. Watch for nonsense speaking or mumbling, slurred speech, confusion, and clumsiness. The victim may exhibit a weak pulse, labored and shallow breathing, or may lose consciousness (Mayo Clinic 2022b).

Victims of hypothermia have been known to irrationally lay down to take a nap during an episode, leading some to freeze to death. With hypothermia, the body redirects blood flow away from the extremities into the core to preserve the heat for essential organs. This lack of blood flow in the fingers, toes, nose, and cheeks can make the tissues appear bluish and cold to the touch. Motor control deteriorates and shivering ensues in an attempt to generate heat.

Frostbite can occur from a short exposure of bare skin to a very cold object. In breweries with liquified CO_2 in cryo-cylinders, frostbite can result from being exposed to a spray of liquified CO_2 or from touching frozen valves, pressure relief valves, or other fixtures. To a lesser degree, anytime a compressed gas is allowed to rapidly expand, such as when using CO_2 under high demand during packaging, fixtures get very cold.

Engineering controls to reduce the risk of hypothermia include safety latches on all cold storage spaces where a worker could conceivably enter that allow a worker inside a cold space to exit under their own accord. These controls are often a large, red button, but it can simply be a lever door handle. Keep a thermometer clearly visible in any such space.

To deal with cold weather, keep sand and road salt on hand for deicing. (Salt may be contraindicated if used near sensitive habitats.) Keep sufficient shovels and warning signs present to make safe passage for customers. In some case, it may be advisable for brewery staff to wear microspikes over their boots during winter maintenance. Walking with small steps, knees bent and toes out, can help avoid slips on icy surfaces.

Workers should be trained in working in cold spaces and advised by signage warning of cold exposure. Recommended warm clothing includes long-sleeve shirt and long pants, close-toed footwear, one or more thermal layers over the torso, insulated gloves, and a warm, knit hat.

As with heat stroke, hypothermia is a 9-1-1 call. Permanent organ damage or death can result from delayed treatment. After calling for assistance, move the victim to a warm, draft-free place. Sit them up and warm their body both externally (with warm towels or blankets) and internally, with warm, sweet, non-alcoholic beverages. If they have damp clothes, remove those and wrap the person in warm, dry materials.

If warming the body, apply heat compresses or a rubber hot water bottle to the chest plate, groin, and neck. To avoid cardiac strain, avoid warming or massaging the extremities. Be prepared to offer CPR if the victim shows no signs of life, such as breathing or coughing.

APPLICABLE STANDARDS

OSHA does not have much in the way of general industry regulations pertaining to burn prevention from contacting hot surfaces or substances, nor for icy conditions or hypothermia. In these cases, the General Duty Clause should prevail: the employer creates a workplace free from recognized hazards (*see* p. 13). Rather than look for compliance standards for kettle boilovers and heat stress, follow the guidelines provided above in this chapter.

Table 14.2 **OSHA and NIOSH standards and guidance pertaining to thermal hazards**

Short title of standard	Citation	Key requirements
OSHA Proposed Rulemaking[a]	n/a	"Heat Injury and Illness Prevention in Outdoor and Indoor Work Settings Rulemaking"
National Emphasis Program – Outdoor and Indoor Heat-Related Hazards[b]	CPL 03-00-024	OSHA-wide instruction issued April 2022 (Parker 2022)
Personal Protective Equipment	29 C.F.R. § 1910.134(d)	Employer conducts hazard assessment to determine PPE
Sanitation	29 C.F.R. § 1910.141	Employer provides potable water
NIOSH guidance	DHHS (NIOSH) Pub. No. 2016-106	*NIOSH criteria for a recommended standard: occupational exposure to heat and hot environments* (National Institute for Occupational Safety and Health 2016)

[a] US Department of Labor (n.d.[e]); OSHA may have a published a final rule after this book went to press.
[b] Giving OSHA authority and priority to protect employees from heat-related hazards and resulting injuries and illnesses in outdoor and indoor workplaces.

15

OTHER HIGH-ENERGY SOURCE HAZARDS

Electrical Hazards: Shock, Electrocution, Arc Flash

	**ELECTRICAL – Electric potential and kinetic energy**	
	Occurs: On contact with electrical current in tools, equipment, lighting; wires, switches, electrical disconnects, breaker panels; batteries and capacitors; lightning strike	**Outcomes:** Pain, confusion, difficulty breathing, tremors, cardiac arrhythmia, deep tissue burns, open wounds, shrapnel; loss of consciousness, death; equipment or property damage

ELECTRICAL HAZARDS

If hazards are the harmful transfer of energy to a person or object, then electrical hazards occur when electricity is released somewhere we don't want it. The most convenient definition for electricity is that it is the flow of electrons. Electrons move in a current along a conductive pathway whenever there is a difference in charge between two points, which is called the electric potential difference, or voltage. Any electrically powered device has potential and causes a demand for current when operated.

Electrical hazards are found in all parts of the brewery operation. Receiving a shock from a piece of equipment can be due to contacting damaged power cords and switches, contacting any unprotected or uninsulated energized component, standing in water that contains a charge for any reason, and commonly, from touching a conductive surface that is charged because of a ground fault. A ground fault occurs when electric current does not pass solely to the device it is intended for, but instead finds its way to the ground conductor. A rapid increase of current will flow through the circuit when a ground

fault occurs—think of it like water escaping a leaking pipe—and the circuit breaker should detect this flow and trip. It can be caused by loose connections, an internal short, accumulation of dust, or moisture. Electricity can also be held in storage in batteries or capacitors. Completing a circuit with these supplies can result in rapid and dangerous electrical injury.

Certain control devices, like the main disconnect, will have bare, energized contacts within reach simply by opening the enclosure. Because these fixtures can result in immediate death, they are labeled with the word "danger" in a black and red coloration, often with the words "high voltage" or a pictogram of a lightning bolt.

Preventing injuries from electricity relies heavily on controls. Engineering controls are built into components so that surfaces with electrical charges are difficult to access. Electrical components need to meet or exceed design ratings established by the National Electrical Manufacturers Association (NEMA) consensus standards accredited by the American National Standards Institute (ANSI). Some devices are made to be safer around wet areas, combustible dusts, or flammable gases or vapors. The National Fire Protection Association's NFPA 70® *National Electrical Code®* (NEC) is the national consensus standard that describes how an entire commercial system is to be designed, installed, and inspected.

Shock and Electrocution

When electricity contacts the body and the injury is non-fatal it is called an *electric shock*. Shocks can be as seemingly harmless as a momentary tingling, but they can be very serious, causing heartbeat irregularities, internal organ damage, entry and exits wounds, and unconsciousness.

Fatal injury from direct contact with electricity is called *electrocution*. An *arc blast* is when electrical energy jumps across airspace from a distribution panel to a worker, causing a damaging shock wave and ejecting molten shrapnel at very high speeds. Some arguably lucky persons have lived through an arc blast, but many are killed.

An indirect way electricity injures us is through the accidental activation of an electric machine. Grist mills, mash mixers, conveyors, packaging line machines, powered hand tools, and kitchen appliances can all cause injury through mechanical force powered by electricity.

CURRENT AFFAIRS

It is often said that current kills. For example, a 50-amp (50 A) 12-volt (12 V) truck battery packs more wallop than a 5 A 120 V outlet. But voltage can also kill. When current and voltage are both high, as with a distribution panel, the risk is multiplied.

0.001 A (1 mA)	You can perceive the flow of current as a tingling
2–10 mA	Shock sensation; may or may not be painful
10–20 mA	Involuntary muscle contractions, person cannot let go
20–50 mA	Respiratory paralysis; can be lethal
50–200 mA	Heart fibrillation or arrest, usually fatal
>200 mA	Tissue burns; can be fatal if affects vital organs
5 mA	Enough for a ground fault circuit interrupter (GFCI) to trip. Yes, you can still be shocked before the GFCI interrupts the circuit.

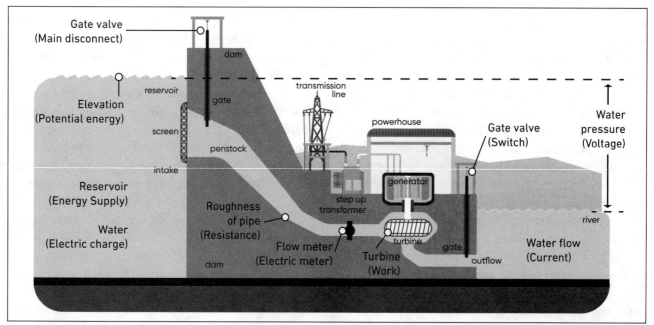

Figure 15.1. Electricity flows like water. A hydroelectric plant as shown as a comparison to an electric distribution system. Terms in parentheses are the electric analog to the hydraulic example. *Illustration © Getty/armckw*

Table 15.1 **Electricity variables and its water analogs**

Electrical Quantity				Aquatic Quantity			
Variable	*Symbol*	*Unit*	*Description*	*Variable*	*Symbol*	*Unit*	*Description*
power	P	W: watt	Volume of electrons	capacity	V	ft³: cubic foot	Volume of water
voltage	E	V: volt	Electric potential	head pressure	Z	h: feet	Height of water
resistance	R	Ω: ohm	Losses due to wiring and appliances	head loss	Δh	hl: feet	Friction and turbulence
current	I	A: ampere	Flow rate of electrons under load	flow rate	Q	cfs: ft³/ sec	Flow rate of water at point of kinetic energy transfer
ground	⏚	gnd: ground	Point in any circuit where electrons naturally flow to Earth ground	sea level	▽	msl: feet above (below) mean sea level	Point in an aqueous system to where water naturally flows

The Nature of Electricity

Although we all use electricity throughout our everyday lives, many people struggle to grasp the core principles of electricity. A helpful analogy is to view electric current like the flow of water (fig. 15.1).

In the scheme depicted in figure 15.1, current is the rate of electron flow and power is the amount, or "volume," of electrons available. Voltage is often represented with *E*, because it represents the amount of potential energy in play; some describe voltage as a "pushing force" behind the current. The atoms of the material that the electric current flows through present collision obstacles for the electrons, essentially offering resistance, R, to the current; this is analogous to textures and obstacles of the riverbed and other obstacles creating resistance for a flowing river. As the river flows, potential energy is converted to kinetic energy, and the same goes for electrons that were at rest being accelerated through a wire. This increase in kinetic energy is a transference of energy, that is, work. These quantifiable variables and their units are outlined in table 15.1.

Electrical variables are interrelated according to Ohm's law. Terms are easily convertible, which is very helpful in many basic electrical calculations. These interrelationships are illustrated in figure 15.2.

Conductors and Insulators

When electrons accumulate in a set place without flow, we call it static electricity (as opposed to current

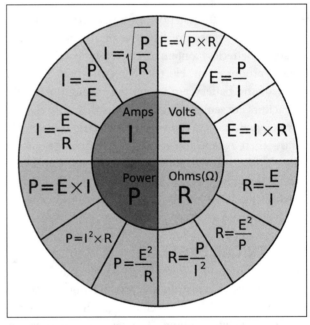

Figure 15.2. Convertibility of electrical terms. Each common electrical measure can be derived from just two other quantities.

electricity described in the previous section). In the water analogy, static electricity is like a puddle or water held behind a dam. It is in a potential energy state, just waiting for a conductor (a channel) to give the electrons a way out. Unlike current electricity, static electricity does not create a magnetic field.

A *conductor* is anything that allows electrons to flow through it. Copper, aluminum, gold, and steel are all good conductors and are used in electrical systems to create efficient pathways for current. A substance that does not conduct an electric current is called an *insulator*. Glass, ceramic, rubber, Teflon®, and some other plastics are used for their insulating capabilities. Materials that have both properties of a conductor and an insulator are called *semiconductors*.

Aqueous solutions are also pretty good conductors, which is why we can be shocked standing in a puddle on the floor. The human body is comprised principally of water with salts and other things dissolved or suspended in it. That makes the human body a decent conductor of electricity, a fact that will advise many of the safety measures we use with electricity. The cardinal rule of all hazard controls is to avoid becoming a conductor of electric current.

Finally, even air can convey electric charge if the potential difference between two points is great enough. This explains how lightning moves through air, but it also allows for a very dangerous and dramatic workplace hazard: the arc flash.

Electric potential energy can be stored in batteries or capacitors. These are like a reservoir in our water analogy. Batteries contain chemical potential energy, which produces an electric current when the two poles of the battery are connected through a circuit. Capacitors are semiconductor devices that store electric charge on plates between which a dielectric substance is sandwiched. A capacitor releases electric charge quickly when a given stored potential is achieved. Capacitors can be thought of as rechargeable batteries that charge quickly, discharge quickly, and usually cannot store as much energy as a battery.

ELECTRICAL SYSTEMS

The most important thing we need to know about a brewery's electrical distribution system is how and where to stop electricity from flowing for both premeditated and emergency situations. Current electricity moves at lightning speed and can easily deliver a fatal shock. Prevention, safe work practices, and a lot of engineering controls are where we will find safety around electricity. It is useful to appreciate how, within the brewery, electricity is modified in terms of voltage, current, and phase and then distributed to all electrified equipment, tools, lighting, and so forth.

Electric Distribution to and into the Brewery

Once it leaves the point of generation and enters the electric grid, electricity is routed to a substation that uses power transformers to increase the voltage. The voltage is stepped up because generating facilities are often far away from where the power will be sold, so large amounts of electricity must travel great distances efficiently through overhead transmission lines. When the power gets close to the community where it will be sold, it comes off the overhead transmission lines and into a step-down or distribution substation where the voltage is lowered. It then goes out across the local grid until it reaches the service drop, the point at which power supplies the brewery. Another step-down transformer will usually be at the service drop, either on a pole or mounted on the ground.

At the point where the electrical service enters the brewery premises, the entire network of breakers, switches, and wiring serve as engineering controls. A properly designed system requires advance knowledge of what is to be powered, how often, and the phase, voltage, and current required at load conditions (*load* being the current that will flow through a circuit when all devices on the circuit are energized). Adding more loads to existing circuits can cause overloads. Adding new circuits as a facility develops requires good record-keeping and must still not overload the overall supply.

Where the power supply connects to the building, there will be a power meter and the main power disconnect. The main disconnect is a special switch with large bus fuses. When turned off, all power to the building ceases. It can be an important emergency component for first responders. It disconnects all of the service energy, whether single phase or 3-phase.

Once power is distributed throughout the building it is called the *feed* and runs in feeder conductors. Next is the main breaker panel, where the feed is sent out into many *branch circuits* connecting to the main building equipment. Each branch is protected by a breaker, or sometimes a pair of breakers or a specialized fuse. Often, equipment like HVAC, chillers, compressors, and beer production machines and outlets are found in the main breaker panel, but these can also be connected through *subpanels*, also called *load centers*.

Two special types of load center are programmable logic control (PLC) panels and large, metal-cladded switchgear panels. Programmable logic control systems can have an important secondary role in safety and are outlined below. A switchgear panel refers to an enclosed collection of circuit protection components such as breakers, fuses, and switches.

Each load center will have a master breaker that protects for the total load through that panel. A load center dedicated to one piece of equipment is called a power disconnect. It comprises a breaker-protected switch in a metal enclosure.

Panels, subpanels, and switchgears are also emergency systems and require easy, unobstructed access. The area to be kept clear in front of all load centers is 36 inches (91 cm) in front of the panel for a width of 30 inches (76 cm). This allows an operator to fully open the panel door, perform emergency de-energization, service the breakers, and connect new branches if compatible with space and load capacity. Load control panels are also where lockout devices are often installed (chapter 19).

Branch circuits each connect through a breaker on the supply end and terminate at one or more places on the load side. Circuits for lighting and wall outlets may supply multiple loads. The anticipated load of the circuit constrains the gauge of wiring, rating of switches, and the setpoint for the circuit breaker.

Panels and switchgears mostly work in the background, whereas we interact frequently with the appliance end of the circuit. We may use either a switch or plug and outlet to select whether a unit is powered on or off. Both interfaces put us close to the flow of current and we should be aware of the risks.

Switch or outlet components may work loose over time; dirt or moisture can accumulate and cause overheating or shorts; or the fixture may have too low a rating for the current passing through it, causing it to heat up. A good preventive maintenance activity is for a qualified person to periodically go through the branch circuits, one at a time, shut them down, then inspect, clean, and tighten each of the switches and outlets on the circuit. Care is required—use proper rubberized gloves, insulated tools, and a multimeter-type circuit tester to avoid contacting stored charge or any bare surfaces.

Thermal imaging devices can depict hot spots in control panels, switches, and outlets. These are helpful in identifying loose or dirty connections that may heat up and catch fire.

Table 15.2 **Types of load control and load control device (LCD)**

Load control	Location	Comments
Main power disconnect	Where service enters the building	High voltage and arc flash risk; turns off everything in the building
Main breaker panel (multiple LCDs)	Typically near main power disconnect	Good emergency shutoff point, may shut down multiple outlets or devices on the same circuit; breaker disconnects all ungrounded conductors
Breaker subpanel (multiple LCDs)	Typically serves a part of the business (plant, offices, front of house)	Good emergency shutoff point, may be closer than main breaker panel, may shut multiple outlets or devices on the same circuit; breaker disconnects all ungrounded conductors; breakers are not switches for routine use
Power disconnect	Accessible and close to powered device	Switches off all ungrounded conductors to a machine; best lockout point
Machine control panel	On or near device	Best immediate on-off control, must be within easy reach; also used on portable devices (e.g., pump cart); may not switch off all conductors
Outlet with GFCI (ground fault circuit interrupter)	General work area	Outlet with a switch that opens both hot and neutral conductors when circuit is overloaded due to appliance demand, a ground fault, or dirty or loose connections; trips at 5 mA of leaked current
Outlet without GFCI	General work area	Does not identify overloads and can overheat and start a fire or transmit a ground fault; using a plug and cord as a control device leads to wear, breakage, and shorting
Common wall switch	Throughout building	Easy to reach, typically only opens "hot" conductor and leaves neutral and ground connected, where they can still result in a shock

Grounding

The grounding of electrical equipment is the deliberate connection of exposed metal surfaces of electrical equipment to the Earth (ground) for personnel safety. Electrical equipment such as transformers, motors, switchgear panels, cables, and busses contain energized and nonenergized components. When an energized component comes into contact with a metal case or structure of a piece of equipment, it is commonly known as a ground fault. This can happen because the equipment is worn, rodents have chewed off insulation, water has been allowed to enter, a loose connection has heated and melted, or a person has touched an energized circuit and become a conductor to the ground.

Ground fault is a very dangerous situation. Without grounding, if a conductive object or person touches the equipment, current will flow through the object or person to ground. This can severely injure or kill the person or will cause extensive equipment damage. A properly designed, installed, and maintained grounding system can prevent this from occurring by sending this errant energy directly to ground.

Table 15.3 Occupational Safety and Health Standards (29 C.F.R. Part 1910) and consensus standards pertaining to electrical hazards

Short title of standard	Title 29 citation	Key requirements
Design safety standards for electrical systems	1910.302-.330	Wiring design, methods, conductors and equipment meet definitions in § 1910.399; special equipment and locations described
Safety-related work practices	1910.331-.360	Work to be performed by qualified and nonqualified persons; training; work practices; use of equipment; personal protection; *see also* NFPA 70E®
Deenergized parts	1910.333(a)(1)	Deenergize whenever feasible, unless the employer can demonstrate that deenergizing introduces additional or increased hazards or is infeasible due to equipment design or operational limitations.
Safety-related maintenance requirements	1910.361-.380	Reserved (currently no standards in these sections). Refer to definitions, current licensed trade practices, and NFPA 70B.
Safety requirements for special equipment	1910.381.-.398	Reserved (currently no standards in these sections). Refer to definitions, current licensed trade practices, and consensus standards.
Definitions for Subpart S	1910.399	Extensive list of definitions for terms used throughout Subpart S
Electrical protective equipment	1910.137	Design requirements for rubber insulating gloves and other insulating PPE, classified according to exposure voltage; numerous ASTM electrical test methods cited
References for further information	1910 App. A to Subpart S	ANSI/ASME, ANSI/ASSE, NFPA consensus standards listed; compliance with these is not a substitute for compliance with 29 C.F.R. Part 1910 Subpart S

National consensus standards from other professional organizations	
Organization	*Relevant code or standard*
National Fire Protection Association[a]	NPFA 70®: *National Electrical Code®*
	NPFA 70E®: *Standard for Electrical Safety in the Workplace®*
	NPFA 70B: *Standard for Electrical Equipment Maintenance*

[a] "Codes & Standards." NFPA. Accessed February 16, 2023. https://www.nfpa.org/Codes-and-Standards.

Load Controls

The term *load control* has two senses: a system that controls the overall load or a collection of loads (such as the loads of an entire building); or any device that controls the load to a specific tool, appliance, or other device. In this book, we use *load control* to refer to controlling a single piece of equipment. Thus, a load control device (LCD) is anything that can direct, turn on, or turn off current. In the simplest case, removing a plug from an outlet is a load control for a powered hand tool. In protected circuits, a breaker or ground fault circuit interrupter (GFCI) outlet are LCDs (table 15.2).

Load control devices turn equipment on and off, or they can be used during emergencies, repairs, or maintenance to isolate the worker from electric shock. It is important to use a device designed for the task. Use commercial quality switches and control panel buttons when turning equipment on or off on a routine basis.

Circuit breakers have a spring and fatigable parts that are not designed for constant use as switches. Power disconnects, circuit breakers, and main disconnects should be thought of as emergency devices. Getting easy access to them in case of emergency is why a three-foot clear area is required to be maintained at all times.

APPLICABLE STANDARDS

The employer will need to comply with a large volume of electrical safety requirements under Subpart S of 29 C.F.R. Part 1910. There are many additional or underlying consensus standards related to component design and approval, public safety, specific industries, and specific work activities. A non-exhaustive list of major supporting standards are listed in the second half of table 15.3.

The employer should take the position from the outset of designing a brewery—through construction, installation, and operation—that commercial electrical requirements in a workplace are fundamentally more demanding, and at times more specialized, than rules applicable to home electrical systems. Not only is the code different, but component ratings, durability, and variety are more sophisticated. Understanding the value of, and maintaining a relationship with, a qualified electrical contractor is fundamental to operating a safe workplace and is a cost of doing business.

Electromagnetic Energy: Non-Ionizing and Ionizing Radiation

	RADIATION – Radiant energy, electromagnetic waves
	Occurs: Sunlight, welding, lighting, disinfection and inspection tools
	Outcomes: Burns, pain; short-term eye injury; degenerative eye disease; genetic damage, cancer

ELECTROMAGNETIC FIELDS

In the physical universe, everything we interact with is either matter or energy. Energy that propagates through space in the form of electric and magnetic fields oscillating perpendicular to each other is known as *electromagnetic radiation*. These oscillating fields act as waves, hence, electromagnetic radiation propagates as a waveform. As with any wave, electromagnetic radiation possesses a frequency and a wavelength; and also like any wave, the greater the frequency, the smaller the wavelength, and vice versa. Since a greater frequency means more wave peaks pass in a unit of time (the distance between peaks is shorter, i.e., the wavelength is smaller), this essentially means the electric and magnetic fields are oscillating more per unit time. These fields exert forces and move charges, thus, they transfer energy. We can see from this that electromagnetic radiation with a greater frequency transfers more energy per unit time than radiation with a lower frequency.

So, if you take anything from this, remember the rule of thumb for high-frequency electromagnetic radiation: it is more energetic than low-frequency electromagnetic radiation, can penetrate matter further, and can do more damage when it interacts with or is absorbed by matter. Higher-frequency forms are said to be *ionizing* electromagnetic radiation. These forms can alter genetic material and cause cancer and other diseases. Mid- to lower-frequency forms are called *non-ionizing* electromagnetic radiation. Non-ionizing electromagnetic radiation causes burns, headaches, nausea, and both reversible and irreversible eye and skin damage. Exposure to non-ionizing electromagnetic radiation has so far been difficult to regulate.

All electromagnetic radiation in the universe lies on the electromagnetic spectrum, which covers an enormous range of wavelengths (fig. 15.3). The human eye can only

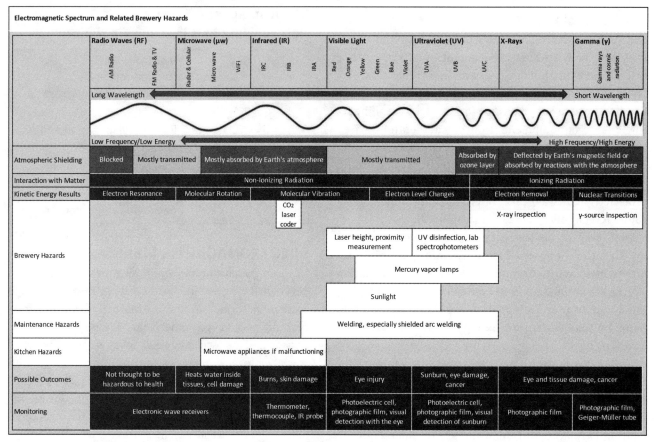

Figure 15.3. Electromagnetic spectrum and related brewery hazards. Brewery and kitchen electromagnetic radiation hazards can result in tissue burns, photokeratitis in the eyes, and, potentially, genetic damage (Madl and Egot-Lemaire 2015, 31; Ranjit and Thakur 2020, 15; Sliney and Stuck 2021; Tuchin 2015; Ulloa, Santiago, and Rueda 2019).

detect a small portion of this spectrum, that is, visible light. In order of increasing energy, the electromagnetic spectrum includes radio waves, microwaves, infrared radiation, visible light, ultraviolet (UV) light, X-rays, and gamma rays. Of these, radio waves are the least hazardous, while ionizing radiation from X-rays and gamma rays are highly energetic and are the most likely to cause permanent cellular damage if human tissue is exposed. All of these forms are summarized in figure 15.3.

Non-Ionizing Radiation

Non-ionizing energy exposure for brewers usually comes from the use of welding systems, lasers, UV instruments, and sunlight. Exposure to high levels of radio wave or microwave radiation is not common, unless in the kitchen with a malfunctioning microwave oven.

Protection with engineering controls and PPE are the preferred controls for hazardous electromagnetic radiation. Device manufacturers should assure that hazardous energy cannot be emitted when machine shields or guards are not in place.

Brewery employees who are working outdoors in customer service, grounds maintenance, or festival staffing should take precautions against sunburn with appropriate loose, comfortable clothing, head covering, and UV eye protection. Clothing is available that is rated for its ability to block UV radiation, sometimes because of chemical pretreatment or simply the materials used. In general, dark, tightly woven, synthetic fabrics are good UV blockers. Thin, translucent, pale, or gauzy fabrics allow more UV to pass through.

Sunburn from outdoors activities can also be reduced using topical sunscreen. Sunscreen UV protection factors are usually printed on the product. A minimum rating of SPF30 is recommended. Ratings higher than that afford only slightly more protection. Under-application is a common problem with sunscreen. Applying only half of the recommended amount of SPF30 lotion only supplies SPF5.5 worth of protection.

Welders are at risk of welding flash, surface burns from hot splatter, and burns from touching hot work. Welders should wear fire-retardant clothing and leather welding gloves, and keep hair behind the head and the skin of hands, arms, and legs covered at all times. Leather work boots are the best footwear, and a leather apron is preferred.

Eye protection is recommended to reduce exposure to hazardous light energies. In general, eye protection will be coded with a letter to indicate the type of light for which it is protective: *U* for UV, *R* for IR, *L* for visible light, and *W* for welding. Proper selection of laser light eye protection requires knowing both the wavelength of light and the power (wattage) of the laser.

Welding shields come with shades of increasing darkness on a logarithmic scale from 1.5 to 14, with 1 being clear glass and 14 the least transmissive to light. Shade numbers 3 to 8 are required for gas torch brazing, cutting, and stick welding. Numbers 10 to 14 are specified for gas-shielded arc welding and carbon arc welding (29 C.F.R. § 1910.252(b)(2)). Welding shades are either made of colored glass or are an autodarkening liquid crystal semiconductor with adjustable shade levels and tunable millisecond response times.

Hazardous Light Energy

In the brewery context, most electromagnetic radiation hazards encountered will fall in the visible light region and on both sides of it, in infrared and UV wavelengths. Even though we cannot see all of these wavelengths of light, they can carry energy sufficient to heat or burn tissues and cause genetic or other cellular injury. Sunburn

usually takes minutes to hours of UV exposure to cause a burn, while UV light from welding, lasers, or certain lamps can injure the skin and eyes more quickly.

Ultraviolet light is used in disinfection systems and in UV-Vis spectrophotometers in laboratories. In these applications the light is isolated by engineering controls, so these must remain in place during equipment usage or locked out during maintenance. Ultraviolet emissions may seem slightly visible, but much more energy is present than can be seen. The same goes for CO_2 laser coders and label cutters used in packaging. These beams are also mostly not visible, but they are strong lasers that operate at mid-infrared wavelengths of 9.3, 10.2, and 10.6 micrometers (μm).

Some packaging lines might have visible light lasers used for measuring bottle height, cleanliness, or checking for foreign objects. These generally operate in red, blue, or green visible light wavelengths and do not require as much power as a laser coder.

Ultraviolet light energy from welding flash, sunburn, and snow blindness can result in an acute eye injury called *photokeratitis*, which is inflammation of the cornea (keratitis) induced by UV radiation. Photokeratitis is usually temporary and causes blurry vision or temporary loss of eyesight. Other symptoms can include eye pain, watering eyes, light sensitivity or seeing halos, headaches, and a gritty feeling in the eyes. Photokeratitis usually disappears in 24 to 48

Table 15.4 **UV protection ratings for classes of products**

Product	Rated for	Protection factor and rating examples
Sunscreen, lip balm, etc.	UVB	SPF: sun protection factor, only rated for UVB, but sunscreens may also protect against UVA depending on formula
		SPF30 – blocks 97% UVB, rec'd minimum
		SPF100 – blocks 99% UVB
Clothing	UVA/UVB	UPF: UV protection factor, but also consider fabric density, color, and material
		UPF30 – blocks 97% UVA/B, recommended minimum
		UPF50+ – blocks 98% UVA/B
Sunglasses, safety glasses	UV	UV380: 99+% protection at 380 nm wavelength
		UV400: 99+% protection at 400 nm, preferred
Welding shades	W No.	Shade 2: soldering
		Shades 3–8: gas torch brazing, cutting, and stick
		Shades 10–14: shielded arc welding

UVA, ultraviolet radiation with wavelengths extending from about 320 to 400 nm and causes tanning and contributes to skin aging; UVB, ultraviolet radiation with wavelengths extending from about 280 to 320 nm and is primarily responsible for sunburn, skin aging, and the development of skin cancer.

hours; long-term injury can result, however, such as after looking directly at the sun or welding without eye protection. Persons who work outdoors at high elevations or near the polar regions are at increased risk due to more UV radiation penetrating the thinner atmosphere. Chronic UV exposure to the eyes can lead to macular degeneration, cataracts, and cancer.

Ionizing Radiation

Occasionally, a brewery may use an X-ray or gamma ray inspection device. X-ray devices generate X-rays from electrical power using an X-ray tube, while gamma ray devices contain a portion of radioactive isotope that emits gamma radiation as the isotope undergoes decay. These devices typically have a mechanical shield that is triggered into an open position during use. Both types of device are focusable to deliver concentrated energy a certain distance from the source. This focus point is where the highest exposure exists.

Gamma radiation requires very thick shielding, on the order of a thick sheet of lead or iron or a concrete wall. This makes shielding a less practical control for things like small electronic instruments. X-rays and gamma rays are similar to light in that they can be focused, concentrated, or dispersed using components that act like lenses and prisms. This means that outside of the specific region where the measurement is made, the radiation energy is usually less concentrated. Nevertheless, prohibiting body parts from entering the exposure region is fundamental to hazard control.

Adequate signage should be mounted around areas of potential exposure. Brewery inspection devices may require licensing, training, and special security considerations, depending on the state or other jurisdiction.

Lower-energy radiation forms, called alpha- and beta-radiation, are in the form of energized particles that can be inhaled or ingested. This makes them actually more hazardous in most cases than X-ray or gamma radiation. Fortunately, these forms of radiation are not encountered in package inspection devices in breweries.

The classic control scheme for ionizing radiation is T-D-S: time, distance, and shielding. The shorter the exposure time and the greater the distance, the less exposure. Also, radiation dose drops off exponentially with distance. Doubling one's distance from the source results in one-fourth the exposure; tripling the distance reduces exposure to one-ninth, and so on.

A safe work practice that originated in radiation safety, and is now used in chemical exposure control as well, is that of ALARA, which stands for *as low as reasonably achievable*. One is naturally exposed to some radiation simply by being on Earth. This radiation comes from cosmic rays, naturally occurring radioactive material in the Earth's crust, radon gas emissions from the ground, medical imaging procedures, and, in some cases, from man-made materials and even the food we eat. The principle of ALARA is that we should keep our overall exposure as low as possible using TDS.

APPLICABLE STANDARDS

OSHA currently has scant standards directed at worker protection against occupational electromagnetic radiation exposure. Health and safety studies continue, but there is little consensus on the hazards of electromagnetic radiation except where specifically seen after acute exposure, such as with photokeratitis, sunburn, and welding flash.

The employer should understand what potential electromagnetic radiation hazards could exist and take action to limit worker exposure through the use of shields, skin and eye protection, and reducing exposure durations and proximity to radiation sources.

Table 15.5 **OSHA and consensus standards pertaining to ionizing and non-ionizing radiation**

Short title of standard	Title 29 citation	Key requirements
Non-ionizing radiation	1910.97	Provides voluntary standard for exposure level and specifications for warning sign.
Control of hazardous energy (lockout/tagout)	1910.147	Energy control program for all forms of hazardous energy.
Welding, cutting, brazing	1910.252	General requirements.
		(b) Protection of personnel.
		(d)(2) X-ray inspection of welds.

National consensus standards or guidelines from other professional organizations	
Area of safety	*Relevant code or standard*
Eye and face protectors	ANSI Z87.1(2020) / CSA Z94.3(2020) Eye and face protection against physical hazards, including ultraviolet, visible, and infrared radiation but excluding lasers, X-rays, gamma rays, or high-energy particulate radiation.
Non-medical X-ray and sealed gamma ray sources	ANSI Z54.1(1963) *Non-Medical X-Ray and Sealed Gamma-Ray Sources Part 1: General*
Threshold Limit Values (TLVs®) and Biological Exposure Indices (BEIs®) (ACGIH 2021)	TLVs for UV exposure (ACGIH 2021), may need revising (Sliney and Stuck 2021). Employers with potential UV exposure from instruments should evaluate worker exposure levels.
Radioactive Source Licensing	Companies using X-ray or gamma source inspection instruments may require a state or local license. Check with jurisdiction's radiation safety agency.

ACGIH, American Conference of Governmental Industrial Hygienists; ANSI, American National Standards Institute; CSA, CSA Group (formerly the Canadian Standards Association)

OXIDIZER

5.1

PART # LP510 / 1-888-801-1515 / PRINTED IN U.S.A.

CORROSIVE

8

BIRK-OX
(ANTIMICROBIAL SOLUTION)

STORAGE AND DISPOSAL

16

CHEMICAL HAZARDS

I f all hazards are an exchange of energy, chemical hazards are those that expose the body or equipment to participating in unwanted chemical reactions. Chemical reactions, for our purposes, are interactions between molecules mediated by electrons that result in a transfer of energy. Chemical bonds arise from a net attractive force between atoms due to a reorganization of the bound atoms' electrons, ranging from an equal sharing of electrons to one or more electrons being transferred completely from one atom to the other. During chemical reactions, chemical bonds can be broken, made (or remade), and rearranged to connect to different atoms. In many chemical reactions, the energy released when new bonds are formed exceeds the energy required to break and rearrange the previously existing bonds. When this excess chemical energy is released and interacts with the human body the result can include chemical burns and injury to the skin and eyes. Chemicals interacting with the human body, whether an existing hazardous chemical or one formed due to an inappropriate or uncontrolled reaction, can also impair biological functions, leading to altered consciousness, allergic sensitivity, and acute and chronic poisoning.

There is a fine distinction between some chemical and physical hazards because chemicals can also cause physical injury. This is the case when a flammable vapor explodes and causes blunt force trauma, or carbon dioxide asphyxiation occurs simply because of the physical displacement of oxygen. Physical hazards involving chemicals that are covered in chapters 8 through 15 include flammability, slipping on chemicals, thermal burns, and radiation. However, simple asphyxiation is included in this present chapter along with chemical asphyxiation, since understanding both hazards has much to do with the descriptions and behaviors of gases and vapors.

HAZARD COMMUNICATION

Hazard communication, known as "HazCom," is all about hazardous chemicals at work. It was originally known as "Worker Right-to-Know," which is perhaps a more accurate description of this OSHA standard. Essentially, workers have a right to know about the hazardous chemicals they work with, how to obtain chemical information, and how to protect themselves from exposure.

HAZCOM REQUIREMENTS

Since 1983, manufacturers have been required to have a written HazCom program and comply with several key provisions (29 C.F.R. § 1910.1200). In 1990, a modified version for laboratories was added (29 C.F.R. § 1910.1450). In a brewery with laboratory facilities, both standards are applicable, although a unified written program may substitute for two separate programs. Between 2013 and 2016, the HazCom standard was revised to make labeling more consistent with international standards and to make safety data sheets (SDSs) consistent with the American National Standards Institute (ANSI) 16-section consensus standard, ANSI Z400.1/Z129.1-2010. This renovation of the HazCom standard was dubbed the Globally Harmonized System of Classification and Labeling of Chemicals (GHS). The terms HazCom and GHS are now often used interchangeably.

The "program" is the policy document that acknowledges the employer's responsibility and describes how it will be met. Other HazCom requirements include the worker notice, use of labels, provision of PPE and training, and acquisition and maintenance of SDSs. Templates for written HazCom programs are widely available from state and federal OSHA On-Site Consultation offices, state labor departments, brewing trade associations, and the internet. The most reliable templates will be those sourced through OSHA Outreach or On-Site Consultation offices.

Worker Notice

What is commonly called the "worker notice" is described in 29 C.F.R. § 1910.1200(h)(2). Employees must be informed that they may come into contact with hazardous chemical products and that information about these materials is available for the employees or their representatives to review. This is usually communicated via the employee handbook, during new worker orientation, or by displaying a poster in the workplace that tells workers they have a right to know about chemical hazards.

Labels

Since the GHS came into effect, hazardous chemical product labels must meet several prescriptive requirements. The label can be thought of as an abbreviated SDS; thus, the information contained on a label must be consistent with that in the SDS. Whenever a hazardous chemical product is shipped, the labeling must have the following components:

- Product identifier, usually the product name
- Signal word, either "Danger" or "Warning"
- Pictogram(s)
- Hazard statement
- Precautionary statement
- Name, address, and telephone number of manufacturer, importer, distributor, or other responsible party

The product identifier should correspond directly to an SDS, so that a worker can identify the product and easily find the SDS for more detailed information.

The signal word is designed to catch the observer's attention and draw them to read the label. The two choices in GHS labeling are "Danger" and "Warning." "Danger" represents a material with significant physical or health hazards. Significant physical hazards can include chemicals that exhibit explosivity, instability, flammability, or oxidative potential that could accentuate a fire or explosion. Significant health hazards can include materials that would cause serious illness or death upon either acute or chronic exposure, acting locally or through systemic action. "Danger" is also used with severe skin or eye irritants, corrosive materials, sensitizers, mutagens, and carcinogens. "Warning" is used when hazards posed by the product are of a more moderate nature.

One or more pictograms should be on the label. Under the GHS standard, OSHA recognizes nine standard pictograms, which are designed to visually depict any sort of physical or chemical hazard posed. Pictograms do not require further explanation and are helpful when the worker may not be fluent in English. The recognized standard pictograms and when they are used are shown in figure 16.1.

Two types of shorthand information on the label are *hazard statements* and *precautionary statements*. Hazard and precautionary statements may be solely narrative in form, but may include standardized codes beginning with *H* or *P*, respectively. Hazard statements describe the expected behaviors of the product and possible outcomes. Hazard statements may be listed starting with the letter *H* followed by a three-digit code. For a corrosive product, these would include "H302: harmful if swallowed" and "H320: causes eye irritation," among others.

Precautionary statements describe the recommended steps that should be performed to prevent or minimize the risk of a hazard occurring or minimize its effects. Precautionary statements may be listed beginning with the letter *P* followed by a three-digit code. A corrosive product might show "P280: Wear protective gloves/protective clothing/eye protection/face protection," for example.

Determining which signal word, pictogram(s), and statements apply to a particular chemical product is a complicated process required of manufacturers, importers, formulators, and distributors. OSHA has published a substantial guidance document on the subject called the *Hazard Classification Guidance for Manufacturers, Importers, and Employers* (OSHA 2016).

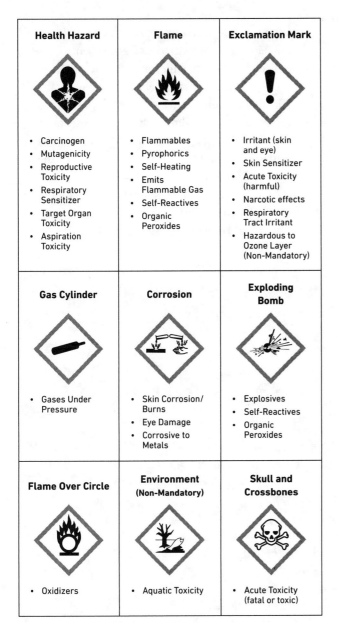

Figure 16.1. OSHA standard pictograms for labels and safety data sheets.

HP-35
(hydrogen peroxide, 35%)
CAS No. 7722-84-1

Danger
May cause fire or explosion; strong oxidizer
Causes severe skin burns and eye damage

Keep away from heat. Keep away from clothing and other combustible materials. Take any precaution to avoid mixing with combustibles. Wear protective neoprene gloves, safety goggles and face shield with chin guard. Wear fire/flame resistant clothing. Do not breathe dust or mists. Wash arms, hands and face thoroughly after handling. Store locked up. Dispose of contents and container in accordance with local, state and federal regulations.

First aid:
IF ON SKIN (or hair) or clothing: Rinse immediately contaminated clothing and skin with plenty of water before removing clothes. Wash contaminated clothing before reuse.

IF IN EYES: Rinse cautiously with water for several minutes. Remove contact lenses, if present and easy to do. Continue rinsing.

IF INHALED: Remove person to fresh air and keep comfortable for breathing. Immediately call a doctor.

IF SWALLOWED: Rinse mouth. Do NOT induce vomiting.
Immediately call poison center.
Specific Treatment: Treat with doctor-prescribed burn cream.

Fire:
In case of fire: Use water spray. In case of major fire and large quantities: Evacuate area. Fight fire remotely due to the risk of explosion.

Great Chemical Company,
55 Main Street, Anywhere,
CT 064XX   Telephone (888) 777-8888

Figure 16.2 Product label with GHS-required parts on the label.

Labels need to be maintained in good condition throughout the life of the product. OSHA requires employers to ensure that labels on incoming containers of hazardous chemicals are not removed or defaced. Labels required by US Department of Transportation (DOT) regulations must remain on the container, even after shipment of the product has concluded. Employees should be provided with appropriate dispensing equipment and training such that they can avoid damaging the labels during product use. If a label is rendered illegible, a replacement label should be created or obtained from the supplier.

When a hazardous chemical is dispensed into another container, and that container is under the direct care of the worker who obtained it, the second container does not require labeling. If the container is handed off to others, or its use extends to another shift, it must be labeled with enough information for a person to track it back to the original container, and hence the SDS.

OSHA permits employers to use alternatives to labels, as long as the contents of the container, vessel, piping, or other equipment is identified. Examples include placards, piping labels, batch tickets, or similar forms of identification attached or written on the container.

Several classes of chemical products are not required to carry GHS labeling since the products are regulated by another governmental agency. These products include the following:

- Pesticides labeled under the Federal Insecticide Fungicide Rodenticide Act, which is overseen by the EPA (40 C.F.R. § 156)
- Chemical substance or mixture regulated under the Toxic Substances Control Act, overseen by the EPA (40 C.F.R. §§ 700–799)
- Any food, food additive, color additive, drug, cosmetic, or medical or veterinary device or product, including as flavors or fragrances in such

products, according to the Federal Food, Drug, and Cosmetic Act (21 U.S.C. 301 et seq.)

- Any distilled spirit, wine, or malt beverage intended for nonindustrial use as defined in the Federal Alcohol Administration Act (27 U.S.C. 201 et seq.)
- Any consumer product or hazardous substance subject to labeling according to the regulations under the Consumer Product Safety Act or the Federal Hazardous Substances Act as overseen by the Consumer Product Safety Commission (16 C.F.R. §§ 1107.1–.30, 1500.121)
- Agricultural or vegetable seed treated with pesticides and labeled or colorized in accordance with

SDS SANITY SUGGESTIONS

- Make an inventory of all chemical products currently in use.
- Obtain the SDSs for all the products in your inventory:
  - Collect them from shipping papers/bills of lading.
  - Find them online at manufacturer websites (manufacturers must provide the SDS by law).
- Index SDSs in one or more binders:
  - Maintain a master SDS file of all chemical products actively in use.
  - It's OK to have different smaller binders for different plant areas, in addition to the master file.
  - Place the binders close to areas where the chemicals are used, not tucked away.
  - Keeping a digital SDS binder on an intranet system or phone app is permitted.
  - Digital archives must be backed up.
- Train all employees who may come into contact with a chemical product
  - when they begin a job, are transferred, or a chemical in use changes;
  - where to find the SDS and how to access SDS databases, if used;
  - how to read and understand SDSs and container labels;
  - how to protect themselves from exposure with PPE and other controls;
  - who to go to with questions or who to notify in case of a spill.
- Remove SDSs for products no longer in use.
  - Archiving removed SDSs is not required by OSHA, but your company policy may differ.
  - **Note:** OSHA requires the employer keep a record of what employees were exposed to for 30 years; either keeping the SDS or the name of the product in the employee's file will satisfy that requirement.
- Contractors should provide evidence of their own written HazCom program and SDSs for all chemical products they bring on site.
- Identify products with severe hazards and look for lower-hazard alternatives that still meet process requirements.
- Remember, some chemicals in the workplace do not have SDSs because they weren't purchased from a supplier. These include the following:
  - Carbon dioxide from fermentation
  - Carbon monoxide from combustion exhaust
  - Hydrogen from forklift battery charging
  - Hydrogen sulfide from anaerobic processes (wastewater and spent grain)

regulations under the Federal Seed Act as overseen by the Department of Agriculture (21 C.F.R. § 2.25)

PPE and Training

Under HazCom requirements, employers must provide employees with training on hazardous chemicals in their work area at time of hiring or whenever a new chemical is introduced to the process or job activity. Training should include where the chemicals are located; how they are used; labeling and SDS information and how to understand these; the hazards of each chemical product; and ways to avoid injury. Employees will be provided with PPE to protect themselves and trained how to use it; employees will also be trained on safe work practices and engineering controls, and how to respond to emergencies. As with any other type of training, a record should be kept of training content, date, trainer, and employee acknowledgement.

Chemical Inventory

The HazCom chemical inventory requirement requires employers to create a list of all hazardous chemicals in the workplace. This is essentially an index of all hazardous chemical products currently in use. It can be the table of contents for the SDS recordkeeping system.

Safety Data Sheet

The heart of HazCom is the safety data sheet (SDS), formerly known as the material safety data sheet (MSDS). The SDS has 16 standardized sections that provide information about the manufacturer, the chemical product's composition and properties, emergency actions, and compliance requirements for transit, workplace, and (often) waste management.

Acquiring and maintaining SDSs for all chemicals in use is a major component of HazCom compliance. Manufacturers, formulators, re-packagers, and distributors provide chemical safety data to commercial end users, who then make the information available to their workforce. Exceptions to this rule include when a chemical product contains no listed hazardous chemical ingredient or when the product is packaged and labeled for home consumer use. It can be a challenge to obtain and collect SDSs, as well as make them convenient for employees to find and reference. The sidebar provides simple suggestions for getting and organizing SDSs.

The standardized SDS has 16 sections. Of these, the ones carrying the most significant information about chemical

identity, composition, behavior, hazards, and controls are sections 1–4, 6, and 8. Sections 11–16 contain chemical reference information, but much of it is less important for day-to-day chemical safety and spill responding.

SECTIONS IN A STANDARDIZED SDS

1. Identification
2. Hazard(s) identification
3. Composition/information on ingredients
4. First aid measures
5. Firefighting measures
6. Accidental release measure
7. Handling and storage
8. Exposure controls/personal protection
9. Physical and chemical properties
10. Stability and reactivity
11. Toxicological information
12. Ecological information
13. Disposal considerations
14. Transport information
15. Regulatory information
16. Other information, including date of preparation or last revision

Exemptions Under HazCom

Certain classes of materials are exempt from coverage under HazCom because they are regulated under a different federal standard, defined as something other than a chemical hazard, or are considered innocuous. These materials include the following:

- Hazardous waste as defined in the regulations under the Resource Conservation Recovery Act as overseen by the EPA (40 C.F.R. § 261.3)
- Hazardous substance as defined under the Comprehensive Environmental Response, Compensation and Liability Act in relation to remedial action being conducted in accordance with Environmental Protection Agency regulations (42 C.F.R. § 304)
- Tobacco or tobacco products
- Wood or wood products, other than those that have been treated with a hazardous chemical and wood which may be subsequently sawed or cut, generating dust

- An article: a manufactured item other than a fluid or particle that under normal conditions of use does not release more than minute or trace amounts of hazardous chemical and does not pose a physical hazard
- Food or alcoholic beverages that are sold, used, or prepared in a retail establishment, and foods intended for personal consumption by employees while in the workplace
- Any solid drug (i.e., tablet or pill) packaged for the consumer, sold over the counter, or intended for employee use in the workplace (e.g., in a first aid kit), as dictated by regulations under the Federal Food, Drug, and Cosmetic Act as overseen by the Food and Drug Administration (21 C.F.R. Subchapter D)
- Cosmetics that are packaged for retail sale and cosmetics intended for personal employee use in the workplace
- Nuisance particulates where the manufacturer or importer can establish that the particulates do not pose any physical or health hazards
- Ionizing and non-ionizing radiation
- Biological hazards

PHYSICAL PROPERTIES OF CHEMICALS

Chemical materials are described in qualitative and quantitative terms. Comprehending SDSs and predicting a chemical's behavior when spilled are two important reasons to know this terminology. Predicting a chemical's behavior during regular use or emergency management advises the worker on appropriate PPE, engineering controls, and monitoring technologies.

If we know a chemical product's physical state at a given temperature, what it looks like and smells like, and how its density compares to air or water, we can be better informed in the case of a spill or unmarked chemical. Many chemicals are hard to detect or identify—the chemical may lack odor, it can be gaseous and colorless, and it can even take away our sense of smell or gradually put us to sleep. It is always a good idea to know as much as possible about a chemical's appearance and behavior in the environment. These characteristics can be found in section 9 of the SDS, which includes physical descriptions.

Physical States of Matter

The physical state of a chemical describes whether the substance is a solid, liquid, or gas at a specified temperature. Gases are said to freely occupy the space they are in, such as a room or a tank. Liquids are said to take on the shape of their container with relation to gravity. Solids have their own shape and do not easily conform to the shape of a container unless made into small particles like powder, kernels, or granules.

If the SDS or other resource does not list a temperature, assume it refers to room temperature, which we will take to be 68°F (20°C). Most materials progress through the expected phase transition of solid–liquid–gas as their heat energy increases. A fourth physical state, plasma, will only be encountered in the brewery setting during certain welding operations, which are discussed in chapter 15 under non-ionizing radiation.

When a solid becomes liquid, we say it has reached its *melting point*. If we are solidifying a liquid by cooling it, it will become solid at the *freezing point*. For most materials, the melting point and the freezing point are the same; which term is used typically depends on the direction of the phase transition, that is, whether the material was being liquified (melted) or solidified (frozen) when the measurement was taken.

When a liquid is heated or the pressure is reduced and the material turns to a gas, we say it has reached its *boiling point*. When a gas is cooled so that the material turns to a liquid, we say it has reached its *condensation temperature*. Many liquids below their boiling point will exist in two physical states at the same time: liquid and vapor. This is because molecules of the substance are moving across the interface between the two phases (the interface, of course, being the surface of the liquid). The pressure exerted by the atmosphere on the liquid's surface "pushes" against the molecules moving between the liquid and gas phases, and this limits the amount of the substance that can transition from liquid to vapor (unless energy is added or the atmospheric pressure is reduced). Hence, at high elevations where atmospheric pressure is lower, vapor production increases.

At a given pressure, when molecules are moving at roughly equal rates in both directions between liquid and vapor so there is no net change, an equilibrium is established. In this balanced state, the molecules bouncing around the vapor portion are in the gas phase and so contribute a force ("push") of their own, which is called the equilibrium vapor pressure, or more typically just *vapor pressure*. The higher a chemical's vapor pressure at a given temperature, the greater the tendency for its molecules to escape from the liquid phase and form a vapor at that temperature. Chemical products prone to producing considerable vapor at a given temperature are said to be *volatile*. There are many substances that are volatile at or near room temperature and can be hazardous as a result; examples include gasoline, paint thinner, peracetic acid, and ammonia-based cleaners.

Gases and Vapors

Chemicals in a gaseous state are particularly challenging because they can migrate easily over great distances. This means they more easily come into contact with workers or sources of ignition and can affect more people than an isolated liquid or solid chemical release.

The fact that a gas expands to fill the volume of its container has important implications. The volume of a gas can vary hugely depending on temperature and pressure. Consider a 20-pound cylinder of compressed carbon dioxide (CO_2): when the contents are released, the CO_2 that moves from the cylinder to the room expands about 63 times in volume. A cylinder of this size holds 175 cubic feet of gas at ambient pressure. Since 7% CO_2 concentration in air can kill a person, the contents of that cylinder released into a 12' × 12' × 8' cooler would present a fatal asphyxiation hazard. Other hazardous scenarios related to rapid changes in a gas's temperature and/or pressure include the following:

- Compressed gas cylinders are exposed to high heat during a building fire; the internal pressure increases and the containers explosively fail.
- A leaking fitting that allows a high-pressure gas to rapidly escape causes the fitting to get cold enough to cause frostbite to unprotected skin.
- Removing a tri-clamp blank end "coin" from a fermentor with a pressure of just 5 psig in it can cause the coin to forcibly eject, causing blunt force trauma if it hits a worker.

In the brewery, gases are regularly used. Gases are also produced daily in the brewery from fermentation, petroleum gas combustion, forklift battery charging, and wastewater processes. It is easy to overlook these gaseous by-products as being workplace hazards because they don't come with an SDS or labeled container.

Vapor Pressure

When a substance, volatile or otherwise, forms any measurable vapor, we say that substance *evaporates*. As we saw in the previous section, a substance's impetus to evaporate is measured by its vapor pressure (*vp*), usually given in Torr, millimeters of mercury (mmHg), or atmospheres (atm). Note that the Torr and mmHg scales are essentially equivalent, so 1 Torr equals 1 mmHg.

A high *vp* means the substance has a strong tendency to evaporate or volatilize. If such a substance were spilled, it will be encountered in both liquid and gaseous forms. In general, substances with a *vp* greater than 760 mmHg (1 atm) at room temperature will automatically be in the gas phase under ambient conditions. Substances with a *vp* less than 760 mmHg will partially volatilize. Be aware that even substances with a low *vp* (between about 5 to 20 mmHg) will still produce vapors sufficient to cause a workplace exposure or even an explosive mixture with air.

Vapor pressure varies according to the purity of the compound evaporating. Grain alcohol (95.5% ethanol/4.5% water) has a *vp* of 45 mmHg at room temperature, while 80-proof/40% spirits have a *vp* of 35 mmHg. The following are examples of other chemicals that may be found in a brewery:

- 100% acetone, *vp* = 266 mmHg, a highly volatile solvent
- 15% peracetic acid, *vp* = 128 mmHg, quite volatile and subsequently irritating to eyes and nose
- 70% isopropyl alcohol, *vp* = 33 mmHg, somewhat volatile
- 100% propylene glycol, *vp* = 0.8 mmHg, non-volatile

Vapor Density

When a volatile liquid is spilled, it is important to know where the vapors will end up. The factor used to define gas and vapor heaviness is vapor density (*vd*). Gaseous substances are compared to the density of air, where vd_{air} = 1.0. Vapor densities greater than 1.0 indicate a gas or vapor that is heavier than air; vapors denser than air will tend to collect in low spots. A vapor density less than 1.0 means the gas or vapor will rise in air.

Only a few common substances are lighter than air, like hydrogen, helium, and methane. Most vapors are at least as dense as air, and the vast majority are heavier. Examples include carbon dioxide (CO_2), chlorine gas, and ethanol and motor fuel vapors.

Since substances in the gas phase can mix freely, *vd* only advises where the gas or vapor would accumulate in a enclosed space. A common misconception about CO_2, (*vd* = 1.53) is that it goes right to the floor and distinctly layers beneath the air one is breathing. The fact is, in a room with even the slightest air disturbance, hazardous gases can be found at any elevation.

Warning Properties

If a substance has a detectable odor in the air, that is evidence of vaporization. Think of a freshly cut onion, chlorinated pool water, or acetone—you are smelling the airborne molecules, not the solid or liquid. Imagine smelling a gasoline spill on the shop floor; even though the *vd* of gasoline is greater than air, enough vapor rises and mixes with air that we can still smell it four or five feet above the floor.

Some substances can be detected by odor, others not. When a substance can be detected by its odor at a concentration below that which would cause harm, we say it has *good warning properties*.

Substances with *poor warning properties* can be especially dangerous since our senses may not detect them even at dangerous concentrations. Carbon dioxide, carbon monoxide, nitrogen, and oxygen-enriched air cannot be reliably detected by smell or sight.

Hydrogen sulfide gas from organic decomposition and the thiol odorants added to natural gas and liquified petroleum gas both smell sulfurous at first, but the nose quickly stops detecting it. This is known as *olfactory desensitization*.

It is a dangerous property of some chemicals in the workplace because we can let ourselves believe that whatever we smelled has dissipated when it really has not.

Brewers are familiar with the phenomenon of olfactory desensitization. For instance, some pilsners will smell burnt or vegetal at first, but after a minute or two they are no longer as offensive.

Appearance of Liquids and Solids

Assessing the color, opacity, viscosity, or solid characteristics of a chemical can be helpful in identifying a spilled substance. Once a chemical is identified, we normally go straight to the SDS to look for health warnings, PPE recommendations, and first aid and spill procedures.

Odor, which was discussed above in relation to gases and vapors, is often listed on the SDS along with appearance data. Be advised that many SDSs use vague

Table 16.1 **Predicting gas and vapor behaviors with vapor pressure and vapor density**

Substance	Vapor pressure (mmHg)[a]	Vapor density (air = 1.0)	Behavior at normal room temperature range and additional comments
Carbon dioxide	42,970	1.53	Gas; denser than air; poor warning
Ammonia	7,500	0.60	Gas; less dense than air; adequate warning
Propane	6,300	1.56	Gas; vapor slightly denser than air; will creep to source of ignition; poor warning by itself, excellent warning once odorant has been added
Air	760	1.00	Note: approximate dividing line between gases and liquids
Ethanol	59.3	1.59	Volatile liquid; vapors slightly denser than air; will creep to source of ignition; poor warning
Isopropyl alcohol (2-propanol)	45.4	2.07	Volatile liquid; vapors slightly denser than air; will creep to source of ignition; poor warning
Water	17.6	0.63	Evaporates slowly, rate dependent on relative humidity; water vapor rises
Acetic acid	11.4	2.07	Low volatility liquid; vapors slightly denser than air; good warning
Kerosene[b]	5.2	5.9	Volatile liquid; vapor significantly denser than air; will creep to source of ignition; poor warning
Mercury	0.0012	6.92	Evaporates very slowly; very dense vapor; poor warning; high toxicity

[a] 760 mmHg = 1 atm = 14.7 psi.

[b] A mixture; typical values shown.

terms for odor like "strong" or "sweet" when "vinegar-like" or "wintergreen" would be more accurate. One should exercise extreme caution in intentionally assessing chemicals by odor, as this can lead to injury. In hazmat jargon, sniffing a chemical is called "the sniff test" or "the nose hair test" and it is not recommended.

Color is usually obvious. The term *colorless* means without any hue, as with water in the case of liquids; for solids, such as table salt, the same property may be listed as *colorless* or *white*. Opacity describes whether a liquid is clear or opaque, or anywhere in between; terms such as "cloudy" or "milky" are often used.

Viscosity describes the thickness of a liquid. An SDS will commonly state a qualitative viscosity, using terms such as "thick," "syrupy," or "water-like." There is a unit of measure for viscosity (η), the centipoise (cP), where the η_{water} equals 1.0 cP at 68°F (20°C). If the SDS lists a product viscosity greater than 1.0 cP, then the substance is thicker than water. Also note that as liquids are heated, viscosity decreases. This thinning due to heat is one reason ill-worn bearings lose lubrication and speed up failure.

Solid chemical products used in the brewery will usually be described as *powder*, *crystalline*, or *granular*. The word "amorphous" may be used to signify powdery with a lack of crystals or crystal structure.

Dry sodium hydroxide (NaOH) or potassium hydroxide (KOH) may come as aspirin-shaped pellets. Calcium chloride ($CaCl_2$) may come as small beads, called prill, or as flakes. *Hygroscopic* solids readily absorb atmospheric moisture to become damp, sticky, sweaty, or rock-like. In brewing, dry malt extract, maltodextrin, and $CaCl_2$ are hygroscopic. A hygroscopic solid that can absorb so much moisture as to liquify is said to be *deliquescent*. Sodium hydroxide pellets are deliquescent and difficult to keep dry; consequently, brewers typically use concentrated solutions of NaOH rather than pellets.

Occasionally, there are substances that transition directly from the solid phase to the gas phase in a process called *sublimation*. Dry ice, which is solid CO_2, is a good example: when dry ice pellets get warm, the solid changes directly into gaseous CO_2. No liquid transition occurs unless the system is under pressure, as it is with

AIRBORNE PHYSICAL STATE MIXTURES

Solid Aerosols
- Dust: fine particles from milling, rubbing, sanding, etc.
- Fume: condensed heated metal
- Soot: carbonaceous particles from combustion
- Fiber: long, narrow particles
- Infectious agents: airborne bacteria and fungal spores
- Allergens: chemical or biological airborne agents, for example, pollen and beer adjuncts

Liquid Aerosols
- Mist: suspended microscopic droplets
- Fog: fine suspended droplets from vapor condensation
- Infectious agents: viruses, bacteria, and spores carried in mist droplets

Gaseous
- Gas: gaseous at ambient conditions, fills available volume
- Vapor: gas phase emanating from liquid surface

Multistate Aerosols
- Smoke: mixture of gases, liquids/vapors, and particulates (e.g., soot) from incomplete combustion
- Smog: photoreactive mixture of smoke and fog

a cylinder of compressed CO_2. If we are responding to a box of dry ice that broke open, we can see from the SDS that dry ice sublimates and use that knowledge to predict the air around us will have increasing CO_2 concentrations even without a monitoring device. This is one example of how we can use knowledge of physical states to predict hazardous conditions.

Since we can only live a few minutes without air, airborne contaminants that gain access through the respiratory route of entry can be a serious health threat. Specific mixtures of chemical substances with air have specific names. These are listed in table 3.2 (pp. 28–29), but a few relevant ones are summarized in the sidebar on the next page.

Liquids and Solids

Liquids, to a lesser degree than gases and vapors, can flow away from a spill; therefore, there is still a risk the spill will contact a source of ignition, cause an unwanted chemical reaction, or expose workers. Oftentimes, we can add an absorbent/adsorbent material to a liquid spill, effectively turning it into a solid spill. This limits migration of the substance and simplifies its collection as part of cleanup.

The main things to avoid when managing spilled liquids and solids are tracking the substance elsewhere, causing the substance to become airborne during handling, and letting it come into contact with incompatible substances. The spilled substance, whether liquid or solid, still represents a slip/trip and fall hazard.

APPLICABLE STANDARDS

The primary standard for chemical safety is HazCom, which encompasses very nearly every hazardous chemical used in beverage manufacturing. OSHA has many other standards that relate to chemical safety, including PPE, respiratory protection, first aid, and spill response (table 16.2).

OSHA has dozens of standards that apply to specific chemical or chemical groups. These are not shown in table 16.2 unless deemed to be relevant to brewing.

Table 16.2 **Occupational Safety and Health Standards (29 C.F.R. Part 1910) relating to chemical hazards**

Short title of standard	Title 29 citation	Key requirements
HazCom	1910.1200	(e) Written HazCom program. (f) Labels and other forms of warning. (g) SDSs. (ANSI Z400.1) (h) Employee information and training.
HazCom Lab Standard	1910.1450	(d) Employee exposure determination. (e) Chemical hygiene plan. (f) Employee information and training. (g) Medical consultation and medical exams. (h) Hazard identification. (j) Recordkeeping.
PPE	1910.132–.136, and .138	.132 Hazard assessment and equipment selection .133 Eye and face protection .134 Respiratory protection .135 Head protection .136 Foot protection .138 Hand protection
Duties Owed	1910.9(a)	Employer shall provide PPE, including respirators as needed

Corrosive Chemicals

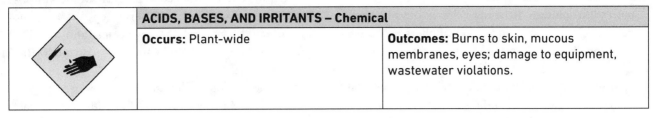

	ACIDS, BASES, AND IRRITANTS – Chemical	
	Occurs: Plant-wide	**Outcomes:** Burns to skin, mucous membranes, eyes; damage to equipment, wastewater violations.

Corrosive chemicals include acids, bases, and oxidizers, with oxidizers having special reactive properties putting them into a separate class. Corrosives, as discussed in this section, are substances with extreme pH. They are hazardous to bodily tissues either by decomposing cells through oxidation reactions or by irritating or inflaming tissues. Corrosives also frequently attack metals and can create violent reactions with a variety of other chemical types.

THE PH SCALE

Since corrosivity often depends on how acidic or basic a substance is, we can use the pH scale as a quantitative measure. For simplicity's sake, the pH scale is named for the "potential of hydrogen" and is the negative $\log_{10}$ of hydrogen ion (H^+) concentration, usually ranging from 0 to 14 pH units. Because it is a base-10 logarithmic scale that inversely indicates the concentration of hydrogen ions in a solution, a solution with a pH of 4.0 has ten times the concentration of hydrogen ions as a solution with a pH of 5.0. A pH of 7.0 is considered *neutral*, less than 7.0 is *acidic*, and greater than 7.0 is *basic*.

Two terms that are used in place of basic are *alkaline* and *caustic*. Technically, alkaline is equivalent to basic, except that the chemical manufacturing industry redefines the term to suggest a substance less corrosive than

a strong base. Caustic specifically refers to bases with the hydroxide ion (OH^-), such as sodium hydroxide ($NaOH$) and potassium hydroxide (KOH).

The pH of a solution is usually expressed as

$$pH = -\log_{10} [H^+],$$

which can be rearranged as

$$pH = \log_{10}\left(\frac{1}{[H^+]}\right)$$

where $[H^+]$ is the concentration of H^+ ions in moles per liter (mol/L). You can also take the antilog (the exponential, or inverse function, of the logarithm) to get

$$1 \times 10^{-pH} = [H^+].$$

A weak acid solution with 1×10^{-4} mol/L H^+ will have a pH of 4.0. Distilled water has a pH of 7.0, which the antilog expression above tells us means [H+] is 1×10^{-7} mol/L. A very basic sodium hydroxide solution at pH 13.2 has $1 \times 10^{-13.2}$ mol/L H^+, which simplifies to 6.3×10^{-14} mol/L H^+. Just remember that solutions with more H^+ will have a lower pH and be more acidic, and a change of one pH unit equals a ten-times change in the concentration of H^+.

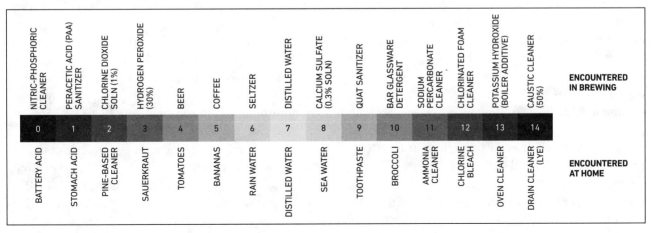

Figure 16.3. The pH of most substances is between pH 0 and 14.0. Shown here are substances commonly encountered in a brewery or the home.

BUFFERS

A buffer is a solution that resists changes in pH when either an acid or base is added to it. Buffers are added to commercial chemical products to maintain the product's pH at the optimal level. Brewery products containing buffers include acid and base chemical cleaners, sanitizers, boiler additives, and pH standards. Buffers also exist in the blood and other body fluids as a means of optimizing biochemical reactions.

In general, a buffer system comprises two related chemicals: a weak acid and its salt analog. Acetic acid (CH_3COOH) and sodium acetate (CH_3COONa) is a common buffer system. Only non-toxic buffers will be used in the food and beverage industry.

ACIDS

As we have seen, acidic solutions have higher concentrations of H^+ ions and pH levels below 7.0. This excess of positively charged ions means the acid will readily react with any molecular structure that has accessible negatively charged groups. In the human body, such vulnerable molecules include proteins, fats, and bone tissue. Acid exposure removes water from tissues, causing brittleness and sores that are difficult to heal.

When acids attack metals, the reaction often produces explosive hydrogen gas (H_2). Even though acids and bases both belong to the corrosives class, they react strongly with each other, often producing excess heat and gaseous by-products.

In breweries, acids are widely used as surface cleaners for stainless steel and glass. They are best for dissolving mineral buildup in draught systems, in brewing vessels, hot liquor tanks, and boilers. Acids are also used in stainless steel passivation, where the H^+ ion reacts with free iron (Fe^0) atoms on the surface of the material. If not removed, Fe^0 chemically reduces beer, resulting in stale flavors and shortened shelf life.

Mineral Acids

Common acids that do not contain carbon are categorized as mineral, or inorganic, acids. In breweries, mineral acids used include primarily nitric (HNO_3), phosphoric (H_3PO_4), and sulfuric (H_2SO_4) acid. They are strong acids that readily donate at least one H^+ when dissolved in water and all can make a solution of pH < 1.0. Phosphoric and sulfuric acids can release additional H^+ ions, but at a rate slower than the first one.

The purpose of these acids is to clean metal brewing equipment, dissolve mineral deposits from hot liquor tanks and draught beer lines, and to adjust mash pH. They may also be used for pH equalization of alkaline wastewater.

Nitric acid is sometimes used in stainless steel passivation. Nitric acid is also a powerful oxidizer that reacts with carbonaceous materials, like wood, cardboard, oily rags, and human tissue. Keep nitric acid away from alcohols, acetone, glycols, petroleum products, peracetic acid (PAA), and all bases. When nitric acid reacts with organic compounds, the mixture may react violently and catch fire or it may heat up and decompose, releasing an orange to red cloud of mono-nitrogen oxide (NO_x) gases, which are extremely poisonous. Any cloud like this requires immediate evacuation.

Organic Acids

With a few exceptions, in chemistry *organic* compounds are defined as those compounds containing carbon; typically, the carbon is bound to hydrogen, oxygen, nitrogen, or other carbon atoms. Therefore, organic acids are those acids containing carbon. For the most part, organic acids are considered weak acids. Strong acids are compounds that give up their hydrogen ions entirely (or almost entirely) when dissolved in water. Weak acids are different in that only a certain proportion of the molecules give up their hydrogen ions in solution; for the remainder, the hydrogens are still bound up and do not change the pH of the solution.

Organic acids commonly encountered in a brewery setting are carboxylic acids. They are so named because these acids contain one or more carboxyl groups, that is, a functional group where a carbon atom forms a bond with a hydroxyl group (OH) and a lone oxygen. The carboxyl group is typically represented in the form R-COOH, where *R* represents the rest of the molecule that is not part of the carboxyl group. The carboxyl group can give up its hydrogen ion, forming a carboxylate moiety (R-COO⁻) and H^+.

You can see that acetic acid (CH_3COOH) is a carboxylic acid. Other carboxylic acids familiar to brewers are lactic acid and citric acid, both commonly used to alter mash pH. Citric acid can also be used as a passivating agent for stainless steel. Lactic acid is an important by-product formed during the fermentation of sour ales.

A particular danger with many organic acids is that they can undergo an exothermic (heat-generating) hazardous reaction with nitric acid or other oxidizers. These reactions can get very hot, produce hazardous gases, and even explode. Store organic acids away from oxidizers and bases.

BASES

Bases are widely used in brewing, especially in the removal of organic residues left after mashing, boiling, fermenting, and packaging processes. Regular draught line cleaning relies on the use of 2%–3% caustic solutions for removing biofilm deposits. Bases can also be used in wastewater pretreatment for pH equalization.

Bases are highly reactive with acids, some soft metals like aluminum and brass, and with fats in the human body. When a strong base, like sodium hydroxide, comes into contact with skin, eyes, mucous membranes, or the respiratory and digestive tracts, it is highly irritating. Where fats or skin oils are present, a strong base instigates a saponification (*lit.* "soap-making") reaction, wherein the tissue fats are dissolved out of the body and converted to waxy or soapy by-products. When fats are removed in this fashion, the affected skin becomes brittle and fissured, being slow to heal and more readily allows contaminants directly into the bloodstream.

There are a few weak bases used in brewing, primarily as water amendments. Sodium bicarbonate, gypsum (calcium sulfate), and calcium chloride make aqueous solutions with pH levels between 7.5 and 8.5. Amines, which may be part of boiler water treatments, are a class of organic bases.

BOILER ADDITIVES

Steam boilers require a variety of different chemical products to control scale, rust, and corrosion pitting. Boiler blowdowns remove accumulated salts and sludges, but they also remove boiler water treatment chemicals. These treatments must be restored, with chemical dosage rates usually determined by portable test kits.

Boiler treatment chemicals are often multicomponent mixtures and may be highly corrosive to skin and eyes. Products have a wide range of pH levels and may have additional hazards. Reviewing the SDS for any boiler treatment chemical is critical prior to making decisions about storage, dispensing, and protective equipment.

One class of boiler treatments are oxygen scavenging chemicals, which remove dissolved oxygen (O_2) from make-up water. Oxygen in boiler water causes pitting of surfaces. Oxygen levels are commonly controlled by maintaining a residual level of sulfite (SO_3^{2-}). The SO_3^{2-} combines with dissolved O_2 to form sulfate (SO_4^{2-}), which is soluble and can be discharged during blowdown. There are many different formulations of oxygen scavengers, some with a pH less than 4.0, some around neutral pH, and others with a pH around 10.0.

Another common boiler additive is an alkalinity building product. These are typically strong caustic solutions containing sodium hydroxide or potassium hydroxide. They pose a serious health risk for skin and eye exposure, as well as respiratory system injury during blowdown.

Amine-based chemicals are used primarily in long steam-line applications. They are volatile, organic compounds that vaporize with the steam and maintain the alkalinity of the steam and condensate beyond the boiler itself. Common amines include cyclohexylamine, diethylaminoethanol, and morpholine (tetrahydro-1,4-oxazine), all examples of organic bases. In addition to the skin and eye irritation expected of bases, many amines have auxiliary hazards like being flammable and producing noxious gases when decomposing.

Scale-preventive chemicals (descalants) are also used as boiler treatments. These precipitate out calcium and magnesium hardness into a sludge that can be removed during blowdown. Descalants may contain caustics, phosphates, polyphosphates, or derivatives of polyglycols or polycarboxylic acids. They are generally moderate to strong bases.

Oxidizers

	**OXIDIZERS – Chemical**	
	Occurs: Sanitation applications, epoxies	**Outcomes:** Burns to skin, mucous membranes, eyes; damage to equipment, adverse chemical reactions.

NOT ALL OXIDIZERS CONTAIN OXYGEN

Any reaction involving oxidation must also involve reduction. What this means is that all oxidation-reduction reactions involve one chemical species transferring or receiving electrons and the other chemical species receiving or transferring electrons, respectively. When looked at this way, oxidation does not necessarily involve the element oxygen. Reactions involved in rusting, respiration, combustion, decay, disinfection, and the staling of food and beverages often are the result of oxygen receiving extra electrons. Oxidizers commonly contain oxygen in an electron-hungry state, this is true, but other elements, ions, and molecules can also be oxidizers.

An oxidizer acquires electrons from a reducing agent; likewise, a reducing agent gives up one or more electrons to an oxidizer. Oxidizers are also known as oxidizing agents, oxidants, electron recipients, or as being electron deficient. Reducing agents are also knowns as reducers, reductants, or electron donors. The reaction between an oxidizer and a reducer is called a *redox reaction*, short for oxidation-reduction reaction. In addition to electrons trading places during a redox reaction, oxygen atoms may join another molecule, sometimes leaving another molecule to do so. Examples of familiar redox equations in the brewery include the following:

$$4Fe_{(s)} + 3O_{2\,(g)} \rightleftharpoons 2Fe_2O_{3\,(s)} \qquad \text{Rusting of iron}$$

$$2SO_{3\ (aq)}^{2-} + O_{2\,(aq)} \rightleftharpoons 2SO_{4\ (aq)}^{2-} \qquad \text{Removing dissolved oxygen in boiler water}$$

The proclivity of oxidizers to attract electrons is what makes them highly reactive with many classes of substances. In the brewery, intentional reactions with oxidizers include sanitizers (which kill microorganisms by oxidizing cell membranes), cleaning chemicals, laboratory tests, and in paints and coatings.

It is often said oxidizers don't play well with anything. They should be kept separate from nearly all other chemical products, but particularly organic compounds such as alcohols, glycols, oils, and materials like human skin, cardboard, and wood. Oxidizing sanitizers can cause deterioration of certain types of rubber and some metals. Follow the manufacturer's recommendations when selecting dispensing equipment.

INORGANIC OXIDIZERS

Inorganic oxygen-based oxidizers include atmospheric oxygen (O_2) and reactive species like ozone (O_3) and hydrogen peroxide (H_2O_2). Oxygen may be part of an oxidizing anion, such as hypochlorite ($HClO^-$), permanganate (MnO_4^-), or nitrate (NO_3^-). Nitric acid, HNO_3, is a strong oxidizer in addition to being a powerful acid. Sulfur dioxide, SO_2, is used to arrest yeast viability and destroy free chlorine in brewing water, usually introduced in the form of sodium thiosulfate.

The halogens chlorine (Cl_2), bromine (Br_2), and iodine (I_2) are strong oxidizers by themselves. In brewery applications they are usually found in a stabilized solution, such as with bleach or iodophor. Bleach is a solution of sodium hypochlorite or calcium hypochlorite. Iodophor is a general term for I_2 in a solution with polyvinylpyrrolidone ("povidone"), and sometimes nitric acid. Chlorine dioxide (ClO_2) is used in yeast washing, but it is unstable. A short working life solution is prepared by combining a dilute solution of Cl_2 with a solution of sodium chlorite ($NaClO_2$).

In breweries, the most common inorganic oxidizers include hydrogen peroxide (H_2O_2) and sodium percarbonate ($2Na_2CO_3 \cdot 3H_2O_2$). Hydrogen peroxide is available in varying concentrations ranging from 3% to 50%. If using H_2O_2 in the brewery, you should only keep on hand the lowest effective concentration, typically 15%. Hydrogen peroxide is not a preferred oxidizing sanitizer because of the poor shelf life of working solutions. If peroxide is to be used for sanitizing, it should have a wetting agent added to increase surface contact. Hydrogen peroxide poses a serious hazard to the skin and eyes.

One use of H_2O_2 in the brewery is as an additive to hot caustic clean-in-place (CIP) chemicals. The reactive oxygen and fizzing decomposition of peroxide increases the effectiveness of caustic in the removal of baked-on residues in equipment like calandrias or kettles.

Hydrogen peroxide is also a weak acid, due to the oxygen radical, $\cdot O$, formed when the peroxide decomposes. When two $\cdot O$ radicals combine, O_2 gas is generated. This is the fizz observed when dilute peroxide is applied to a laceration:

$$2H_2O_2 \rightleftharpoons 2H_2O + 2\cdot O$$

$$\cdot O + \cdot O \rightleftharpoons O_{2(g)}$$

When H_2O_2 oxidizes dissolved metals that exist in the body, such as $Fe^{2+} \rightarrow Fe^{3+}$ or $Cu^+ \rightarrow Cu^{2+}$, the hydroxyl radical, $\cdot OH$, is formed. The $\cdot O$ and $\cdot OH$ radicals are two of many reactive oxygen species (ROS) that are damaging to bodily tissues. The $\cdot OH$ radical can react with lipids and proteins and cause structural damage in DNA similar to that caused by ionizing radiation (Ahsan, Ali, and Ali 2003).

Sodium percarbonate may be called sodium peroxycarbonate or sodium carbonate peroxide. Sodium percarbonate comes as a white granular solid. It is the primary ingredient in trademarked products such as Oxyclean® and PBW® and has good sanitizing and detergent properties. Granular percarbonate products usually contain an additional cleaning compound, such as sodium carbonate ("washing soda"), sodium metasilicate, or trisodium phosphate.

From a health and safety standpoint, granular oxidizing sanitizers are less hazardous to skin and form solutions with a lower pH than caustics. Powders can become airborne during dispensing and represent more of an eye or respiratory injury hazard than liquids. Sodium percarbonate cleaners may require more contact time and/or temperature to give equivalent cleaning to caustics.

ORGANIC PEROXIDES

Organic peroxides are carbon-based chemicals that carry the reactive peroxide radical (O_2^{2-}). In breweries, peracetic acid (PAA) is preferred because of its shelf stability, dose accuracy, and wettability.

The other case where organic peroxides are found in breweries is with epoxy coatings and adhesives. These products typically consist of two chemical products: the resin and the hardener. The hardener is the component containing organic peroxides and, thus, the more hazardous of the two components.

Non-Oxidizer Disinfectants, Sanitizers, and Cleaners

	FLAMMABLE, POISON, IRRITANT – Chemical
	Occurs: Sanitation applications, laboratory
	Outcomes: Irritation to skin, mucous membranes, eyes; digestive upset; slips, trips, and falls; solvent narcosis; damage to equipment; fire and explosion

The peroxide initiates and catalyzes a polymerization reaction in the resin that results in a hard finished mixture. Some resins can be catalyzed by UV light or heat instead.

DISINFECTANTS VERSUS SANITIZERS

A disinfectant is a substance that can be used to treat a surface and render it almost entirely free from undesired bacteria and viruses, whereas a sanitizer does not target viruses. (Note that disinfecting and sanitizing are not as absolute as sterilizing.) Disinfectants and santizers may have some antifungal properties. In beer production, sanitizers are commonly used to control undesirable bacteria that could affect the beer's flavor, pH, or shelf life. Spoilage organisms can exist in the product, in production equipment, and throughout the brewery environment on external surfaces of vessels, chillers, drains, and walls.

UNDERSTANDING AND SELECTING CLEANERS

Cleaners covers a diverse range of chemicals. All of the chemical cleaners categorized here may be encountered in a brewery setting. The categories are summarized in table 16.3. We will go on to look at detergents and solvents in more detail.

Table 16.3 **Types of chemical cleaner commonly used in breweries for cleansing, disinfecting, and sanitizing**

Category	Description	Common applications	Safety notes
DETERGENTS			
Cleaners			
All-purpose soaps and detergents	Chemicals that can dissolve in both water and oils/fats/greases at the same time.	Useful for food and beverage service items, normal kitchen cleanup, hands and face.	Often have an alkaline pH but are designed to not harm the skin.
Alkaline cleaners	Inorganic cleaners with alkaline pH, and sometimes oxidizer behavior. Dissolves in water, reduces surface tension to lift and suspend soils.	Cleaning wood and concrete surfaces, removing stubborn brewing deposits.	Hazardous to skin and eyes and corrodes aluminum, copper, brass, chrome. Damage to painted surfaces.
Quaternary ammonium cation compounds ("quats")	A family of detergents that have antimicrobial properties.	Commonly used for bar tops, bathroom contact surfaces, and mold control on floors and beer production and refrigeration exteriors.	Typically gentle on skin, but can cause digestive discomfort or diarrhea if traces are ingested.
Degreasers			
Strong caustic	Products that are primarily sodium or potassium hydroxide that chemically break down organic soils. These usually have added chelating agents and surfactants to help remove inorganic deposits and improve rinsing, respectively.	It is the most common type of cleaner for beer production equipment and draught system maintenance. Foaming formulations may be used for oven and hood cleaning.	Hazardous to skin and eyes and corrodes aluminum, copper, brass.
Citrus or pine oil cleaners	These are mixtures that usually contain plant-derived terpenes, alcohol-derived non-ionic surfactants, and wetting and chelating agents. Some of these products have limited antimicrobial properties.	Commonly used for heavily greasy surfaces, like kitchen or shop floors.	Eye, skin, and respiratory irritant; can cause digestive discomfort or diarrhea if traces are ingested.
ACID CLEANERS			
Inorganic and organic acids	Common acids in brewing include phosphoric, nitric, and citric acids.	Acid cleaners primarily work to break down mineral deposits, like hard water deposits, beer stone, and rust. Acids are used for periodic deep cleaning cycles of beer production equipment and draught systems, and also for passivation of stainless steel.	Acids can be very damaging to body tissues and metal surfaces. Nitric acid is also a strong oxidizer and will react with organic materials to form combustible or explosive mixtures.
ENZYMATIC CLEANERS			
	Enzymes are proteins that facilitate the breakdown of organic soils.	At time of writing, enzymatic cleaners are not commonly used in breweries, but they do hold promise for draught line cleaning, ice makers, chiller coils, and other areas where it is difficult to get mechanical action or there are parts that can be damaged by caustic cleaners.	Respiratory sensitization or allergic response; skin or eye irritation.

Table 16.3 **Types of chemical cleaner commonly used in breweries for cleansing, disinfecting, and sanitizing** (cont.)

Category	Description	Common applications	Safety notes
SOLVENTS			
Polar[a] *aprotic solvents*	These organic compounds possess some solubility in water.	Acetone is sometimes used in laboratory work as a surface cleaner.	Flammable. Drying and defatting of skin; eye and respiratory system irritant; headache, racing pulse.
Inorganic polar protic solvents	The most common example is water-based ammonia cleaner. Water is the simplest polar protic solvent.	Light duty cleaners of glass, porcelain, and plastic contact surfaces.	Skin and respiratory irritation; acute or delayed eye damage.
Organic polar protic solvents	Solvents miscible in water and can dissolve or decompose soils consisting of polar molecules. Some may have disinfecting properties. They are characterized by having one or more alcohol functional groups (R–OH) in the molecule.	Ethyl and isopropyl alcohols are commonly used in brewery microbiological sampling and laboratory work. Graffiti removal products are often polar protic solvents with added aprotic or non-polar solvents added.	Flammable. Central nervous system effects: depression, headaches, confusion, and intoxication if consumed or inhaled.
Non-polar[b] *solvents*	Petroleum-derived solvents that are either alkanes (e.g., hexane, heptane) or aromatic compounds (also called arenes; e.g., toluene, xylene, ethylbenzene). Common solvent mixtures include mineral spirits, lacquer thinner, and turpentine (which is plant-derived).	Mostly limited to brewery repair and maintenance tasks, these products are useful in removing petroleum lubricants and adhesives and in cleaning painting supplies.	Flammable. Drying and defatting of skin; respiratory irritation. Central nervous system effects: depression, headaches, confusion, euphoria, and fatigue.
ABRASIVES			
	Abrasives are insoluble mineral, fine metal, or hard plastic particles that enhance the physical action of cleaning. Abrasive tools are commonly made of plastic or natural fibers with embedded abrasive particles or are in the form of a fine metal mesh made of steel or copper.	Dry abrasives, also called powdered cleansers, use mineral particles combined with surfactants, bleaching agents, and/or mild acids.	Eye, skin, and respiratory system irritation; eye and skin irritation. May contain oxidizers and react strongly with other cleaners or solvents.

[a] Polar: electric charge is unevenly distributed across the molecule, making it water soluble to varying degrees.

[b] Non-polar: electric charge in the solvent molecule is evenly distributed; generally insoluble in water, but readily miscible with other non-polar solvents.

DETERGENTS AND INORGANIC CLEANERS

The term detergent applies to surfactants used for cleaning. A surfactant changes the properties of a surface that lies between an oil and water phase.

Detergents come in organic and inorganic forms. In breweries, common inorganic detergents include sodium carbonate (soda ash or washing soda), sodium metasilicate, and trisodium phosphate, which primarily work as water-softening and saponification agents. These tend to lift soils away from surfaces and suspend them in solution by lowering surface tension. They may not actually dissolve soils into their structure. Inorganic cleaners have a relatively high pH. Some in the brewing industry refer to cleaners with a pH≥12.5 as corrosive or caustic, and those pH

<12.5 as alkaline, but this is marketing parlance, not a technical definition. Many of these products have oxidizing capabilities, either from their own chemical behavior or because of other ingredients like sodium percarbonate. One or more of these is the basis for powdered brewery cleaners and powdered glassware washing products.

Inorganic detergents pose a hazard for dust getting into the eyes or lungs. They form slippery solutions that should be cleaned up quickly, if spilled. Some are also hygroscopic and will absorb moisture from the air. Containers of these products should be kept tightly closed.

Ammonia gas dissolves readily in water to form a solution of dissociated ammonium hydroxide (NH_4^+ + OH^-). The resulting liquid has a strongly alkaline pH and is often effective in removing organic and oily soils. Ammonia by itself is not considered a suitable disinfectant or sanitizer. Quats, which are also based on the NH_4^+ ion, fall more appropriately into both detergents and disinfectants.

Ammonia solutions are odorous and readily burn the eyes as ammonia gas is given off. Ammonia reacts readily with chlorine to produce a toxic gas, chloramine. It is important that all workers who may have to use ammonia and bleach solutions in their work know this hazard and never combine these products.

Sodium hydroxide and potassium hydroxide are also inorganic detergents; they are usually classed as strong caustic due to their extreme alkalinity and were discussed in the "Corrosive Chemicals" part of this chapter.

CLEANERS CONTAINING OXIDIZERS

If the product includes an oxidizer, like a peroxide or hypochlorite, the formula will act as a sanitizing cleaner. These products are susceptible to hazardous reactions with organic materials, which can in some cases spontaneously erupt into flames. It is best to store inorganic cleaners away from all oils, alcohols, and solvents. Storing solids on upper shelves and liquids on lower shelves is a good practice to avoid cross-contamination and possible hazardous reactions.

Organic detergents include what we generally call soaps. Soaps are the oldest known detergents, being formed by treating fats with a very basic solution (the process of saponification that we saw on p. 207). These may or may not foam, but all have the structure of one or more carbon-containing molecules attached to a polar end. Each molecule possesses a water soluble (hydrophilic) head and a fat soluble (hydrophobic) tail. Detergents are effective because they can be easily dispersed into water, yet they still mix with oily and fatty substances. When a group of detergent molecules surrounds a greasy deposit, the oil molecules in the grease interact with the hydrophobic tails of the detergent while the hydrophilic heads point outward into the surrounding water (fig. 16.5). The resulting structure is a microscopic, emulsified bead called a *micelle*, which is often described like a dried dandelion flower, with the organic soil particle at the center and the outer

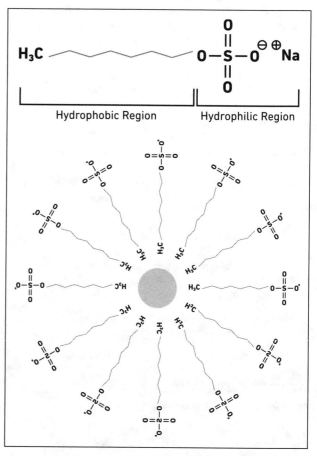

Figure 16.4. Molecular structure of sodium dodecyl sulfate detergent (*top*), showing the non-polar hydrophobic tail and the polar hydrophobic head. A micelle formed by sodium dodecyl sulfate molecules surrounding an organic soil particle (*bottom*). The hydrophilic heads are dissolved in the aqueous medium while the tails interact with the soil.

edges dissolved in the watery environment. This allows the grease to be carried off with the detergent when it is rinsed away.

Dishwashing detergents, hand soaps, and floor cleaning products are all organic detergents. Most organic detergents are mild irritants and can dry the skin out upon repeated contact. They are slippery. They are usually of slightly alkaline pH and are not very reactive with other chemicals.

Some detergents have disinfecting capabilities. These include citrus or pine oil detergents, and quats. "Quat" is short for quaternary ammonium cation compound. There are many variations of quats, depending on the organic (carbon-containing) functional groups attached to the ammonium ion (NH_4^+) at its polar end.

SOLVENTS

When talking about molecular structure, a molecule is said to be *polar* when its electrical charge is unevenly distributed across the whole molecule. Molecules that contain atoms which tend to "pull" on electrons, like oxygen and nitrogen, are usually polar to some degree. Water (H_2O) is a polar molecule and liquid water is the most prevalent polar solvent found in nature. Molecules that are primarily carbon and hydrogen tend to have an evenly distributed charge and are considered *non-polar*. Hydrocarbons like alkanes and arenes (aromatic compounds) that are liquid at ambient temperatures are typically non-polar solvents. Polar and non-polar liquids are immiscible—hence why water and oil do not mix—unless a substance like a surfactant is added.

Polar Solvents

Polar solvents include alcohols, ketones, and aldehydes. In the context of breweries, the most likely to be used are ethanol, isopropyl alcohol, methanol, and acetone. Ethanol and isopropyl alcohol (a.k.a. isopropanol) are fairly good disinfectants if the concentration and surface contact time are adequate. Methanol may be used in brewery laboratories as a carrier solvent for cellular dyes. Acetone is widely used for drying glassware in laboratory settings.

Alcohols and acetone are flammable, readily form ignitable vapors, and have flashpoints close to room temperature. They represent a fire hazard if vapors are allowed to accumulate and come into contact with a source of ignition. Flames from these fires can be colorless to light blue and difficult to see, especially in bright light. These chemicals also cause eye and respiratory irritation and will dry the skin upon repeated contact.

Laboratory-grade ethanol may be denatured with toxic or bad-tasting additives to preclude its consumption for recreation. Denaturants used include methanol, isopropyl alcohol, acetone, benzene, or pyridine. While most are toxic, alcohols and ketones also elicit symptoms of drunkenness with sufficient exposure.

Non-polar Solvents

Non-polar solvents have little to no water solubility, and are called hydrophobic. Their molecules do not have charged ends that combine well with water's polar nature. Non-polar solvents have fewer applications in a brewery than polar solvents and detergents. Non-polar solvents are usually associated with servicing machines, since they dissolve oils, machine greases, lacquers, and non-polar adhesive residues.

Non-polar solvents are often highly volatile, and some are notoriously hazardous due to flammability or their potential to cause illness upon exposure. Fortunately, with the range of products available, the employee can select the least hazardous solvent adequate for the job by reviewing the SDS and other manufacturer literature.

Some common non-polar solvents on the safer side of things include hexane, cyclohexane, D-limonene, and the mixture mineral spirits. Solvents derived from pine tar or citrus peels contain D-limonene. These are best described as "from renewable sources" not "natural."

Aromatic solvents such as toluene, benzene, and xylene, and products like paint thinner and lacquer thinner that often contain them, are thought to be more hazardous to health with repeated exposure, exhibiting greater dermal transmissivity than hexane. Benzene is a known human carcinogen. Using gasoline or other motor fuel as a cleaning solvent is not recommended. These are composed of a variable mixture of chemicals, including toxic aromatic chemicals and specialized fuel additives. All of the non-polar solvents mentioned above are particularly flammable.

Special non-polar solvents that contain chlorine, fluorine, or bromine are called halogenated solvents. Halogenated solvents are found in carburetor cleaners and closed loop refrigeration and HVAC coils. The most common are methylene chloride, 1,1,1-trichloroethane (TCA), trichloroethylene, perchloroethylene (PERC, a dry cleaning solvent), and trichlorotrifluoroethylene (CFC-113). While halogenated solvents are often not flammable, they will give off toxic and corrosive gases if involved in a fire. They are long-lasting water pollutants that should not be released into the sewer system or surface waters. Some affect ozone levels and some are known or probable carcinogens. It is best to review the SDSs and find safer alternatives.

SELECTION AND CONTROLS

Whenever product labeling is nonspecific about when and where to use a cleaner, disinfectant, or solvent, first knowing what types of materials it will be interacting with will guide correct selection. Controls for both non-polar and polar solvents start with substitution and elimination whenever possible.

During use, good ventilation and isolation for sources of ignition are essential. Store flammable substances in designated flammable storage cabinets. If maintained in bulk containers, be sure these are above secondary spill containment. Keep absorbent materials on hand in case of a spill.

Inhaling solvents may cause narcotic effects, headaches, dizziness, and nausea. Respirators, if used, need to be rated for organic vapors, but even these can fail in short order as some solvents—acetone being a notable example—will "break through" the charcoal filtering material in minutes. Avoid skin and eye contact, as many solvents can pass through the skin, potentially carrying other substances into the bloodstream with them. Check with glove manufacturers for the most compatible materials because solvents can penetrate many glove materials.

Other Chemical Hazards

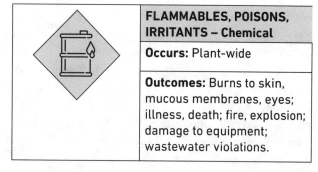

	FLAMMABLES, POISONS, IRRITANTS – Chemical
	Occurs: Plant-wide
	Outcomes: Burns to skin, mucous membranes, eyes; illness, death; fire, explosion; damage to equipment; wastewater violations.

Beyond the chemicals used in the most common brewing and cleaning tasks, breweries will have a wide range of other chemical products and hazards present. These chemicals are often purchased and used in small quantities or in non-routine applications. They may be labeled for commercial or domestic use. As a result, it is important to get into the habit of checking product labels to confirm recommended uses, observe application rates, and ensure that relevant hazards are understood and effective controls are used.

FRONT OF HOUSE AND KITCHEN CHEMICALS

Specialized products for cleaning, polishing, and sanitizing surfaces are used in bar and kitchen areas. If these are detergents or sanitizers, they have been discussed in the previous section. Oven cleaners and some metal polishes fall into either corrosive or corrosive-oxidizer classes and have been discussed earlier in this chapter. Glassware cleaners and automated dishwashing products are also alkaline corrosives and relevant guidance can also be found in earlier sections of this chapter. Floor cleaning and waxing products may be either mildly corrosive or neutral, and contain polar or non-polar solvents and other specialized ingredients. See the sections on corrosive chemicals and solvents for these.

Safety controls for these types of everyday products should include maintaining legible labels, marking intermediate containers, and dedicating certain types or colors of rags, buckets, and mops to certain products. Store these products away from foodstuffs.

PEST CONTROL AND GROUNDSKEEPING CHEMICALS

Pest Control

Pest control is discussed in chapter 17 with an emphasis on using a designated commercial provider for pest management systems (p. 221). If the business is deploying its own pest control products, preference should be given to non-chemical methods. If poisons or baits are used, they should only be in the context of closed baiting systems that leave the poisoned pest unable to roam after contacting the poison. Keeping such products away from beer and food operations is of the greatest importance, because many pest control chemicals affect humans as well. Local jurisdictions may have prohibitions against non-licensed pest control applications in a commercial setting.

Groundskeeping Chemicals

If the brewery has outdoor spaces with plantings, grassy areas, or adjacent forested areas, it may either use a contract groundskeeper who provides their own fertilizers, soil amendments, and pesticides, or it may elect to perform those activities with brewery-employed staff. Keep grounds chemicals stored away from beer and food production areas. In particular, aim to purchase insecticides, herbicides, and fungicides in small, less cost-efficient containers, and use them in their entirety. Dispose of the empty container promptly; do not ever reuse these. Improperly used, stored, or discarded groundskeeping chemicals can pollute the environment. Salt or sand used for icy surfaces may be bought and stored in greater quantities.

Hazard controls for grounds chemicals should be advised by the label and, if available, the SDS. At a minimum, physical protection should include eye protection, work gloves or rubber coated gloves, and sturdy footwear. If using groundskeeping machinery, a face shield, hearing protection, long-legged pants, a long-sleeved work shirt, and a weather-appropriate hat are also warranted. Wearing a respirator for dispensing pesticides may trigger OSHA to intervene and insist the employer maintain a complete respiratory protection program, which is another reason it may be better to contract out the work.

Conduct periodical evaluations of groundskeeping products for substitutes or alternative methods. Engineering controls include using compatible storage, mixing, and dispensing equipment in an adequately ventilated area. Administrative controls include training and hygiene after chemical use.

17

BIOLOGICAL HAZARDS

Biological hazards are generally environmental hazards in the workplace that originate from microbiological, animal, or plant sources. Outcomes can include sickness of the respiratory or digestive system, allergic response, skin irritation, and, occasionally, blindness or death. If you have jumped directly to this chapter, it is recommended you be familiar with the routes of entry discussed in chapter 6.

Human and Food Transmitted Illnesses

HUMAN AND FOOD TRANSMITTED ILLNESSES – Biochemical	
Occurs: Plant-wide	**Outcomes:** Sickness, disease, death; spreading of disease to others.

Human illnesses arise from contact with contaminated air, contaminated food or water, spilled or shared bodily fluids, human excrement, or close contact with an infected individual. The main hazard in these exposure incidents is coming into contact with *pathogenic* (Gk. disease-causing) microorganisms.

Typically, pathogenic microorganisms require adequate moisture and warmth to remain viable on their journey from one human to another, but they may live for hours to years outside of a host (Gerba 2015, 510). Microorganisms may be contained in aerosol emissions from sneezing, coughing, or even simply exhaling. Some are able to persist on bodily surfaces for some time, while others are borne along in food, water, or feces. Those present in the blood or other bodily fluids may persist at least as long as the host organism lives.

When living organisms transmit a viable pathogen, the transmitting organism is called a *vector*. For example, humans, mosquitos, ticks, and bats can be vectors. When pathogens are transmitted from an inanimate object, the object is called a *vehicle*. Vehicles include food, water, biologic products, and *fomites*. Fomites are inanimate surfaces, like tables, linens, and doorknobs, that house a pathogen. (Gerba 2013; Centers for Disease Control and Prevention 2006, pp. 1-64–1-65).

In a brewery or taproom setting, fomites can include bar surfaces, touch screens, menus, condiment bottles, glassware, and toilet handles. Smooth, non-porous fomites may afford pathogens shorter viability but transmit the pathogens more easily than porous surfaces such as clothing and paper currency. Fomites typically form when aerosols land on a surface, but they can arise from detritus shed from organisms, like hair, dead skin cells, and saliva.

With the exception of viruses, many pathogenic micro-organisms can proliferate outside of a human host—in animal vectors or on fomites (if conditions allow)—as well as within it.

Viruses, which are technically not alive and are not considered cellular organisms, must use the molecular machinery of other living cells to replicate their genetic material and must therefore be within a living host to proliferate. However, viruses can typically remain viable as infectious agents on fomites longer than bacteria or the spores of parasitic organisms.

Some microorganisms can produce spores, which are long-lived, resistant cellular forms containing genetic material that can survive difficult environmental conditions. Many fungi produce spores as part of their asexual or sexual reproductive life cycle, and certain bacterial species form what are called endospores, essentially a dormant phase in response to unfavorable conditions such as dehydration or lack of nutrients. Spores and endospores require more extreme control measures than the viable organisms, requiring sterilization rather than disinfecting or sanitization (National Library of Medicine n.d.).

Parasites are organisms that live and often can reproduce within the body. They may be microscopic (protozoa), barely visible (head lice), or macroscopic (intestinal worms). Parasites can enter the body through ingestion of contaminated food or water, broken skin, bites from insects or spiders, or simply by landing on the skin. While probably quite uncommon in the United States, parasitic infections in the workplace can be controlled by the same methods of surface sanitation, food safety, and good personal hygiene used to fight bacterial and viral hazards.

HUMAN- AND FOODBORNE ILLNESSES IN THE WORKPLACE

Just as we do not wish to be injured in a manufacturing process, most of us are loathe to become sickened by another individual or by consuming contaminated food or drink. Since most beers are of sufficiently low pH and were boiled during their manufacture, beer is usually the least concerning from an infectious disease standpoint. That leaves other people and food as the primary vector or vehicle of disease in the brewery workplace.

 HUMAN AND FOOD TRANSMITTED ILLNESSES BY PATHWAY AND VECTOR

Airborne, Direct Contact, or Fomite
- Viruses: influenza (flu), common cold, SARS-CoV-2 (COVID-19), varicella (chicken pox), measles, hepatitis A
- Bacteria: *Streptococcus* species (strep throat), *Neisseria gonorrhoeae* (gonorrhea), *Chlamydia trachomatis* (chlamydia)
- Parasites: body/head lice, pubic lice (crabs)

Bloodborne Pathogens
- Viruses: hepatitis B, hepatitis C, human immunodeficiency virus (HIV)

Food, Water, and Feces Transmitted
- Viruses: norovirus, hepatitis A, the common cold
- Bacteria: *Salmonella* species, *Clostridium perfringens*, *Campylobacter* species, *Listeria monocytogenes*, *Escherichia coli* (*E. coli*), *Vibrio* species
- Parasites: giardia, pinworm, hookworm
- Latent toxins:[a] botulinum toxin, *Staphylococcus aureus* toxins

[a] These are toxic substances produced by food-contaminating bacteria. The toxins persist even if the bacteria are destroyed. Latent toxins are not communicable between human individuals.

Airborne and Direct Contact Illnesses

Many notorious illnesses are caused by breathing aerosols of exhaled air carrying viable, pathogenic microorganisms, such as common cold viruses, flu viruses, coronaviruses, measles viruses, and streptococci. Coughing and sneezing increase the chances of transmission. Given time, these aerosols will settle on surfaces where the pathogens can remain viable long enough to enter a new host. This usually involves being picked up on the skin by contact with a contaminated surface, and then later entering the body through actions like rubbing one's eyes, smoking, eating, or biting nails. It is easy to imagine coworkers and visitors carrying a disease into the workplace, then spreading it by exhalation.

Pathogens and parasites can also be transmitted through close person-to-person contact, like hepatitis A or lice. Sexually transmitted diseases (STDs) rely on intimate physical contact between individuals.

Illnesses from Blood and Bodily Fluids

Contact with contaminated bodily fluids typically occurs while rendering aid to a coworker or patron with a bloody injury, or cleaning up after such an incident has happened. While the principle of universal precautions (discussed below) would have you act as if all bodily fluids contain disease, this should never preclude you acting with care and respect for the injured party.

Illnesses from Contaminated Food, Water, or Feces

Inadequate sanitation, improper food handling, and poor personal hygiene can all contribute to the risk of transmitting disease by food or water. Food, not water, is most likely to be the source of illness in a brewpub or taproom setting. Most drinking water in the US is disinfected with chlorine, chloramine, or ultraviolet light prior to distribution, and is unlikely be to a source of disease transmission.

Food-borne illness arises from contaminants in the food itself, from food contacting contaminated work surfaces and kitchen implements, and from direct human contact with food. The duration of time that food is kept above refrigeration temperature (i.e., above 40°F or 4°C) and below a food-safe holding temperature (below 140°F or 60°C) directly influences the growth rate of spoilage microorganisms. Pantry foods require sterile processing or a sufficiently low pH or low moisture content to remain shelf stable.

When kitchen staff do not maintain good personal hygiene, there is a possibility of transferring enteric bacteria and viruses from toilet activities. Lacerations in the kitchen can also introduce the chance of a blood-borne infection.

Toxins Resulting from Contaminated Food

In rare cases, food-borne illness can result from toxins created by bacteria or fungi. These toxins exist even after the organism that produced them has been killed by pasteurization. Botulinum toxin, formed by *Clostridium* bacteria, affects the nervous system. It is the same substance used in small, local doses to prevent facial wrinkles.

The toxins made by the bacteria *Staphylococcus aureus* ("Staph") can cause fast-acting, violent food poisoning characterized by nausea, vomiting, and diarrhea. Food poisoning from *S. aureus* is typically short lived, with the toxins usually being metabolized and eliminated from the body. Food-borne bacterial toxins are not communicable between individuals, but the bacteria that produce them are transmissible.

The extensive use of antibiotics has resulted in antibiotic-resistant strains of *S. aureus* called MRSA, meaning that some individuals may not respond to treatment with antibiotics. It is estimated that one third of the population carries systemic *S. aureus* without evident sickness (National Institute for Occupational Safety and Health 2023). Systemic exposure to *S. aureus* toxins can lead to more severe symptoms, damaging red blood cells and white blood cells and, in rare cases, causing toxic shock syndrome.

APPLICABLE STANDARDS

OSHA standards relating to biological hazards either address sanitation in very general terms or focus on blood-borne pathogens (table 17.1). A business's comprehensive program to control biological hazards should take a broader view. Specifically, care should be given to preventing transmission of all diseases, whether they be spread by bodily fluids, aerosols, direct contact, or food and water.

HAZARD CONTROLS FOR HUMAN DISEASE HAZARDS

The starting point for any infection control program is something called *universal precautions*. Essentially, one assumes that any blood or other bodily fluid is a potentially infectious material and one should take

Table 17.1 **Occupational Safety and Health Standards (29 C.F.R. Part 1910) relating to biological hazards**

Short title of standard	Title 29 citation	Key requirements
Sanitation	1910.141(a)(3)	Place of employment kept clean to the extent that the nature of the work allows (§ 1910.141(a)(3)(i)); every floor, working place, and passageway shall be kept free from protruding nails, splinters, loose boards, and unnecessary holes and openings (§ 1910.141(a)(3)(iii)).
First aid	1910.151, app. A	"If it is reasonably anticipated that employees will be exposed to blood or other potentially infectious materials while using first aid supplies, employers are required to provide appropriate PPE in compliance with 1910.1030(d)(3), such as gloves, gowns, face shields, masks, and eye protection."
Bloodborne pathogens[a]	1910.1030(b)	This standard applies to those with "reasonably anticipated skin, eye, mucous membrane, or parenteral contact with blood or other potentially infectious materials that may result from the performance of an employee's duties," e.g., employees trained in first aid or those cleaning up after an injury.
	1910.1030(c)	Employer provides written Exposure Control Plan, exposure determination, lists job classifications, tasks and procedures; employees to use universal precautions and engineering controls.
	1910.1030(d)(3)	Employer provides PPE, responsible for use, issuance, cleaning, disposal, repair and replacement; includes gloves, face protection, apron, or other protective clothing.
	1910.1030(d)(4)	Worksite, equipment, and surfaces are clean and sanitary; waste and laundry managed.
	1910.1030(f)(2)	Hepatitis B vaccination available at no cost to an employee who may have occupational exposure.
	1910.1030(g)	Hazard communication includes labels, signs, information, and training. Inform employees how medical evaluation and training records maintained, made available upon request to the employee, employee's representative, or regulatory agency.

[a] While this standard applies strictly to bloodborne disease, most of the principles also apply to safely managing illnesses transmitted by air, food, water, feces, and direct contact.

actions designed to limit potential contact with infected substances or surfaces. Universal precautions are most often discussed with regard to bloodborne pathogens, specifically hepatitis B and human immunodeficiency virus (HIV).

According to the National Institutes of Health, universal precautions do not apply to "sputum, feces, sweat, vomit, tears, urine, or nasal secretions unless they are visibly contaminated with blood because their transmission of hepatitis B or HIV is extremely low or non-existent" (Broussard and Kahwaji 2022). However, the underlying principles can certainly help in preventing the spread of illnesses transmitted through air, food, water, or contact with non-bloody human fluids, surfaces, or waste.

Bloodborne Pathogen Controls

Bloodborne pathogen controls usually consist of PPE and simple engineering controls to contain, collect, and manage potentially contaminated materials. The PPE is designed to protect the surface of the hands, face, and eyes using rubber gloves and a face shield. Respiratory protection is in the form of a medical-grade disposable face mask. The engineering controls consist of a disinfectant, absorbent granules or pads, a trowel or dustpan for collecting waste, and a red biohazard trash bag.

Bloodborne pathogen spill kits are commonly available from industrial and first supply houses or can be easily put together from the elements listed. Disinfectant supplied in store-bought kits is likely to be a quaternary ammonium cation (quat) compound. Homemade kits can use chlorine bleach applied at a rate of 1 tablespoon per gallon of water (4 mL per liter). Keeping a plastic spray bottle in the kit makes it easy to get the disinfectant onto all surfaces. Diluted bleach solutions will lose their effectiveness over time and should only be made up on an as-needed basis. In cases where there is significant trauma to the injured, or where coworkers are emotionally unable to clean up after the accident, a crime scene cleanup company may be used.

Surface Disinfection Controls

The primary control of biological contamination on fomites is through application of disinfectants. While most microorganisms do not live long on fomites, they often live long enough in a bar or restaurant environment to be transmitted to a new host. It is important to clean a surface well before applying the disinfectant, since the chemical activity of the disinfectant may be used up reacting with soil that is sheltering the target organisms.

Chemical agents used to *mostly* eliminate microorganisms on a surface are called sanitizers or disinfectants. This is different from sterilants, which destroy *all* viable cells and spores. If products designed for commercial application are used, the SDS should be kept on hand and reviewed prior to use. Products packaged for domestic use will not have SDSs, so relevant use and storage instructions will be found only on the label. Correct disinfectant selection requires consideration of the target microorganism, the surface, and whether direct human contact will occur. Most common disinfectants fall into four chemical categories:

- Disinfectant cleaners and quats
- Oxidizers
- Alcohols and phenols
- Skin disinfectants

The following discussion focuses on where these products might be used in a brewery. These common brewery and front of house disinfectants are discussed in further detail in chapter 16 as workplace chemical hazards.

Disinfectant cleaners possess both disinfecting and detergent properties. The most common of these are quaternary ammonium cation compounds, called quats or QACs. They are diluted to a working solution in water and have a long shelf life. Quats form positively charged ions in solution, which bind to negatively charged cell walls of bacteria, viruses, and fungi, resulting in death or inactivation of the cells (American Cleaning Insitute

2022). Because these disinfectant molecules possess both water-soluble charged ends and hydrocarbon (alkyl) ends, they can also act as detergents to remove organic soils.

Oxidizing disinfectants provide reactive electron-scavenging compounds that break down intracellular proteins. This class of compounds may be based on reactive oxygen or weak acids containing halogens. Reactive oxygen disinfectants include hydrogen peroxide, sodium percarbonate, peroxymonosulfates, peracetic acid, and ozone. Halogen containing compounds include chlorine bleach (sodium or calcium hypochlorite), chlorine dioxide, chloramine, and iodophor. These agents are generally short-lived at working strength and require longer contact time than quats or alcohol. Because they are mild acids with oxidizing behavior, these halogen compounds may irritate the skin or damage metallic surfaces over time.

Alcohol disinfectants, including ethanol, methanol, isopropanol, thymol (extracted from thyme), and pine oil (contains cyclic and acyclic terpene alcohols, like α-terpinol and linalool, respectively) all have fast disinfecting properties if used at an appropriate concentration. Use concentrations, however, are high enough that the substance may be considered flammable. These compounds also cause drying of the skin if used topically.

Skin disinfectants have to meet two additional challenges: they must work on porous and rough skin surfaces and they must not be injurious to the individual. In addition to the alcohols, these products include chlorhexidine, chlorhexidine gluconate, and povidone-iodine. Their use in a brewery or taproom environment would likely be limited to certain first aid products.

Food Contamination Controls

There are four factors to controlling food contamination: cleaning, separation, cooking, and chilling (Centers for Disease Control and Prevention 2020c; Partnership for Food Safety Education 2022; Food Safety and Inspection Service 2008).

Cleaning, or sanitation, includes both worker personal hygiene and the cleanliness of working surfaces and implements. Personal hygiene requires administrative controls, training, and regular reinforcement. Often, signs are posted near sinks and in restrooms to remind workers to wash well. The types of products used for cleaning and sanitizing working surfaces and implements are those mentioned above in "Surface Disinfection Controls."

Separation includes avoiding cross-contamination during food preparation and the proper storage of foods. It is an important control that should not be overlooked. For instance, storing raw poultry and meats on the lowest shelf in a refrigerator keeps any leaked juices from contaminating food that will be eaten raw, like fruits and vegetables.

Cooking food to the correct internal temperature is an important part of making food safe. There is a hierarchy based on minimum safe cooking temperatures (MSCT) for meat and eggs. This hierarchy can be combined with separation guidelines to inform how the raw ingredients are stored in a refrigerator (StateFoodSafety n.d.):

- Red meat, fish and seafood, and whole eggs (MSCT = 145°F/63°C)

Are stored above

- Ground or marinating meats and pooled eggs (155°/68°C)

Are stored above

- Raw poultry (165°F/74°C)

Chilling involves making sure perishable food (raw or cooked) is never left long enough at a temperature that allows microorganisms to multiply and reach harmful levels. By limiting the time that food is kept between 40°F and 140°F (4°C and 60°C) to less than two hours, one can prevent the proliferation of foodborne microorganisms, if they are present.

The National Restaurant Association has developed a widely recognized training series called ServSafe. Most state or local health departments will require certification for cooks, servers, and food managers.

Destruction of Bacterial and Fungal Spores

Spores are inherently robust and designed to survive in adverse environments where there are extremes of temperature, moisture, and chemical exposure. Destroying spores requires sterilization (National Library of Medicine n.d.). Common means of sterilization include the following:

- High temperature, e.g., direct flame or steam
- High temperature and pressure, e.g., autoclaving
- Radiation, e.g., UV or gamma radiation

- Oxidizing chemicals: ozone, hydrogen peroxide, peracetic acid, chlorine dioxide
- Sterile filtration

Most of these methods cannot safely be used on humans, nor on many types of fomites in beer manufacturing and front of house activities. So, instead of high-energy destructive techniques, effective controls rely on thorough surface cleaning with appropriate detergents, followed by sanitizing with an appropriate sanitizer or disinfectant, or by using an agent like a quat that has both detergent and sanitizing properties.

Illnesses and Allergies Caused by Animals, Plants, and Consumables

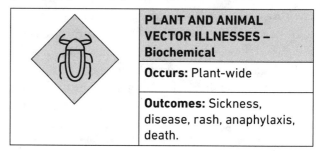

	PLANT AND ANIMAL VECTOR ILLNESSES – Biochemical
	Occurs: Plant-wide
	Outcomes: Sickness, disease, rash, anaphylaxis, death.

Many substances originating from plants, animals, and synthetic compounds can make us sick or cause us to have an allergic reaction. These may be substances we knowingly avoid or substances we intentionally use.

ILLNESSES RELATED TO PESTS

Diseases which can be transmitted from an animal vector to a human, or vice versa, are called *zoonotic* diseases. While many are rare, there are some that can occur in the context of beverage manufacturing or food handling.

Vermin

In a brewery setting, grain, grain dust, and beer residue are powerful attractants for rodents, birds, and invertebrate pests. An animal may transmit disease or cause injury through a bite or sting, but diseases can also be transmitted by workers inhaling an aerosol containing viral particles from animal waste.

Venomous Native Animals

Irritating vertebrate and invertebrate species may be encountered during facility construction, in untrampled crevices, or in outside seating and festival spaces. The invertebrates of most common concern, depending on geographic location, are insects and arachnids, such as hornets, wasps, biting spiders, and scorpions. Injury usually involves a painful bite or sting, but for some individuals a sting can cause an allergic reaction up to and including life-threatening anaphylaxis. Allergens are discussed below.

Spider bites can elicit an immune response, as well as cause a region of ulceration and dead tissue called necrosis. Necrotic wounds may need to be treated by a medical professional.

Mosquitos and ticks are important disease vectors carrying microorganisms that are pathogenic to humans. Most well-known among these are Lyme disease carried by the deer tick, and various flus caused by mosquito-borne viruses.

Venomous snakes are native to most of the US, though they are more common in number and variety in southern and western states. You are unlikely to encounter a venomous snake in a brewery, but snakes can be present where there are outdoor storage areas or crawlspaces. If a snake bite occurs, identifying the species of snake may be critical for the administration of the correct antivenom.

Irritating Native Plants

Poison ivy (most common), poison oak, and poison sumac all cause a bothersome skin rash in most individuals who contact these plants. Oils in the leaves cause itching, blistering, and, ultimately, open sores that are slow to heal and prone to infection.

The stinging nettle is a common perennial in moist soil throughout North America. Skin contact causes a painful itch due to tiny hairs along the stems. These hairs contain formic acid, the same compound that gives ant bites their sting. A small percentage of persons will experience an allergic reaction, in addition to the acid sting.

Table 17.2 **Examples of zoonotic and fungal diseases**

Zoonotic disease	Transmission	Symptoms and treatment
Rabies	Virus from animal bite	A bite from an infected mammal. Affects central nervous system. Fatal if not treated, although effective vaccination exists if administered prior to onset of symptoms.
Hantavirus	Inhalation of animal waste	Inhaling dust containing dried rodent excrement or urine can cause this potentially fatal respiratory illness. In the US, most common in the desert and mountain areas west of the Mississippi River.
Lyme disease	Bacteria from a tick bite	Infection by *Borrelia burgdorferi* bacteria. Can be treated with antibiotics, but some individuals will experience a long-lasting syndrome affecting their skeletal, circulatory, and nervous systems.
West Nile virus	Virus from mosquito bite	A mosquito-borne disease that affects humans, horse, and birds. Infection can result in a sometimes fatal, flu-like illness, primarily in older individuals.
Equine encephalitis	Virus from mosquito bite	There are several subtypes of this disease, mostly affecting the very young and the old. In advanced cases, flu-like symptoms may be followed by seizures, coma, and death.
Fungal Disease	**Transmission**	**Symptoms and Treatment**
Histoplasmosis	Inhalation of fungal spores	A respiratory disease caused by *Histoplasma capsulatum* associated with exposure include soil, soil enriched with bird or bat guano, remodeling or demolition of old buildings, and clearing trees or brush in which birds have roosted
Valley fever (coccidioidomycosis)	Inhalation of fungal spores	A respiratory ailment that may require antifungal treatment. Common in the soils of the US Southwest.
Blastomycosis	Inhalation of fungal spores	Caused by the spores of soil and wood fungal species in the genus *Blastomyces*, which live in moist, forested environments. Causes a potentially fatal respiratory disease that may require antifungal treatment.

Fungal Spores

Like some viral particles, fungal spores can also be present due to animal waste. Bat or bird droppings are a source of *Histoplasma capsulatum* spores, which can cause severe respiratory disease if inhaled. Fungal species are also ever present in soil. Do not think dry areas are free from fungal spores, as diseases like Valley fever (coccidioidomycosis) are caused by a fungus endemic to the American southwest and Mexico (table 17.2).

ALLERGENS

Allergies are a common condition in which the body's immune system launches defensive countermeasures to an otherwise non-harmful substance. We call these substances *allergens* (Gk: different and activity). Allergens can enter the body through inhalation, ingestion, skin/eye contact, and injection.

For non-allergic people, exposure to an allergen causes no noticeable symptoms. An allergic individual's immune system, however, will recognize a specific allergen, consider it a threat, and unleash a series of biochemical reactions resulting in runny nose, watery or itchy eyes, tingling in the mouth, wheezing, coughing, or hives. This reaction begins within minutes to up to a few hours of exposure. In those with a non-severe allergic reaction, the normal allergic response does not significantly compromise the cells of the body. In persons with hypersensitivity, the immune and inflammatory responses of the body are more damaging.

Some allergic individuals experience a life-threatening allergic response called *anaphylaxis*. During anaphylaxis, blood pressure rapidly drops, the upper respiratory tract swells, and breathing becomes difficult. During this time, the body releases histamines, which not only drain the nasal cavity, but cause inflammation and swelling in the throat. This is accompanied by a sense of panic. Other symptoms can include an itchy skin rash, vomiting, or diarrhea.

Although allergies are known to have a genetic component and can be inherited, they can also arise from environmental factors through the process of *allergic sensitization*.

Food Allergens

Food allergies typically occur when the body wrongly identify certain proteins as being harmful. There are over 160 known food allergens (US Food and Drug Administration 2023), but the federal government currently requires warning labels for nine of them: milk, eggs, peanuts, tree nuts, wheat, soybean, sesame, shellfish, and finned fish (21 U.S.C. § 321(qq)(1)).[1] Of these, allergies to cow's milk, eggs, wheat, and soybean are most common in children, although these childhood allergies are often outgrown by adulthood. Shellfish, peanuts, tree nuts, and similar food allergies are generally lifelong (Mayo Clinic 2021).

Of particular interest to beer makers and sellers is the practice of adding flavorings to beer that are known allergens. Table 17.3 includes common allergens, including a grouping for allergens used as beer additives, but this should not be considered a comprehensive list.

Table 17.3 **Common allergens**

Plant and animal	Facility and household
Cockroaches	Household chemicals
Dust mites	Latex
Insect stings (e.g., bees, hornets, wasps)	Medications
Mold spores	Perfumes
Pet dander	
Plant pollen	
Allergenic beer ingredients	**Foods per 21 U.S.C. § 321(qq)(1) (2023)**
Aniseed	Eggs, especially egg white
Banana	
Chamomile	Fish (e.g., bass, flounder, cod)
Isinglass (fish-based flocculant)	Milk (incl. lactose)
Kiwi fruit	Peanuts
Lactose	Sesame (e.g., tahini)
Mustard seeds	Shellfish (e.g., crab, lobster, shrimp)
Oyster shells	
Passion fruit	Soybeans (incl. soy products)
Peach	Tree nuts (e.g., almond, pecans, walnuts)
Peanuts	
Wheat	Wheat
Yogurt (as sour culture starter)	

---

[1]   Sesame was added as part of the Food Allergy Safety, Treatment, Education, and Research Act of 2021 (Pub. L. No. 117–11, § 2, 135 Stat. 262), effective January 1, 2023.

Medications and Cosmetics

In the context of a brewery or brewery laboratory, the natural rubber in latex gloves is a common allergen. Other substances that can affect employees and customers who have developed an allergic sensitization include perfumes and cleaning compounds with scent added, cosmetic preservatives, and dyes (US Food and Drug Administration 2022).

Some medications can cause allergic reactions. This includes important classes of drugs, such as penicillin and non-steroidal anti-inflammatory drugs (NSAIDs) like aspirin and ibuprofen.

DRUGS AND ALCOHOL

Over-the-counter drugs, recreational drugs, tobacco, and alcohol all cause changes in the human body. Exposure is not caused by contact with a vector or fomite, but rather by intentional ingestion, inhalation, or injection. These changes can alter reaction time, attention span, and rational thought. Being under the influence of such substances relates to a higher chance of industrial accident or unintentional overdose (National Research Council 1994; Bureau of Labor Statistics 2020a).

HAZARD CONTROLS

The control of illnesses and allergies caused by animals, plants, and consumables emphasizes administrative controls such as policies, training, warning signage, and emergency planning. In addition, some situations may require PPE and engineering controls.

Hazard Controls for Animal and Plant Pests

Animal and plant pest detection and control is usually accomplished with an integrated pest management (IPM) program. An integrated pest management program typically has four parts: action thresholds, monitoring and identification, prevention methods, and specified controls (Environmental Protection Agency 2022).

Action thresholds are statements about what frequency or population of pest requires the integrated pest management program to engage. An occasional bird in the brewery may not be a problem, whereas roosting or nesting birds might be too much. A sighting of a single rat, however, would be cause for immediate action. The program should list the types of pests that are likely at the business and decide what action threshold is right for the business. The threshold usually is set to the point where an economic or quality threat is perceived.

The IPM should focus on identifying only those pests that truly pose a problem for employees, customers, or product integrity. It is not necessary or desired to list every weed species in the back lot or stray hornets around the dumpster. Identifying specific threats allows the business to develop monitoring, prevention, and control strategies that are pertinent. This will reduce the use of pesticides and other costly control regimens.

Just as with any workplace hazard, we start with prevention first. Prevention methods against rats and cockroaches involve robust housekeeping to eliminate their access to food sources. Prevention against mosquitos is to eliminate standing water, where practicable. Prevention against irritating plants in the beer garden might be the use of landscaping cloth during construction. Prevention costs less in the long run and does not have the human and environmental health risks of pesticides.

If you are establishing a brewery in a location that previously had no food or beverage production nearby, you may not know what pests to expect. You also want to make sure not to give pests a chance to get established. Maintaining the absence of a pest problem is much easier than trying to control an existing one. You should consult a pest control contractor before you encounter any pests.

When pests do exist and they threaten the livelihood of the business, some engineering controls may be warranted. The hierarchy is to first choose methods that are species-specific and do not involve the use of pesticides. This can include using beneficial predatory insects, trapping, weeding, bug zappers, mechanical rodent traps, or using pheromones to disrupt pest reproduction.

Where poisons are warranted, they should be the last resort. Poisons should be applied on a schedule by a licensed pest control technician. The contractor will be able to choose the best products and application rates for your situation. Bug control sprays are liquids that quickly dry, reducing tracking and uptake by non-target populations. Rodent traps that use bait are enclosed and use solid bait blocks. As a general rule, never use loose bait or poisons where they could conceivably enter the production process or cause harm to non-target populations like customers, their kids, dogs, or beneficial predators.

Avoiding and Responding to Allergic Response

Every brewery should establish a policy on the presence and use of allergens in their beer and other consumables. If you are not allergic to the sorts of substances, it can be easy to overlook how significant food allergens are to your customer base. Remember that allergic reactions to food can cause serious health effects, up to and including death.

A policy on allergens can look a lot like an integrated pest management program, in that you will list the allergens of interest, how to identify them, what processes or beverages they can occur in, and then prevention and protection methods to control hazardous exposure. In terms of prevention, a brewery should consider each and every case of using a food allergen and ask themselves whether it is essential to use. In cases where a known allergen is used, provisions for notifying customers of its presence are important. This can be done with labeling, menu listings, the company website, and social media. Consumers who know they have a food allergy are very conscious about their purchases and the brewery needs to make

this information easily available to them. Keep in mind, not every consumer knows that a pastry stout will likely have lactose, or that an oyster stout could have actual oyster shells in it.

If prevention and informational controls fall short, the possibility exists for a product recall or a lawsuit from an affected customer. These are both very costly scenarios. They can create financial hardship and a public relations disaster for the company.

Hazard Controls for Alcohol and Drugs in the Workplace

Preventing and controlling workplace safety incidents from the effects of drug or alcohol use are as much about human behavior and social norms as they are about policies and biological responses. We address drug and alcohol use in chapter 18 on psychosocial hazards.

18
PSYCHOSOCIAL HAZARDS

Circumstances in workplace structure and communication that can result in emotional, mental, and physical stress are called *psychosocial hazards*. As with any common type of safety hazard you might imagine, psychosocial hazards have individual properties and hazard control solutions. They may show up in a specific, situational context, but are often systemic throughout the workplace.

Psychosocial hazards most often involve worker interactions with the workplace organizational structure, interactions between coworkers, conflict or toxicity in the workplace, or harmful societal messaging. In this chapter, we also consider the worker's individual setting, that is, what's in their own head.

Stressful situations can include events that are ambiguous, random, unpredictable, uncertain, or perceived to be out of a person's own control. They may involve conflict between a worker and other individuals, workplace equity disparity, job insecurity, or unattainable work expectations. The worker may be internally conflicted between their own experience and the way jobs are performed at the brewery. Stresses outside the workplace might include family issues, transportation to work, or socioeconomic forces.

Employers may be under the impression that their employees' mental and emotional state is not under their care. This view needs revising for several reasons. First, the employer has the legal responsibility and obligation to create a safe and healthful workplace. If an employee's behavior is impinging on the well-being of others, the remedy falls mainly on the employer. Secondly, improving organizational culture has been shown repeatedly to be superior to risk management on a case-by-case basis (Way 2020; Michie 2002; IOSH 2018; Akiomi et al. 2021). There are obvious financial implications: a dysfunctional workforce is less productive and will cost the employer in loss of sales and damage to corporate image. Workplace stress is measurable and is discussed more in chapter 21 with regard to inclusion and wellness.

Instances of workplace psychological injury are increasing in both number and severity. They affect the well-being and safety of individuals and threaten negative outcomes for businesses (Way 2020). Research into cumulative risk assessment has shown that psychosocial hazards and personal issues can combine to trigger outcomes like post-traumatic stress disorder (PTSD) or diabetes, and that psychosocial hazards combined with physical stresses can result in heart disease or biochemical diseases or imbalances that can cause disabilities (Fox et al. 2021).

While there has been a more concerted study of psychosocial hazard in the past two decades (Way 2020), the underlying psychology is nothing new. Raymond Gastil described four behavior determinants sixty years ago: situational events between people, the basic nature of the individual, behaviors learned from the setting, and biosocial factors of cultures and societies (Gastil 1961, 1286). Each of these has a role in creating occupational stress, which can lead to physical, mental, emotional, and psychological outcomes.

PSYCHOSOCIAL SETTINGS ARE LIKE CLASSIC DRAMATIC CONFLICTS

In literature, theater, and song, all cultures understand the same classic conflicts facing a protagonist. We have the protagonist against another person, against themselves, against society, and against nature. Sometimes the protagonist is challenged by technological or supernatural forces.

In the brewery, we have analogous conflicts: between a worker and one or more coworkers, the worker dealing with internal conflict, the worker against societal conventions, and the worker against their workplace environment.

Going to work at the brewery should not have to feel like you are in a Greek tragedy, Wagnerian opera, or *Game of Thrones*. Learning how to confront these conflicts, which come with being human, is critical to your mental and physical well-being.

ALARM AND ADAPTATION

A person's relationship with stress is usually a two-step process. First is the alarm reaction, where the body reacts physiologically and psychologically to a threat. This is followed by some sort of adaptation, which can involve getting away from the situation or otherwise coping with it.

Alarm Reaction and Outcomes

The alarm reaction may be acute and physical, with symptoms including tense muscles, headache, and increased heart rate, blood pressure, and respiration. Acute responses may include feelings of anxiety or irritability, or a diminution of thought processes involving concentration and problem solving. What is usually most apparent to others are behavioral manifestations, such as aggression, withdrawal, or fragile emotional state.

Serious chronic health consequences can result when a person remains in the alarm state for an extended period, which can happen with ongoing workplace stress. These can include cardiovascular disease, immune deficiency, neuroendocrine disorders, poor physical health, and lasting mental states like anxiety, depression, and PTSD (Michie 2002).

Table 18.1 **Occupational stress outcomes**

Physiological outcomes	
• Gastrointestinal dysfunction	• Physical or sexual assault
• Elevated blood pressure, or hypertension	• Substance abuse, injuries under the influence
• Headache, neck ache, jaw clenching	• Type 2 diabetes
• Fatigue, sleep dysfunction	• Cardiovascular disease
• Musculoskeletal injury	• Premature mortality
• Increased chance of workplace injury	
Mental/emotional outcomes	
• Anxiety, uncertainty, fear	• Distraction
• Unhappiness, volatility	• Disengagement, self-concealment
• Job dissatisfaction, burnout	• Bullying, verbal abuse, shutting out
• Disempowerment	• Substance abuse, addiction
• Depression, fatalism, suicide	• High turnover, absenteeism

Adaptation

Alarm reactions are not pleasant, so we naturally try to understand them and adapt. Adaptation gets us to a safer place, or a place where we understand the alarm but are no longer threatened by it.

We often think of adaptation as the "fight-or-flight" mechanism. This evolutionary conserved response can still be relevant if one were confronted with an extreme event, like a robbery or active shooter. But in the day-to-day operations of a business, we tend to use transactional means of coping. These are learned behaviors and we do not always act in our own best long-term interest or choose the healthiest adaptation. Owners and employees may not have arrived at the company able to recognize occupational stress, nor possess the skills to listen, observe, check in with others, resolve conflicts, and manage workplace time and role expectations.

In an effective safety culture, communication skills are continuously modeled. Management is actively working to create a workplace with fewer stress alarms. The culture's value system reinforces the importance of working together toward common goals of productivity, collaboration, inclusivity, quality, and safety. New learners may benefit from outside experts in cognitive behavioral techniques, whether through books, training, or counseling.

In the same spirit as chapters on physical, chemical, and biological hazards, this chapter describes common workplace human behaviors and management constructs that increase the risk of accident, injury, or illness. This treatment goes beyond injury to the body's physical integrity—with psychosocial hazards, we are viewing mental well-being as if it is a body part that can be injured or sickened.

Biosocial and Situational Interactions

	**INTERPERSONAL – Psychosocial**
	Occurs: Plant-wide
	Outcomes: Stress, anxiety, social isolation, burnout, substance abuse and addiction, suicide; metabolic stress, nausea, musculoskeletal disorders, headache, elevated blood pressure; physical or sexual assault.

Interactions between coworkers sets the tone of a workspace. It may be friendly, supportive, collaborative, and satisfying for all. It may be homophobic, chauvinistic, hostile, or racist. Often it is a mélange of positive and negative messaging combined, which can result in internal conflict for the worker. Regardless of whether these interactions are consciously or unintentionally antagonistic, individual employees may exhibit the sort of outcomes shown in table 18.1.

SOCIAL MODELING BEHAVIORS

Since humans are social creatures, we necessarily observe, copy, and set examples for acceptable and unacceptable behaviors. These cultural norms both evolve and devolve with time and intercultural interaction. There are several common behaviors that often work against a safer workplace, and these can spread, almost like a communicable disease, through a workforce. The result might be labeled dangerous, unsafe, hostile, abusive, disruptive, noninclusive, narrow-minded, or toxic.

Normalization and Habituation

One of the most common behavioral defects is called *normalization*. It can also be called cutting corners, taking shortcuts, or bad habits. It is a "phenomenon demonstrated by the gradual reduction of safety standards to a new normal after a period of absence from negative outcomes" (Wright et al. 2021). Normalization increases the likelihood of performing a task using unsafe methods or actions, introducing risks that would otherwise be easy to avoid.

Normalization is a sign of complacency or indifference, and it is all too common in breweries. Examples include jumping off of a loading dock instead of using the stairs, wearing personal listening devices, ignoring CO_2 alarms, not wearing PPE, or sliding down brew deck railings. When workers are heard saying things like "I know I shouldn't do that" or "It's never hurt me yet," normalization has probably set in. A good toolbox talk topic would be to have coworkers identify as many normalized behaviors as possible. Other ways to reduce normalization include leading by example, conducting task reviews and modifying procedures, encouraging worker reporting, and establishing and maintaining ongoing training.

Normalized behaviors are repeated and become habit. However, safe ways of doing things can also become reflexive through *habituation*. Though habituation can have positive or negative implications, the term is often used in safety jargon to describe positive habits. In this usage, habituation can be considered the obverse behavior to normalization.

Nearly all of us have become habituated to putting on a seat belt when getting in a car. When we forget to do that, we feel awkward, naked, like we are missing something. That sensation is the feedback generated by good habits. In a brewery, examples include the wearing of PPE for certain activities, walking like a duck in slippery areas, and maintaining good housekeeping as a matter of course.

Habituation is limited by the frequency of reinforcement opportunities. We may get in and out of a car several times a day. This makes it easier to instill seatbelt use. Uncommon activities, like erecting brewery tanks or mopping up bodily fluids, don't lend themselves to habituation. When the activity is not routine enough to develop safe work habits, we have to make a mindful assessment of the steps in the task, consider hazards, and act consciously to avoid or reduce those hazards. In other words, we will use the hazard assessment process in the moment.

Perceived and Assumed Risk

Somewhat related to the habits we form are the notions of *perceived risk* and *assumed risk*. These are most often personal value judgments made in respect to the activity, though there is a potential for the sense of risk to become communally adopted and normalized in a workplace.

Perceived risk is exactly as it sounds. How much risk do we *think* is involved in an activity? If the hazard is obvious and severe, we will hopefully perceive the risk to be great and take appropriate preventive and protective measures. However, our perceived level of risk may be less than the actual risk, such as when we underestimate a hazard due to inexperience, or, contrarily, when we believe we have sufficient experience with the task.

Perceived risk can often be seen in physiological and behavioral ways, such as sweating and blood flow

WE CAN NOW PERCEIVE YOUR PERCEIVED RISK

It used to be that researchers would assess perceived risk using surveys, which was not very reliable. Lately, researchers have been successfully measuring the stress associated with perceived risk by using wearable bio-monitors. These send signals through a small wireless transmitter, after which they can be processed and analyzed. Sensors include the photoplethysmograph and electrodermal activity (EDA) sensor.

Once processed, photoplethysmogram (PPG) data reports blood pulse volume, heart rate, interbeat interval, vasodilation/constriction data, and heart rate variability. Electrodermal activity is synonymous with galvanic skin response, the old-school lie detector test that monitors sweating.

Researchers may also use video monitoring of workers to understand how certain activities correlate with observed biometric changes. Another cross-reference technique is to measure cortisol in saliva samples during the testing cycle. Cortisol is a hormone that increases when the body is experiencing stress.

Now, if only there was a way to measure our assumed risk . . .

changes, and nervousness or panic. Heart rate, sweating, respiration, and cortisol levels all increase when we experience fear (*see* sidebar).

Assumed risk is how much risk we decide to allow ourselves, knowing already that there is an existing hazard. On one end of the spectrum a worker may be risk averse and assume the job is very unsafe; on the other end, there are risk-tolerant individuals performing the job as it is laid out and dealing with the risk as a matter of probability. The problem with a "taking your chances" approach is that, sooner or later, you are bound to lose.

Any workforce will benefit from a frank discussion about perceived and assumed risk that is put into the context of strong messaging from management that taking chances is not part of the job. Signs that an employer is truly concerned with risk reduction include institutionalized performance of hazard assessments, providing safety equipment and systems, and regular training on the safe way to do things.

Social Proof and Peer Influence

When an individual is in a work, social, or societal setting and they do not yet know the correct or expected behaviors to exhibit, they often look to others in the group to learn acceptable behaviors. This observed action is described in *social proof* theory. Almost all humans innately want to belong to a social group.

Peer pressure, peer influence, or setting examples are behaviors where existing members of a culture demonstrate commonly accepted or preferred behaviors. Social proof is the other side of the coin, where a person observes and then mimics behaviors of those they associate with the group, particularly the leaders within an organization.

Social proofing is most fertile when there is uncertainty about how to act. It also works best when there are similarities between the individual and the group members (e.g. age, gender, political ideology, appearance, etc.). If many people in a group think or act similarly, social proof will have more grip (Psychology Notes HQ 2018).

Social proof can lead to a kind of group one-track-mindedness if opportunities for diversity of opinions are scarce. The resulting singularity of condoned behaviors might be called herd mentality or mob psychology. The *bystander effect* is a negative social proof outcome of individuals taking cues from uninformed or under-engaged individuals. Research has shown that the more witnesses there are to a situation, the less likely any one of them is to engage with a solution (Darley and Latané 1968).

In the context of brewery safety, it follows that a group of employees who are of like age, gender, race, and preferences will be more prone to adopt normalized unsafe behaviors from each other compared with a diverse team, which will possess more variety of experience. A diverse workforce will likely use more than one style of communication, have different societal origins, work histories, and socioeconomic status.

The best way to defeat negative social proof is to openly develop consensus and set examples that are directed at safe working behaviors or revised social and behavioral customs. Examples can also be set with safety campaigns or endorsements by a trusted personality.

If the ownership has set the example or permitted it, a press release with an apology will not correct the problem. Increasingly, specialized business consultants may be employed to help plot a new course in social norms. This approach has been seen in cases of company-wide misogynistic behavior. Those individuals who behaved in such a way will need to adopt new social norms that can spread through improved communication, habituation, social proofing, and recognition and reward.

When the majority of the group is exhibiting desired, safe behavior, it becomes the new social norm from which new employees will form their social proof. Safety culture can improve over time through iterative improvements in modeling desired behaviors.

Cognitive Dissonance

Cognitive dissonance is a mental conflict that occurs when a person's beliefs and actions are not in agreement, or when two beliefs conflict. It often results in unease or stress, though individuals vary in their tolerance to this stress.

Stress of this kind may manifest as anxiety, guilt, or shame. Individuals may try to cope by avoiding conflicting facts, concealing their actions or beliefs, or avoiding discussions about particular subjects. Individuals experiencing cognitive dissonance will attempt to reconcile the stress. They may seek out additional support from like-minded persons or go on the offensive, trying to convince others that conflicting information is not true or relevant (Leonard 2019).

Cognitive dissonance conflicts can cut two ways. One is an outgrowth of social proofing, where an individual adopts group behaviors that are not in alignment with their own beliefs, values, or presented facts. A person might adopt an unsafe behavior even when they know it is unsafe. By way of example, consider a brewery worker who knows it unsafe to drink beer while on shift, though the rest of staff routinely drink while brewing. The worker may follow the same behavior as others to avoid interpersonal conflict, leaving the worker with elevated stress levels due to the internal conflict, in addition to increasing the potential for physical injury. Employees need to know they will be heard by safety managers and owners without reprisal and supported by management through policy changes when hazards are uncovered.

Cognitive dissonance can work to the contrary, however. A person may be forced to adopt a safe behavior they feel challenges their belief system. One's sense of individualism or libertarianism may lead them to internal conflict if they are required to take certain safety precautions. This is commonly seen in breweries, for example, where shorts are worn or personal listening devices are in use. Strong messaging from management about the employer's responsibility to create a safe workplace will help. The General Duty Clause states the employee has to abide by safety rules and use hazard controls. Make the argument black and white when improving the safety mission statement: this is the required way from now on.

Recognizing cognitive dissonance is often easy, though reconciling it may require supportive training, procedure development, changes in hiring practices, employee discipline, and management reinforcement. It requires an individual to undergo personal growth and change. Behavioral change is best facilitated by creating rational experiences and examples that resolve the dissonance, and by supporting social interaction with persons who amplify the factual basis rather than ideological beliefs.

WORKER-TO-WORKER INTERACTIONS

In addition to the insidious behavioral modeling discussed above, there is also a wide range of overt interactions in the workplace. These include the way workers communicate to and about others, the formation of cliques, bullying, avoidance, exercising prejudices, and the playing of practical jokes. Many of these undesired behaviors are constituents of a toxic workplace, and though mentioned here in the worker-to-worker context, the primary discussion is later in this chapter under "Employment Setting."

IDENTIFYING AND RESOLVING BIOSOCIAL AND SITUATIONAL INTERACTIONS

To address unhealthy worker-to-worker interactions, first they have to be made known. This can occur because an employee moved up the management hierarchy with a complaint or concern, or it may be that an astute manager spotted unwanted behaviors early. Either way, identifying and resolving these harmful interactions requires mature communication, expressed management values, and accountability.

Employees benefit from knowing what behaviors are preferred and those that are unwanted. Holding individuals to account is done consistently in relation to expressed behaviors. Managers model effective interpersonal communication, continuously improve, and openly address shortcomings of their own.

Table 18.2 Biosocial and situational stressors and interventions

Stressors	
• Interpersonal conflict	• Cognitive dissonance
• Lack of coworker or supervisor support	• Bullying
• Normalized unsafe actions	• Poor social support, shutting out
• Unsafe risk perception or acceptance	• Prejudice based on gender, race, religion, etc.
• Bystander effect; non-supportive environment	• Physical abuse, assault, violence

Interventions or controls	
• Management commitment	• Conflict management systems
• Improve communications	• Peer support groups and coaching
• Expressed code of conduct	• Sensitivity training
• Policies and procedures	• Reassignment; team restructuring
• Improve manager skills	• Didactic "fact reset" counseling

Biosocial and Situational Interactions

	INTRAPERSONAL – Psychosocial
	Occurs: Plant-wide
	Outcomes: Stress, anxiety, discontent, insecurity, burnout, self-harm, substance abuse and addiction, suicide; metabolic stress, nausea, musculoskeletal disorder, headache, elevated blood pressure.

In this part we look at the employee by themselves, largely apart from encounters with other people or their workspace. Individuals can engage in thoughts and feelings that are counterproductive to safety, they can be challenged by the structure of their brain or behavioral proclivities, and they can resort to substance abuse.

HARMFUL SELF-NARRATIVES, ANXIETIES, AND DISCONTENT

The specific relationship between an employee and stressful coworkers or work setting can result in a variety of detrimental intrapersonal narratives. These can be harmful by virtue of not making known and resolving workplace hazards, by increasing chronic worker stress over safety concerns, and through disengagement from others. The following are some descriptive examples:

- "If I report an incident, I could be punished or shunned." (Fear of reprisal, job insecurity, cognitive dissonance)
- "Safety is someone else's responsibility." (Disempowerment, abdication)
- "Injuries happen. They're just part of the job." (Fatalism)
- "Nobody will listen. What's the point." (Fatalism)
- "Safety takes time" and "Safety makes me less productive." (Trade-offs, cognitive dissonance)
- "I can't afford to lose this job." (Socioeconomic pressure)
- "I'll keep my head down and do like others." (Complacency, normalization)
- "It will take forever for this culture to improve." (Complacency, fatalism)
- "Should I stay and put up with this or quit and face uncertainty?" (Job insecurity, anxiety)

The best means of mending a dysfunctional self-narrative is through engagement with coworkers who have a positive outlook. Setting examples with positive communication and making sure ongoing culture improvement is visible are essential. When a disengagement feeling or behavior is identified, it needs to be matched with action in the workplace that provides factual experience and a social context to disprove the negative thought cycle.

Identifying these counterproductive thoughts requires conversation with the involved employee. Frequent "check-ins" from colleagues or managers will help, if they are done with objective inquiry, good listing skills, and empathetic response. If one uncovers suicidal thoughts in another individual, early steps include expressing how valued the person is and recommending either calling a hotline or considering medical/psychiatric intervention.

NEURODIVERSITY

Although all individuals are neurologically unique, we use the non-medical term *neurodivergent* to refer to individuals whose brain developed or works differently than most people for some reason. Those who are not neurodivergent are called *neurotypical*. These terms are generally preferred over words such as normal, abnormal, different, or disabled.

Neurodivergent persons may have behaviors that are thought of as limitations, such as struggling with communication or possessing tactile super-sensation, extreme social anxiety, or have a learning disability. They may also possess exceptional abilities, like extraordinary memory, the ability to do complex math or 3-D visualization in their head, or excellent focus and concentration on specific tasks. Some neurodivergent persons require medical care due to related physical or psychiatric dysfunction, but many individuals function well in an accepting environment, either with or without accommodations.

Examples of diagnosable neurodivergent conditions include autism spectrum disorder (ASD), attention deficit/hyperactivity disorder (ADHD), dyslexia, dyspraxia (congenital clumsiness), intellectual disability, sensory processing disorders, phobias, and social anxiety. Coincidence of neurodivergent conditions is common and also associated with mental health states. Among those diagnosed with ADHD in childhood, approximately 25% are also dyslexic (Bates n.d.). Among those with dyslexia, 10%–20% will have an anxiety disorder and 2%–14% will experience depression.

Mental health conditions like obsessive-compulsive disorder (OCD), bipolar disorder, and PTSD broadly fall into the category of neurodiversity (Cleveland Clinic 2022). Even a small workforce is likely to have neurodivergent employees (table 18.3).

Table 18.3 **Prevalence of some common neurodivergent conditions diagnosed in childhood**

Condition	Prevalence
Attention deficit/ hyperactivity disorder (ADHD)	Worldwide: 5% United States: 8%
Autism spectrum disorder (ASD)	Worldwide: 1.0%–1.6%
Dyspraxia	Worldwide: up to 6%
Dyslexia	Worldwide: approx. 9%–12%; varies with native tongue

Sources: ADHD, world and US (Doyle 2020; Danielson et al. 2018); ASD (Elsabbagh et al. 2012); dyspraxia (Blank et al. 2019); dyslexia (European Dyslexia Association n.d.)

For neurodivergent employees, job communication and the methods by which training is imparted may be different than that used with neurotypical employees. A dyslexic worker may benefit from an audio recording of a meeting, while someone on the autism spectrum may find a captioned video meeting helpful in complex conversations. A human resources (HR) professional should be involved when hiring and assigning a neurodivergent employee, both to establish a professional connection between the employee and HR resources, and to ensure compliance with the Americans with Disabilities Act (ADA), when applicable.

From the perspective of workplace safety, neurodivergent individuals may benefit from assignment to certain work activities over others and from the provision of adapted protective equipment. Other steps include diversifying job candidate assessments and using training or internship programs that are sensitive to neurodivergent workers. Leveraging educational resources from state and local agencies and non-profit organizations can help breweries become better employers for a wider range of employees. Increasingly, HR officers and consultants are being trained to recognize and place neurodivergent workers into satisfying job roles.

ACCOMMODATING NEURODIVERSITY IN THE WORKPLACE

Sensory accommodations

- Sound sensitivity
  - Avoid loud workplace noise (e.g., blaring music), communicate expected loud noises (like fire drills), supply noise-cancelling headphones
- Tactile sensitivity
  - Modify work uniform
  - Learn how and if the individual prefers touch
- Agitated movement
  - Allow for fidgeting toys, flexible seating, movement breaks

Communication

- Provide concise task instructions (verbal or written), outline tasks in small steps
  - Speak clearly, avoid implied messages, check for comprehension
  - Avoid sarcasm or euphemism
- Ask a person about their preferences, needs, and goals and let them know you are available to help
- Provide advance notice of changes whenever possible and an explanation for the change
- Inform coworkers of preferred communication practices with the neurodivergent worker
- Offer alternative records of meetings with audio or video recordings or subtitled video
- Practice kindness and patience

Source: adapted from Baumer and Frueh (2021).

SUBSTANCE USE

The use of chemical substances to alter perception, mood, and behavior is common in almost every society worldwide. Breweries are in the business of producing the commonest of these mind-altering substances: alcohol. Other substances include nicotine, marijuana, prescription drugs, and "street drugs." Street drugs, also called illicit drugs, include heroin and other opiates, amphetamines, cocaine, MDMA (ecstasy), and the psychotropic alkaloids LSD, psilocybin, and mescaline.

Casual Substance Use

Occasional and moderated use of substances is commonplace. The concern is whether safety and worker safety awareness has been compromised in any way. Common effects of alcohol or drugs to be on watch for include poor motor control, reduced reaction time, mood swings, conflict with others, and difficulty focusing.

As with other psychosocial hazards, substance use is best tackled from the top of the organization. Expressed policies need to exist in writing. They can include under what conditions substances can be consumed, shift drinks policy, sensory tasting limits, marijuana policies, and drug testing triggers and protocols.

Substance Use Disorder

Substance use disorder (SUD) and alcohol use disorder (AUD) are the current preferred terms to refer to problematic drug or alcohol use, respectively. Here we will use substance use disorder to include alcohol use disorder. Substance use disorder applies to either patterns of substance misuse or abuse, or addiction; in other words, it refers to a spectrum ranging from abuse to addiction. Addiction can occur rapidly or over a long period of time, depending on the substance, the individual, and use patterns.

In general, when substance use becomes recurring and specific symptoms appear, the user has potentially moved from causal substance use to substance use disorder. High absenteeism, workplace conflicts, and injuries could be signs of a problem. Substance use disorder can be diagnosed with or without the patient's acknowledgement of the problem. Substance use disorders are progressive and result in a variety of mental and emotional disruptions, organ failure, reduced longevity, domestic violence, workplace safety issues, suicide, business losses, and other unfortunate outcomes.

Substance Use Disorder Interventions

Intervention has a specific meaning in this context. It is when family, friends, or caregivers confront a person about their dependance and urge or force them into detox and recovery. Interventions for acute abuse or addiction usually involve both psychiatric care and medication assistance. It is common for sufferers of substance use disorder to have coexisting mental health conditions, either due to neurodiversity, emotional or physical trauma, or genetic predisposition.

Medically assisted therapy may be used during detoxification or withdrawal, with therapeutic drugs used to ease the unpleasant symptoms. Detoxification with medically assisted therapy is followed up with intensive therapy, usually some form of cognitive behavioral therapy (CBT). Post-treatment care can involve personal counseling sessions, involvement in peer-to-peer recovery groups, and careful reintegration into the workforce.

Peer-to-peer recovery groups include Alcoholics Anonymous (AA), Narcotic Anonymous (NA), or various local programs that pair peer recovery coaches with a recovering person. A peer in this context is a person who is themself in recovery from substance use disorder. Al-Anon is a support group for family and friends of persons recovering from alcohol use.

Management and human resources should have protocols developed in case an employee is taken away from work due to a substance use problem. These should include defined sick leave, reassessment after recovery, and the conditions under which reentering the workforce would be permitted.

IDENTIFYING AND MANAGING INDIVIDUAL PSYCHOLOGICAL STRESSORS

Of the various relationships in psychosocial dynamics, individual/personal issues are the most delicate. Addressing these issues will likely involve private conversations between a manager or HR and an employee, which may contain protected medical information if the worker is disclosing a neurodivergent condition or conditions. Also, the worker may feel quite vulnerable about disclosing something about themselves that they feel is an extra challenge or even a personal weakness. In identifying individual psychological stressors, managers should be good listeners, offer genuinely workable solutions, and remain extremely careful with confidential information.

Table 18.4 **Individual psychological stressors and interventions**

Stressors	
• Time pressure; role overload	• Environmental conditions
• Emotional work demands	• Lack of recognition and reward
• Job control and autonomy	• Physical demands
• Clarity of role	• Work-life imbalance
• Individual's neurodiversity	• Lack of socialization at work
Interventions or Controls	
• Management commitment and competency	• Exercise, healthy diet, adequate sleep
• Supervisor or coworker check-ins	• Time management training
• Meditation, biofeedback	• Rehabilitation after sick leave
• Individual counseling:[a] CBT, EAC, CISD, GT	• Disability management/case management
• Adaptive provisions for neurodiversity	• Energy management training

CBT, cognitive behavioral training; CISD, critical incident stress debriefing; EAC, employee assistance counseling; GT, group therapy.
[a] These public or private services vary with locality.

Employment Setting

WORKER-WORKPLACE INTERACTION – Psychosocial
Occurs: Plant-wide
Outcomes: Stress, anxiety, discontent, insecurity, burnout, self-harm, substance abuse and addiction, suicide; metabolic stress, nausea, musculoskeletal disorder, headache, elevated blood pressure.

Psychosocial stress can result from the way brewery jobs are structured and how work is managed or mismanaged. The composition of the workforce is also significant and will relate to advancement potential, lines of communication, and inequities in social position, pay, hours of work, and task assignments.

EMPLOYMENT SETTING

The employment setting includes the content and context of the work, and any other environmental characteristics of the workplace, such as coworkers, the difficulty of the work, and the compensation for work performed. In an effective safety culture, expressed management values and good communication are driving the conditions of the workplace. When culture is in trouble, occupational stress will be magnified due to poor communication, poor leadership, inconsistency, lack of behavioral rules and expectations, and confusion about job roles, company processes, and equity.

Work Design

Work design encompasses the division of labor into specialized roles and how those roles interact. It also includes expectations of what skills or knowledge are required, and specifies pay rates and work hours.

Looking at the tasks themselves, are they diverse, interesting, stimulating, achievable, and to some degree under the control of the worker? Or are they monotonous, repetitive, isolated or solitary, meaningless, and either too hard or too boring? Is there pressure to complete the assignment in the allowable time? In breweries, there are certainly jobs that fall into the categories of repetitive, physically demanding, and difficult to accomplish in the available time. Add to this a feeling of not

being in control and treatment inequities and you have a formula for worker burnout and occupational injuries.

Every worker has their own particular value equation. When conditions are met to satisfy the equation, the employee is likely to thrive in the work environment and will wish to continue being employed there. Variables in this value equation include compensation, work schedule, role in the company, influence or learning opportunities, promotion potential, and social good, among others. One employee might be almost entirely motivated by the pay rate, while another might be in it mostly for career advancement learning opportunities. Good work design transmits essential information that allows people with different value equations to know what is expected of them.

Management Behaviors

In designing work, employers should consider equitable pay schemes, shift assignments, hours to be worked, work pace and performance expectations, compensation, and reward and recognition. Work design should endeavor to reduce job insecurity, obstacles to promotion, avoid conflicting roles, and make clear the value of each employee's contributions. Most significantly, owners and managers need to model safe and responsible behaviors. Earning respect serves to also set expectations for group normative behavior.

Management will need to objectively look at itself and identify weaknesses. Is management being supportive, considerate, and listening? Are owners and managers continuously evaluating the function of the parts that make up the company, or are they using the company as a personal playground at the expense of employee morale? When managers uncover occupational stressors like work-life balance issues, poorly explained job responsibilities, or toxic behaviors in the workplace, employees expect them to take action.

Toxic Workplace Culture

In brewery safety, "toxic" usually means we are talking about a hazardous chemical or dangerous disease. The other type of toxic encounter is a psychosocial hazard. In a toxic workplace culture, people or work design problems become so negative and disruptive that the company cannot properly function. Employees may withdraw, suffer ill health effects, lose focus on safety, or resign from work. Toxic workplaces have low morale, are less productive, and experience higher than normal turnover rates. Examples of behaviors that contribute to a toxic workplace include:

- Verbal or physical aggression from one person against another
- Yelling, throwing items, operating machines unsafely, losing control
- Gossiping, talking behind someone's back, spreading rumors, sabotaging careers
- Bullying an individual or subgroup
- Harassing others with sexual inferences, unwanted touching, or lewd gestures
- Ignoring, isolating, or refusing to work with a coworker; forming cliques
- Withholding information
- Outwardly expressing prejudices against persons because of skin color, appearance, race, religion, national identity, language, expressed gender, diet, religion, political perspective, or physical disability
- Taking credit for another's work; stealing ideas or intellectual property
- Out of touch management that is passive toward or unwilling to manage toxic behaviors in the workplace
- Selective micromanagement
- Chronic absenteeism requiring others to pick up the slack
- Communicating by text or social media to avoid direct conversation
- Preoccupation with social media feeds and other digital distractions
- Systematic failure to listen, leading to lack of problem resolution
- Triangulation: adding a third person to problem resolution to avoid direct communication or responsibility for resolution
- Narcissistic behavior by company leaders
- Management inaction or excuse-making
- Discrimination in hiring, job assignment, and advancement
- Atmospheres of pessimism, fatalism, and rejection

This list represents examples of the wide range of behaviors that can define a toxic workplace culture. Admittedly, every workplace will experience some amount of these behaviors. It is when they become prevalent enough to effect productivity, safety, and workers' emotional and mental well-being that they need to be addressed.

Leading a Non-Toxic Workplace

Toxic workplaces are created by, condoned by, or ignored by senior management and ownership. This is where prevention or course correction necessarily takes place. Management shapes the work design and has the role of hiring, reassigning, and firing.

Senior management should seek to avoid the creation of a toxic workplace in the first place. It begins by setting examples for direct communication, work-play balance, and fair treatment of all people. Ideally, policies, procedures, and examples for acceptable behavior are modeled from the day the company opens. Managers actively observe and listen to all members of the workforce and proactively seek solutions for workplace conflicts and inequities. Policies are reviewed and evolve as the business grows.

Hiring for a Non-Toxic Workplace

In hiring, employers should do more than assess whether the employee has the technical skills for the position. They should include behavioral assessment questions with no right or wrong answer. Interview questions should be consistent for all candidates and they should be chosen with the goal of assessing particular strengths and weaknesses.

Behavioral questions should be designed to get at the person's personality. They assess communication, teamwork, time management, response to change, relationship to others, and more. Interview questions should have a purpose, a particular trait to evaluate. Here are some examples for a packaging tech interview with the traits being assessed shown in parentheses:

- "What positive effect do you have on people?" (Sociability)
- "Making beer is fun and it's hard work. What would be an example of fun you might have with your coworkers during the shift? What would be an example of coworker fun that is better outside of work hours?" (Safety, Inclusivity)
- "What's an example of a goal you failed to meet. How did you handle the situation?" (Humility, Resilience, Problem-solving)
- "Based on your past experience in packaging, tell me three safety issues you are always watching out for." (Hazard awareness, Safety competency)
- "Some of the packaging staff wear personal listening devices while working. Tell me why we should or should not allow that." (Autonomy, Safety)

- "Packaging is a pretty repetitive job. How do you imagine you will remain engaged with the job a year from now?" (Value equation, Adaptability, Advancement)
- "Give me an example of when you took a risky decision with your own health or safety and it didn't pay off. It could be an example from work or outside of work. What did you learn from this?" (Autonomy, Hazard awareness, Teachability)
- "Tell me about a time where you found out there was more to the job than you thought or were comfortable with. How did you deal with this? What did you learn from the way you approached these changes?" (Adaptability, Initiative)
- "Can you describe a time where your supervisor gave you too much work to accomplish in the available time? What did you do?" (Time management, Resilience)
- "Think of a time when you had to work with someone completely different from you. How did you adapt to effectively collaborate?" (Communication, Teamwork, Diversity, Inclusivity)
- "Have you ever seen a coworker isolated or bullied by another employee? Was there anything you could do, and if so, how did you get involved?" (Hazard awareness, Communication, Initiative, Inclusivity)
- "Tell me about a conflict you experienced at work. What was the conflict and were you able to overcome it? How did you resolve the conflict?" (Self-management, Communication, Temperament)

Having the candidate meet with future coworkers or be compensated to shadow work for a day will add other voices to the decision process. It is important to hire for cultural growth rather than cultural fit.

Repairing a Toxic Workplace

We normally consider prevention the superior hazard control strategy. One example of prevention would be to manage weekly hours to prevent job stress, burnout, or suicide. Another might be to discontinue employment of an employee who is disrupting workplace safety, communication, or inclusivity objectives. These types of solutions necessarily derive from intentional actions of management.

When a workplace becomes toxic, ownership is faced with a day of reckoning. If the status quo continues, the business will earn a bad reputation in the industry and with consumers. High-quality employees will leave for better positions and only the more desperate will remain. Turnover costs the business in several ways. It requires more recruitment and more training. It can cause lost productivity and overtime pay expenses while shorthanded. Turnover due to a toxic workplace also causes insidious damage to worker morale and brand appreciation, which can result in even more turnover. The other path is one of serious redesign and repair. How does one "rewire" workplace–worker interactions to lower the frequency and potency of emotional stressors and the physical and mental outcomes that result?

Lamontagne et al. proposed a prioritized systems approach to psychosocial hazard control: (1) prevention of psychosocial hazards at their source, (2) amelioration (protection) by educating employees to cope with stress, and (3) reaction (protection) by treating and reintegrating employees following the adverse effects of psychosocial hazards (Lamontagne et al. 2007, 269).

Studies have shown the most significant contributors to a healthy psychosocial setting are management commitment, management priority, communication, and employee participation. These factors are associated with lowered perceived emotional demands and lowered distress and emotional exhaustion (Akiomi et al. 2021). Modeling desired behaviors from the top down and providing training on maintaining social, substance, and technology boundaries is a good start (Takahashi 2017). Management values, communication, and inclusion are three of five key cultural drivers discussed in chapters 20 and 21.

After management accepts responsibility, it must ensure that the entire workforce follows in the new path. This will likely require revised policies and procedures, an improved employee handbook, monitoring and disciplinary action, and perhaps increased accountability from human resources or outside assistance from occupational psychologists. Table 18.5, at the end of this chapter, lists a wide variety of possible interventions or improvements that can help curb toxic behaviors in the business.

WORKFORCE CHARACTERISTICS

Individuals and groups of employees can become stressed by obstacles to advancement or by their perceived weakness due to age, gender, or socioeconomic background. Progressive management of the company can help break down these stigmas. To reduce employee stress over their position, employees should be evaluated and promoted in their roles based on aptitude, demonstrated communication skills, active participation in safe operations, and work ethic. These criteria, or other non-prejudicial core values the company espouses, need to be communicated to the workforce regularly and implemented consistently.

Job Position and Tenure

Brewing has traditionally been an industry where employees may work themselves up the ranks to positions of higher status. In the US, a brewer—one who produces wort using brew deck equipment—is often considered senior to cellar workers, who are considered more senior than packaging and warehouse personnel. (In contrast, in Europe, cellar workers are prized and brewers might be labeled "wort boilers.") In front of house, bartenders may have seniority over servers, who in turn lord over barbacks and bussers. Positions of status may not be based entirely on skills or experience, but by how close the individual is to the group norm, how long they have been with the company, their relationship to the owners, or simply by blindly following industry traditions. In all cases, a healthy, progressive brewery culture may have avoided these typical hierarchies and their associated drama.

Senior or tenured employees, like head brewer or bar manager, may be treated with more tolerance by ownership because of a perception of their relative worth to the organization. If such employees create a stressful or toxic workplace for others and they are viewed as harder to replace than wage laborers, these tenured employees might not be reprimanded or replaced. The resulting message to the workforce is that the exhibited behaviors are acceptable. Because a functional safety culture depends on active participation of all employees, giving some individuals special treatment while fostering stress and anxiety in the rest will stand as a significant obstacle to achieving an effective and inclusive culture.

Age, Gender, and Experience

While not a hard-and-fast rule, numerous studies have identified that younger people, particularly younger males, often operate from a lower perceived risk/ higher assumed risk posture in occupational and social settings (Reid and Konrad 2004; Halkitis and Parsons 2000; Rhodes and Pivik 2011). Gaining experience through aging is borne out in workplace injury data. Workers aged 16–19 years have the greatest likelihood for occupational injuries due to contact with moving equipment, cuts, punctures, and injuries involving the extremities (National Safety Council n.d.[b]). By adulthood, most people have learned the best balance of risk tolerance and experience with hazardous energies. Mature workers tend to be more accurate in their work and make more correct decisions than less experienced workers (Canadian Centre for Occupational Health and Safety 2022).

As we enter later life, the body gradually becomes less resilient to injury and an uptick in injury frequency also occurs. Musculoskeletal injury frequency peaks over the middle span of life, typically in a person's 40s and 50s. In general, the older the injured person, the more substantial the long-term effects of ergonomic injury (Smith and Berecki-Gisolf 2014). In senior adults, increased susceptibility to falls (especially among women), heat disorders, and COVID-19 are well established (Talbot et al. 2005; Centers for Disease Control and Prevention n.d.; National Safety Council n.d.[b]). With age also comes

GROWING INTO MORE ENERGY

Recall that injuries, illnesses, and property damage are caused by a transfer of energy to a person or piece of equipment. As toddlers, we fell down often but we were close to the ground, weighed little, and were pretty soft and pliable. As we grew older, we climbed trees and swing sets, and maybe jumped from roofs. The force of gravity increased with distance from the ground, momentum, and our own body weight.

In our teen years we began experimenting with higher energy things, like driving cars and setting off fireworks. We likely did higher energy activities: played contact sports, used motorized recreation, or experimented with drugs or alcohol. We began leveraging more energy, but still did not have a lot of experience with which to form reasonable values of perceived and assumed risk.

Stop for a moment and think about your own life. During what period in your life did you experience your most worrisome event, one where you almost died or someone else did die? What kind of energy was involved? What psychosocial components were present? Did this event change your perspective on risk or safety?

an increased chance of debilitation due to arthritis, hypertension, and diabetes (National Institute for Occupational Safety and Health 2015).

To reduce the incidence of negative perceptions of age and gender in the workplace, understanding and tolerating differences is a good place to begin. Establishing multigenerational teams within the workforce can provide mentoring for younger workers and physical strength and mobility support for older workers. Improving communication skills within the workforce will allow workers of different ages and genders to interact positively.

Socioeconomic Status

We all bring with us a legacy of who we are, where we live, how much we earn, our education, the particulars of our family life, and our health. In other words, what defines us in the rest of our lives will spill into work life, and work life will contribute to our life outside of work. Nowhere is this crossover more apparent than in terms of socioeconomic status. Socioeconomic status "can encompass quality of life attributes as well as the opportunities and privileges afforded to people within society" and is "a consistent and reliable predictor of a vast array of outcomes across the life span, including physical and psychological health" (American Psychological Association 2022).

Research shows a direct correlation between lower socioeconomic status and reduced advancement possibilities, further compounding stress and prejudice in the workplace. Lower socioeconomic status puts stresses on the worker that their coworkers may not appreciate, such as commuting, housing, and clothing issues. Persons with low socioeconomic status may have multiple roles to fill, such as holding two or more jobs and being a single caregiver. Oftentimes, workers of lower socioeconomic status are the most likely to be let go during quiet periods (e.g., in seasonal work) or during austerity measures. Workers of lower socioeconomic status are more likely to be placed in positions requiring greater physical exertion, repetitive motion, and poorer environmental controls (Aittomäki, Lahelma, and Roos 2003).

Physiological outcomes of reduced socioeconomic status include increased blood pressure, chronic back injury and musculoskeletal disorders, fatigue, and even male infertility; workers of lower socioeconomic status are also more likely to smoke cigarettes. To compound these effects, workers with lower socioeconomic status are more likely to work for small businesses where benefits like health insurance and paid time off are less common (American Psychological Association 2022).

Another disparate outcome is the difference between workers with children and those without. A worker with children may face more pressures from outside of work, notably, childcare, childhood illnesses, school and extracurricular activities, and the stress of needing to provide for a larger household. Workers with children may spend less post-shift social time with colleagues, which can lead to poorer awareness of workplace news, less social involvement, and, possibly, reduced opportunities for advancement.

IDENTIFYING AND MANAGING EMPLOYMENT SETTING STRESSORS

As with other psychosocial contexts, management needs to take the leading role to create and sustain a workplace with few stress alarms:

> Workers are less likely to experience work-related stress when demands and pressures of work are matched to their knowledge and abilities, control can be exercised over their work and the way they do it, support is received from supervisors and colleagues, and participation in decisions that concern their jobs is provided. (World Health Organization 2020)

Desired behavior is predicated on beneficial experiences, learning, and socialization. Solving a specific employee–workplace conflict will not address systemic problems in the organization, but it may provide a template for a broader solution to wider, endemic stresses.

Table 18.5 **Employment setting stressors and interventions**

Stressors	
• Environmental conditions	• Demands exceeding ability
• Changes in management	• Non-participation in decision-making
• Lack of organizational justice or equity	• Shortage of advancement possibilities
• Poor inclusion or diversity	• Wage or benefits inequities
• Lack of recognition and reward	• Unclear performance evaluation criteria
• Lack of job control and autonomy	• Work-life imbalance and lack of mutual appreciation of responsibilities
• Time pressures, role overload	• Toxic culture behaviors: aggression, gossip, bullying, narcissism, alienation, sexual harassment, procrastination, negativity, avoiding engagement, passivity, taking credit for other's work, unclear or incapable management
• Role ambiguity or conflict	
• Physical demands	
• Poor task design: monotony, no variety, busywork, low workload	
Interventions or Controls	
• Management commitment	• Adaptation to personal life issues
• Management competency	• Periodic job demand/satisfaction assessment
• Communication improvements	• Special counseling:[a] CBT, EAC, CISD, GT
• Environmental improvements	• Pre-employment medical assessment
• Better job role design for work-life balance and workload	• Aptitude-based activity assignment
• Restructured roles, hours, and rosters	• Exercise facilities
• Support participation and autonomy	• Mentoring: interpersonal skills, trades
• Time management training	• Terminating employees with toxic behavior
• Sensitivity training	• Use behavior questions and check references during candidate consideration
• Ergonomic training and warm-ups	• Engage passive workers in work design and problem solving
• Expressed code of conduct	
• Policies and procedures	

CBT, cognitive behavioral training; CISD, critical incident stress debriefing; EAC, employee assistance counseling; GT, group therapy.
[a] These public or private services vary with locality.

CAUTION

CONFINED SPACE USE
LOCKOUT AND ENTRY
PROCEDURES PRIOR
TO ENTRY

CUIDADO

ESPACIO CONFINADO
SIGA LOS PROCEDIMIENTOS
DE ENTRADA Y SEGURIDAD
ANTES DE ENTRAR

ESC444PB

www.nationalmarker.com

NMC

16

19

COMBINED HAZARDS

OSHA has two standards important to breweries that each relate to the control of multiple types of hazards simultaneously. Confined spaces can have mechanical, chemical, thermal, and atmospheric hazards, while energized systems can possess mechanical, electrical, gravitational, chemical, pressure, light, and sound hazards. For confined spaces, the standard in 29 C.F.R. § 1910.146, "Permit-required confined spaces," is applicable. For the management of hazardous energy, the standard in 29 C.F.R. § 1910.147, "The control of hazardous energy (lockout/tagout)," applies. These two standards often apply concurrently, like when performing service or maintenance inside a brewery vessel.

Confined space and energy control rules are complicated and nuanced. They require workers to take extra steps in the performance of common activities, extra steps that may seem superfluous or time-consuming. Applying these rules on a daily basis makes nearly impossible safety incidents become assuredly impossible. Breweries that abide by these standards will ensure worker safety and be well positioned in the event of an OSHA inspection.

This chapter provides examples of inventories, floorplans, and procedures related to confined spaces and energy control. Happy Hour Brewing Co. (HHBC) is our fictitious brewery that you will see referenced. No relationship to any existing brewery is intended or implied.

Confined Spaces

	CONFINED SPACES – Rotatory Motion, Falling, Caught Between, Asphyxiation, Chemical, Thermal
	Occurs: Plant-wide
	Outcomes: Asphyxiation, engulfment, heat stress, fatigue, chemical or thermal burns, falling from height, blunt force trauma, laceration, amputation.

WHAT IS A CONFINED SPACE?

Confined spaces are prevalent throughout breweries. Entering these spaces risks asphyxiation, burns, mangling, and chemical injury. The conventional definition of a confined space is a place that meets three basic criteria:

- Is large enough that a person can get inside to perform work
- Has limited access and egress
- Is not designed for continuous occupancy

Based on this definition, grain silos, brewhouse vessels, cellar vessels, bright beer tanks, and serving vessels are all confined spaces. Less common confined spaces are large boilers, large grist cases, and wastewater vaults and tanks. Walk-in coolers, basements, tractor trailers, and the restroom after a chili festival are not confined spaces.

PERMIT-REQUIRED CONFINED SPACE

If your facility has anything that looks remotely like a confined space, an OSHA representative will assume you have a permit-required confined space (PRCS). This is in accordance with the rebuttable presumption that we looked at in chapter 2: confined space regulations are written to presume hazards exist unless or until they are shown not to exist. To say guilty until proven innocent may seem a harsh comparison, but if one mischaracterizes a confined space with a serious hazard in it, workers can die. This follows the adage that people don't get injured in confined spaces (hint: because they die there).

If an area has met the three basic criteria for a confined space, OSHA requires the employer to evaluate the space for additional hazards (29 C.F.R. § 1910.146(c)). Unless a brewery system is so small that no vessel could ever be bodily entered by someone, a brewery will inevitably need to conduct this assessment. A confined space that is determined to have one or more of the following exceptional hazards in addition to the ordinary physical constraints will be termed a PRCS:

- The space contains, or has potential to contain, a hazardous atmosphere.
- There is the potential for an entrant to be engulfed.
- The space is internally configured such that an entrant could be trapped or asphyxiated by inwardly converging walls or by a floor that slopes downward and tapers to a smaller cross-section.
- The space contains any other recognized serious safety or health hazard.

Many confined spaces will end up having one or more of these serious hazards and will be classified as a PRCS. The good news is that PRCSs with limited additional hazards can sometimes be worked down into lesser classifications by controlling these additional hazards. These classifications are called *reclassified spaces* and *alternate procedures spaces*. They are depicted in figure 19.1 and discussed further in the following section on special requirements for PRCSs.

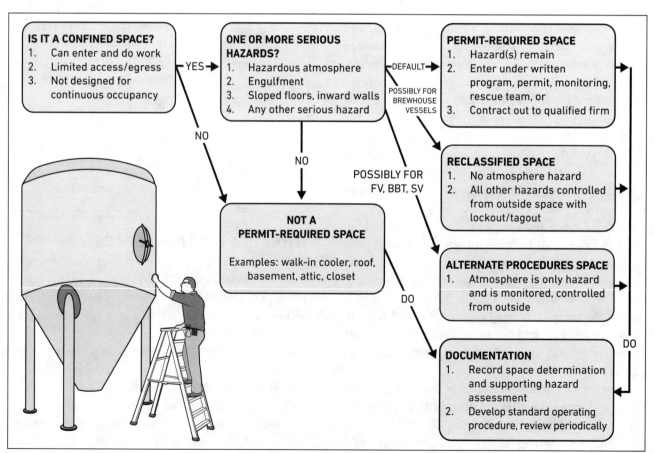

Figure 19.1. Confined space definitions and decision logic. BBT, bright beer tank; FV, fermenting vessel; SV, serving vessel.

An individual space can end up with any of five possible classifications: PRCS, reclassified, alternate procedures, confined space (no permit required), and an enclosed place that is not a confined space. Knowing the classification of a space may influence you to not ever enter the space, when to hire out the work to a team with the equipment and expertise to do the work, or when to enter if only under specific conditions. In general, if you can do what you need to do without physically entering the space, you have prevented the possibility of worker injury or fatality.

Actual or Potential Hazardous Atmosphere

In the context of a confined space, an *actual or potential hazardous atmosphere* is any atmospheric composition that could injure or impair the worker in any way. In a brewery, excess carbon dioxide (CO_2) is the most common atmospheric hazard found in fermentors, bright beer tanks, and serving vessels; excess CO_2 displaces oxygen (O_2) in the surrounding air. Carbon dioxide can also be toxic at high enough concentrations, but even gas that is inert, like nitrogen (N_2) or argon (Ar), will displace O_2 to workers' detriment. Oxygen levels in the air need to remain between 19.5% and 23.5% to be healthy and to comply with OSHA's respiratory protection standard (29 C.F.R. § 1910.134(b)).

Although far less common, oxygen-enriched environments can exist in a brewery. Oxygen cylinders are used for yeast nutrition and in oxyacetylene welding. If a cylinder leaks O_2 into a restricted space and the enriched air contacts a source of ignition, normally combustible materials like clothing, PPE, and hair can burst into flames, causing horrible injuries or worse.

To alleviate this hazard, confined spaces can be rendered non-explosive by displacing the enriched air with either a non-oxygen gas, like CO_2 or N_2, or with regular air. The air in the confined space or vessel becomes inert as a result and will not support rapid combustion or explosion. A tragic accident occurred in a brewery when a welder needed to do hot work inside a fermentor. The tank was CO_2-rich and O_2-deficient. In order to make it "safe" to enter, the welder inserted oxygen into the vessel, unintentionally causing O_2 enrichment. After entry, the worker commenced welding and his clothes and hair combusted. He died of burn injuries and organ failure after a protracted stay in a burn treatment unit.

Table 19.1 **Atmospheric oxygen (O_2) concentrations and effects**

Concentration (% O_2)	Definition	Bodily and physical hazard effects
20.9	Average	Normal physiological function
19.5–23.5	Acceptable[a]	Normal physiological function
>23.5[a]	Oxygen enriched	Chest pain, coughing, and difficulty breathing (dyspnea); central nervous system effects, which can include twitching of the hand, tinnitus, dysphoria, nausea, convulsions (Cooper, Phuyal, and Shah 2022); simple combustibles (clothing, hair, cardboard, solvents) become highly flammable or explosive
19.5	Oxygen IDLH	Immediately detrimental to life or health—poses an immediate threat to life, would cause irreversible adverse health effects, or would impair an individual's ability to escape from a dangerous atmosphere[b]
12–19.5	Oxygen deficient	Increased respiration (tachypnea), increased heart rate (tachycardia), faulty judgment, coordination difficulties, tiredness
10–14	Oxygen deficient	Faulty judgment, intermittent respiration, exhaustion
6–10	Oxygen deficient	Nausea, vomiting, lethargy, asphyxiation, unconsciousness
<6	Oxygen deficient	Convulsions, cessation of breath (apnea), cardiac arrest, permanent organ damage, death

[a] Applicable OSHA General Industry and Construction standards define oxygen-enriched as greater than 23.5% O_2 (29 C.F.R. §§ 1910.134, 1926.1202). OSHA Shipyard Employment standard (29 C.F.R. § 1915.11(b)) cites 22.0% as O_2 enriched. The National Fire Protection Association reviewed industry standards during revisions to NFPA 53—values for "oxygen-enriched atmosphere" published by trade groups and standardizing agencies ranged from 21% to 50% (National Fire Protection Association 2020, p. 6/23).

[b] As defined in 29 C.F.R. § 1910.134(b).

Engulfment

The second qualifying hazard is that of engulfment potential. Breweries with grain silos or spent grain silos or vaults must avoid engulfment. This is mostly done by never entering silos. If an inspection is needed, it can often be done by looking down through a manway. The same is true for piles of grain or bulk grain in rail cars or tractor trailers.

If grain gets damp and clumped up, a long pole can be used to break the clumps. If entry is required to get to the feed mechanism at the bottom, grain should first be vacuumed out with a vacuum truck (to ankle deep or a whatever depth where a worker can no longer be engulfed) to alleviate the engulfment hazard. Entry into a silo is a permit-required entry and will necessitate trained individuals use harnesses, a retrieval system, air monitoring, and a designated attendant.

Grain can flow deceptively fast and bury a worker in only seconds. The force to remove oneself from this predicament is more than a person or even several people can apply. Workers engulfed in grain often die. The three most common mechanisms for grain movement in a silo are flowing grain during normal operations, bridged grain breaking through and filling the cavity below, and a grain wall dislodging and causing a grain avalanche (fig. 19.2).

The vast majority of engulfment incidents occur during the normal operations of filling or emptying silos or bins. A suggested approach for most breweries would be to perform a hazard assessment on any silos, document them as PRCSs, and never enter them without the full permit approach, a qualified attendant, and fall protection and retrieval systems.

Sloping or Converging Walls or Floors

Another qualifying hazard for a PRCS is that of inwardly sloping or converging walls or floors. These can make self-escape impossible and, in some situations, can restrict a stuck person's breathing. This hazard is common in breweries that use cylindroconical vessels for primary fermentation and beer conditioning. Some silos and large grain hoppers also have sloping bottoms, as do some brew kettles. The shallow, dish-shaped bottoms of water tanks, bright beer tanks, and serving vessels do not pose the same hazard in terms of poor egress and bodily restriction.

A special case for potentially entering fermentors under the alternate procedures rules hinges on the sloping surfaces hazard. This is discussed below in the sidebar "The Case for Fermentor Inspection Without a Permit" on page 244.

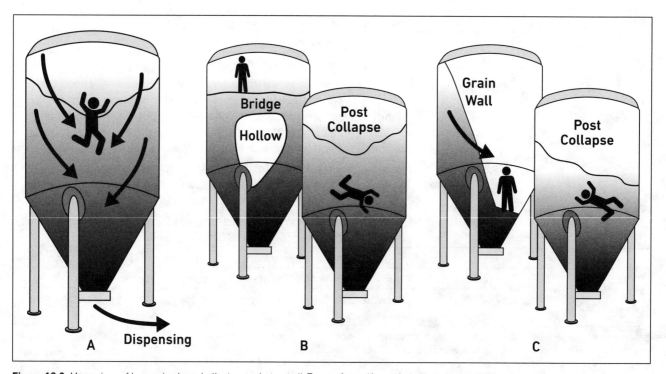

Figure 19.2. Many sizes of breweries have bulk storage, but not all. Types of engulfment in bulk grain are *A*, flowing grain; *B*, bridged grain; and *C*, grain avalanche. *Adapted from Great Plains Center for Agricultural Health (2015).*

Any Other Serious Safety or Health Hazard

The catch-all phrase "any other recognized serious safety or health hazard" (29 C.F.R. § 1910.146(b) "Permit required confined space") composes the final qualifying hazard for PRCSs. Two operative terms in this definition are *safety* and *health*. Safety concerns are generally those that might result in an injury outcome. In breweries, confined space hazards that could injure a worker include moving rakes, agitators, grinders, and pumps, all of which might cause physical trauma. Chemical and thermal burns are also possible, as are slips, falls, and being hit with falling objects.

Health concerns are those that relate to disease, or in some other outcome that isn't simply due to physical energy sources. In breweries there is the potential for asphyxiation by CO_2 and lack of O_2, illness from welding fume or chemical cleaners, and heat stress, to give three examples. The "any . . . serious hazard" category puts a lot of brewery confined spaces into the ranks of PRCSs. However, many of these hazards can be mitigated and monitored from outside the space, opening up the possibility for reclassification or alternate procedures.

SPECIAL REQUIREMENTS FOR PERMIT-REQUIRED CONFINED SPACES

If any confined space is found to be a PRCS during assessment, the employer must inform employees who could be exposed to the attendant hazards. This is done initially through the posting of danger signs. The employer must also implement a written confined space program, which includes hazard assessment, permit entry, monitoring, equipment and rescue gear, designated individuals, and training.

Reclassified Confined Space

OSHA provides for two ways out of a PRCS determination. Both are based on controlling hazards from outside the space prior to entry. For spaces where all potential hazards can be mitigated from outside the space, and there is no possibility of the atmosphere becoming hazardous, the space can be redefined as a *reclassified* space. In breweries, this is commonly done with brewhouse vessels that have thermal, chemical, and mechanical hazards that can be locked out prior to manual inspection, cleaning, and repair. Reclassified spaces should have written documentation of the hazard assessment and an accompanying standard operating procedure (SOP) describing how to control all hazards prior to entry.

Alternate Procedures Confined Space

A space where the existence or potential of a hazardous atmosphere is the only additional hazard can be redefined as an *alternate procedures* space, provided the atmospheric hazard is monitored and mitigated

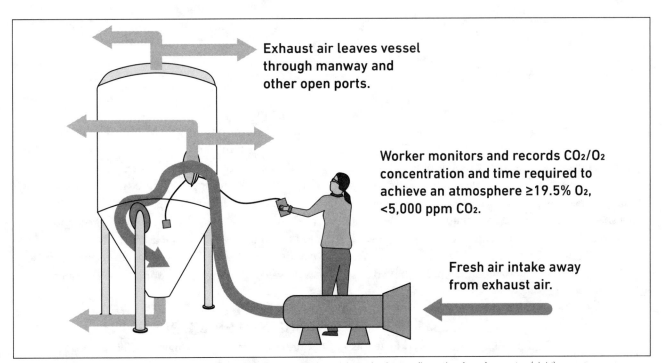

Figure 19.3. Continuous forced air ventilation system and air ventilation and monitoring configuration for a fermentor (*right*).

Exhaust air leaves vessel through manway and other open ports.

Worker monitors and records CO₂/O₂ concentration and time required to achieve an atmosphere ≥19.5% O₂, <5,000 ppm CO₂.

Fresh air intake away from exhaust air.

THE CASE FOR FERMENTOR INSPECTION WITHOUT A PERMIT

A special argument can be made for using alternate procedures when inspecting or servicing cylindroconical vessels without full bodily entry. OSHA's definition of entering a confined space is when any part of the body breaks the plane of the opening. To put one's face through the opening of the manway is to have entered the space.

Visual inspection of the inside of a fermentor is a ubiquitous task in brewing and is necessary for quality reasons. It is performed after a CIP cleaning cycle and the tank has been left open for a period of time, typically 10 to 20 minutes. Brewers also reach through the manway to remove the manway hatch gasket for cleaning. To "enter" the tank with only hands and head is not to be exposed to any hazards, since CO_2 is no longer accumulated and the sloping sides are only a hazard if one is inside the tank and can't climb out.

To perform this routine activity as an alternate procedures space instead of a PRCS, assemble hazard assessment and operational procedures documentation:

- Using a four-function gas monitor, and ideally a CO_2 air monitor as well, determine what passive or active ventilation process and duration reliably alleviates the CO_2 hazard that was in the space. Document results. Note that OSHA regulations require a four-function monitor for alternate procedure safety.
- Identify any other subordinate hazards, like process lines, and establish an energy control procedure for those.
- Create an SOP for the tank inspection task and make sure it explicitly states that it applies only to hand and head entry following purging, and that full-body entry is prohibited and is a permit-required activity.
- Seek regulatory approval from your local OSHA authority.

Some wash fermentors used in distilleries have a mixer installed. This may present a mechanical hazard to anyone reaching into the vessel. This mechanism requires lockout/tagout and may invalidate this approach to using alternate procedures for cylindroconical vessels.

prior to entry. As with reclassified spaces, any alternate procedures space should have written documentation of the hazard assessment and an accompanying SOP describing how to control all hazards prior to entry.

Alternate procedures were developed with utility vault workers in mind. A designated alternate procedures space can be entered by a single worker without an attendant. Most often this involves the use of forced air ventilation while continuously monitoring the air. If conditions change inside the space, the worker immediately exits.

Continuous forced air ventilation means that a space is ventilated by blowing fresh air inward to displace a potentially unhealthy atmosphere. This procedure works best when there is flow in through one end and out

through the other. When there is only one way for air to flow in and out, best results are obtained by placing the blower hose in the far or deep end of the space and forcing the air out the hole at the other end. Getting in and out around the hose can require the use of special flat tube called a saddle vent (fig. 19.3). The operator must also consider where the contaminated air is moving to, and whether it will become a hazard there or be sufficiently diluted. Avoid placing the blower intake close to exhausted air or any other poor quality air source.

WRITTEN PERMIT-REQUIRED CONFINED SPACE PROGRAM

A brewery with even a single PRCS will need to operate with a written permit-required confined space program (a.k.a. permit space program). The permit space program should cover all different classes of spaces, regardless of whether a permit is required.

One can either write a permit space program by following the program requirements line by line, or by acquiring a template and filling in your particulars. Developing your own program from scratch will instill a deeper understanding of the regulations but will take a lot of effort. Most breweries will find a template online and go from there. Since these templates vary greatly and are easy to find, it is best to search state and federal agency resources, institutions with public records, or pay a safety professional to develop a permit space program for your brewery.

Control Entry to Spaces

First and foremost, control entry to confined spaces. As soon as the classification of a space is known, mount a sign on the entrance point(s) indicating a signal word and a major message (*see also* fig. 8.2, p. 84). The signal word for PRCSs will be "Danger" due to the imminent risk of life safety or serious injury. Hazardous spaces not categorized PRCS may be marked with either "Danger" or "Warning." If entry into a particular space could only ever result in minor injury, then "Caution" as the signal word is appropriate. The Spanish equivalents for danger, warning, and caution are, respectively, *Peligro*, *Advertencia*, and *Precaución*. Figure 19.4 shows some examples.

The major message should indicate the primary action desired of the worker. For a PRCS vessel that no employees are to enter, the major message can be "Do Not Enter." For enterable spaces, the message may indicate what must be done prior to entry, such as "permit required," "apply LOTO before entry," or "enter under continuous forced air ventilation only." OSHA does not prescribe specific language for confined space entry control. It is up to the employer, having conducted a hazard assessment, to determine the wording. Sign language should be unambiguous, so an untrained person would clearly know not to enter the space.

Develop Entry Procedures

Every confined space should have a well-defined entry procedure. The procedure may be as simple as the instruction not to enter under any circumstance. For spaces that may be entered, provide an SOP specifically

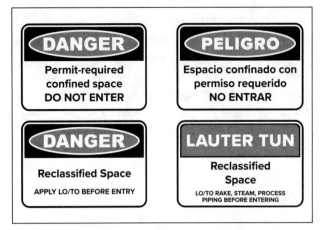

Figure 19.4. Confined space signage examples. Permit-required confined spaces require the "Danger" signal word (*top*). Other spaces may have signal words and major messages appropriate for the class of space (*bottom*).

for the space. This SOP will most likely include energy control methods (lockout/tagout), and possibly atmospheric measurements and limits. When a space is entered under permit, the written entry procedure goes along with the permit as a description of work.

When a one-time, unique entry to a space is required, there may be no previously written procedure. It is okay to thoughtfully develop just a permit, as long as it is thorough enough to assess all of the possible hazards, specify monitoring and controls, and provide needed people and equipment. In developing procedures, be sure to provide detail on the following elements of safe work:

- Specify acceptable entry conditions, that is, when it is okay or not okay to enter a space.
- Provide each authorized entrant the opportunity to observe any monitoring or testing of permit spaces.
- Specify how to isolate the permit space.
- Specify procedures for purging, making inert, flushing, or ventilating the permit space as necessary to eliminate or control atmospheric hazards.
- Provide barriers and signs to protect entrants from external hazards involving pedestrians, vehicles, or other activities or external hazards.
- Verify that conditions in the permit space are acceptable for entry throughout the duration of an authorized entry.

In summary, a good procedure will explicitly name which spaces it was created for and will define exactly how a safe entry should be performed.

Inventory and Site Plan

Procedural development is best preceded by making a comprehensive inventory of spaces (fig. 19.5). This allows procedures, specific vessels, and a site map to be linked (fig. 19.6). You should always know at a moment's notice the classification of every space and what is and is not allowed under your permit space program.

Provide Equipment

The equipment required for confined space work varies from basic for some non-PRCS entries to highly specialized and expensive for PRCS entries. OSHA requires the employer to provide necessary equipment at its own expense, to properly maintain it, and to train workers in its proper use. The equipment required will vary, but can include any or all of the following:

- Testing and monitoring equipment necessary to adequately evaluate the space during entry operations. For breweries, this generally means CO_2 and O_2 monitoring equipment. If bulk spirits are used for products, or the brewery is combined with a distillery, then explosivity or LEL (see p. 159) monitoring should be provided. Other monitoring could include temperature, hydrogen sulfide (H_2S), and

	Confined Space Inventory for Happy Hour Brewing Co.						Revision No. 23.1
SPACE NO.	**SHORT/NICK-NAME**	**ID**	**CONFINED SPACE CLASS**	**LOCATION**	**ENTRANCE FREQUENCY**	**ALLOWABLE ACTIVITY**	**NOT ALLOWABLE / USE CONTRACTOR**
1	silo	GS	PRCS	BACK LOT	NEVER	Visual inspection ONLY; no entry, no hot work	Any entry to, e.g., dislodge grain, hand clean, or hot work, requires contractor under permit with atmospheric monitoring
2	grist case	GC	PRCS	BREW DECK	NEVER		
3	mash tun	MT	RECLASS	BREW DECK	WEEKLY	Full entry for inspection, manual cleaning following SOP No. 29; for CIP without entry, use SOP No. 8	Any non-routine work, e.g., tasks with heavy mechanical tools or power tools, is subject to hazard assessment; if determined to be PRCS without atmosphere hazard, HHBC employees may perform or hire contractor; if atmospheric hazard present, a contractor must be hired
4	lauter tun	LT	RECLASS	BREW DECK	DAILY	Full entry for grain out, inspection, manual cleaning, rake adjustment and maintenance following SOP No. 31; for CIP without entry use SOP No. 8	
5	kettle	BK	RECLASS	BREW DECK	MONTHLY	Full entry for inspection, manual cleaning following SOP No. 29; CIP without entry use SOP No. 8	
6	whirlpool	WP	RECLASS	BREW DECK	MONTHLY		
7	hot liquor	HLT	RECLASS	BTWN BREW DECK & BOILER RM	ANNUALLY	Visual inspection ONLY; no entry, no hot work; acid wash with CIP if required, no manual cleaning; for CIP without entry, use SOP No. 8	Any non-routine work, e.g., manual scale removal, is subject to hazard assessment; if determined to be PRCS without atmospheric hazard, HHBC employees may perform or hire contractor; if atmospheric hazard present, a contractor must be hired
8	cold liquor	CLT	RECLASS	BTWN BREW DECK & BOILER RM	ANNUALLY		
9	Alice (FV1)	FV	PRCS	CELLAR	NEVER	Visual inspection, CIP, entry of hands and face ONLY, following SOP No. 32	Any entry with more than hands and face, e.g., for parts retrieval, hand cleaning, internal fixture access, repair, or hot work, requires contractor under permit due to CO_2/O_2-deficiency hazards
10	Jiggy (FV2)	FV	PRCS	CELLAR	NEVER		
11	Dankster (FV3)	FV	PRCS	CELLAR	NEVER		
12	Muffy (FV4)	FV	PRCS	CELLAR	NEVER		
13	Leonard (FV5)	FV	PRCS	CELLAR	NEVER		
14	Pollywog (FV6)	FV	PRCS	CELLAR	NEVER		
15	Gordo	BBT	PRCS	CELLAR	NEVER		
16	TDV1	SV	PRCS	TAP ROOM	NEVER		
17	TDV2	SV	PRCS	TAP ROOM	NEVER		
18	TDV3	SV	PRCS	TAP ROOM	NEVER		
19	TDV4	SV	PRCS	TAP ROOM	NEVER		
20	TDV5	SV	PRCS	TAP ROOM	NEVER		
21	TDV6	SV	PRCS	TAP ROOM	NEVER		
24	pH equalization tank	WWV	PRCS	BTWN BREW DECK & BOILER RM	NEVER	Visual inspection, no entry; sample from side valve	Any entry, e.g., to remove sludge, hand clean, or repair, requires contractor under permit
25	the tunnel	UV	ALT PROC	UNDER-GROUND BTWN BLDGS	WEEKLY	Inspection, fitting assembly and gasket maintenance, pest control placement; must measure CO_2/O_2 and ventilate, following SOP No. 33	Electrical service, if req'd, to be performed by licensed electrician after HHBC has cleared space and provided ventilation

Figure 19.5. An example of a confined spaces inventory worksheet.

carbon monoxide (CO) levels.

- OSHA specifically requires a four-function monitor (CO_2, CO, H_2S, and explosivity) be used when entering a space with alternate procedures (29 C.F.R. § 1910.146(c)(5)(ii)(C)). Refer to the permit-required confined space sidebar.
- Ventilating equipment needed to obtain acceptable entry conditions. For entry into cellar vessels this usually means an axial forced air blower, flexible ducting, and a saddle vent for allowing the worker through manways.
- Communications equipment as needed for the entrant to communicate with the attendant or entry supervisor. Empty vessels echo, magnify, and distort sound, making communication difficult. If power tools are being used, sounds inside a space can prevent voice communication. Additionally, hearing protection may be advised. Prearranged hand signals, rope pulls, or radio communications should be established.
- Personal protective equipment to supplement best available engineering controls and work practices.
- Adequate and intrinsically safe (i.e., unbreakable, explosion-proof) lighting to enable employee

to see their work and to exit quickly in an emergency.

- Barriers and shields to protect entrants from external hazards and people.
- Ladders as needed for safe ingress and egress.
- Rescue and emergency equipment needed for rescuing entrant, except to the extent that the equipment is provided by rescue services.
- Any other equipment necessary for safe entry into and rescue from permit spaces.

Designated Roles and Responsibilities

In cases of entry into non-PRCSs, reclassified, or alternate procedures spaces, a single worker may conduct the work and enter the space, provided that they have implemented all necessary controls as established in the written entry procedure. For this discussion, that worker will be called the *entrant*. To be clear, it is always a good idea for another worker to be nearby and to periodically check in with the entrant.

Entry into PRCSs is a different matter. OSHA specifies three specific roles for individuals: the authorized entrant, the attendant, and the entry supervisor. There should also be a collective rescue team. The authorized entrant

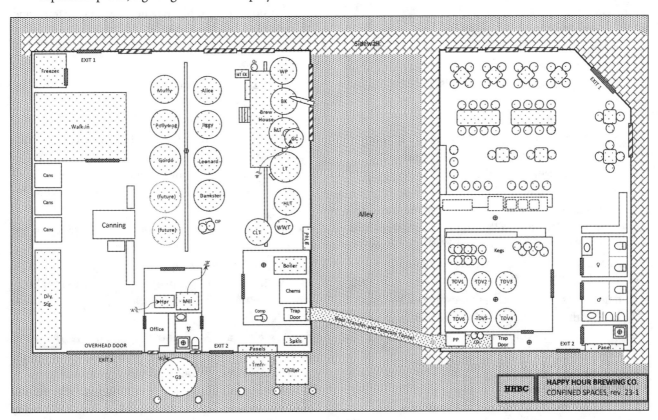

Figure 19.6. Confined space site plan. Confined space identifiers agree with tabulated confined space inventory and SOPs.

is an individual who has been trained in confined space work and is in good condition for the entry. It is often hot, cramped, and physically demanding inside a confined space. An entrant should be fit, well-hydrated, sturdy enough for the work at hand, small enough to enter and exit the space, not subject to claustrophobia, and able to clearly communicate with the attendant. Those suffering from cardiac conditions, high blood pressure, obesity, or a propensity for heat stress should avoid PRCS entry.

Specifically, OSHA states that the employer must assure that the authorized entrant(s) meets the following requirements:

- Knows the possible hazards during entry, including information on the mode, signs or symptoms, and consequences of the exposure.
- Be capable of properly using the equipment provided.
- Be able to communicate with the attendant to allow the attendant to monitor entrant status and to enable the attendant to alert entrant(s) of the need to evacuate.
- Be able to alert the attendant whenever
  - the entrant recognizes any warning sign or symptom of exposure to a dangerous situation, or
  - the entrant detects a prohibited condition.
- Be able to exit from the permit space as quickly as possible whenever
  - the attendant or entry supervisor orders evacuation;
  - the entrant recognizes any warning sign or symptom of exposure to a dangerous situation;
  - the entrant detects a prohibited condition; or
  - an evacuation alarm is activated.

The *attendant* is the person outside the confined space that observes the well-being of the entrant and may offer assistance such as passing tools and supplies, communicating changing conditions, and issuing evacuation orders. The employer must be sure that each attendant meets the following requirements:

- Knows the hazards, including information on the mode, signs or symptoms, and consequences of exposure.
- Knows behavioral effects of hazard exposure in authorized entrants.
- Remains aware of who is in the space at any moment.

- Remains outside the permit space during entry operations unless relieved by another attendant. In some written programs, the attendant is allowed to enter for rescue, but only if they have been trained in rescue and if there is another attendant to take their place.
- Be able to communicate with authorized entrants to monitor entrant status and to alert entrants of the need to evacuate the space.
- Be capable of monitoring activities inside and outside the space to determine if it is safe for entrants to remain in the space
- Be able to order the authorized entrants to evacuate the permit space immediately under any of the following conditions:
  - Attendant detects a prohibited condition.
  - Attendant detects the behavioral effects of hazard exposure in the authorized entrant.
  - Attendant detects a situation outside the space that could endanger the authorized entrants.
  - Attendant cannot effectively and safely perform all their duties.
- Can summon rescue and other emergency services as soon as the attendant determines that authorized entrants may need assistance to escape from permit space hazards.
- Is able to keep unauthorized persons away, including
  - warning the unauthorized persons that they must stay away from the permit space;
  - advising the unauthorized persons that they must exit immediately if they have entered the permit space; and
  - informing entrants and the entry supervisor if unauthorized persons have entered the permit space.
- Can perform non-entry rescues as specified by the employer's rescue procedure.
- Will not perform any duty that might interfere with the attendant's primary duty to monitor and protect the authorized entrants.

Overseeing a PRCS entry operation is the *entry supervisor*. The supervisor can double as an attendant or an authorized entrant, provided someone remains as an attendant during the execution of the permit. The entry supervisor is responsible for the following:

- Knowing the hazards faced during entry, including the mode, signs or symptoms, and consequences of the exposure.
- Verifying that the permit lists results for all testing specified by the permit, and that all procedures and equipment specified by the permit are in place before endorsing the permit and allowing entry to begin.
- Terminating the entry and canceling the permit when the work is complete or the space has been evacuated due to changing conditions.
- Verifying that rescue services are available, timely, and can be summoned.
- Removing unauthorized workers or customers who enter or who attempt to enter the permit space during entry operations.
- Determining whenever responsibility for a permit space entry operation is transferred and, at intervals dictated by the hazards and operations performed within the space, that entry operations remain consistent with terms of the entry permit and that acceptable entry conditions are maintained.

Finally, there is the rescue team. In a well-characterized confined space task being performed by experienced workers with the right equipment and monitoring, the chance a rescue team would ever be needed is low. However, missing a lockout/tagout (LOTO) control point, a medical emergency, or a rare event such as an earthquake could conceivably require active rescue of the authorized entrant.

The rescue team may be formed from the employer's own workers, a private hazmat team, or a public response team like the fire department. It is vitally important to familiarize potential responders with plant layout, known PRCS vessels, and on-site emergency systems long before a confined space incident occurs.

In selecting the team, the employer should evaluate their ability to respond to a rescue summons in a timely manner, depending on the hazards identified. For O_2 deficiency, such as might occur within a cellar vessel, the rescue team must be available for "immediate action" while wearing respiratory protection adequate for immediately dangerous to life or health (IDLH) atmospheres (29 C.F.R. § 1910.134). In such cases, calling 9-1-1 would not be considered timely enough. The rescuer must have proficiency with rescue-related tasks and equipment. The employer must inform each rescuer of the hazards they may encounter when called upon and provide the rescuers with access to all permit spaces from which rescue may be necessary so that the rescue service can develop appropriate rescue plans and practice rescue operations.

When an employer employs its own workers as permit space rescuers, it must ensure the following:

- Appropriate PPE is provided and workers are trained in its proper use.
- Rescuers are trained to perform assigned rescue duties; in addition,
  - rescuers are also trained as authorized entrants,
  - at least one member of the rescue team is trained in basic first aid and CPR, or a trained individual is available nearby, and
  - rescuers practice making permit space rescues at least annually by conducting simulated rescue operations (i.e., removing a dummy or actual person from an actual permit space or from a representative permit space).
- That retrieval systems shall be used whenever an authorized entrant enters a permit space so that there is a plan for non-entry rescue, unless the retrieval equipment would increase the overall risk of entry or would not contribute to the rescue.
- Retrieval systems must include the following features:
  - A chest or full-body harness, with a retrieval line attached at the center of the entrant's back near shoulder level, above the entrant's head, or at another point that the employer can establish presents a profile small enough for the successful removal of the entrant.
  - The other end of the retrieval line shall be attached to a mechanical device or fixed point outside the permit space in such a manner that rescue can begin as soon as the rescuer becomes aware that rescue is necessary.
  - The mechanical retrieval device shall be able to retrieve an entrant from vertical type permit spaces more than 5 ft. (1.5 m) deep.
- If an injured entrant is exposed to a substance for which an SDS is required to be kept at the worksite, that SDS shall be made available to the medical facility treating the exposed entrant.

Entry Permit

A written entry permit is required prior to entering a PRCS. This is not a permit issued by the local fire marshal or other entity. It is a checklist or form that is completed by the employer's staff or by a contractor. The permit is not submitted to any agency, instead it is kept in-hand during entry operations. It is a mini-safety and health plan for a single activity.

The entry permit is an essential tool to make sure that all hazards have been considered and, if present, measured or qualified in the permit. It is best to look at the confined space permit not as an obligation, but as a useful tool in guiding the entire entry operation to a safe and proficient outcome for all. For this reason, reprinting completed permits and hastily filling out permits is dangerous and reflects substandard practice.

The OSHA standard (29 C.F.R. § 1910.146(f)) provides a list of essential parts to a confined space permit. The list does not go into great detail about what hazards could exist and how to monitor and control them—that is up to the employer to decide. The entry permit follows the same logic as any hazard assessment, namely, outline the task, identify hazards, select hazard controls, and write down the safe way to do the work. The following list covers the minimum requirements for the entry permit:

- Identifies the permit space to be entered (a good reason to inventory and ID all confined spaces on the premises; see figs. 19.5 and 19.6).
- States the purpose of the entry, i.e., the work activity.
- States the date and authorized timeframe of the entry permit.
- Lists authorized entrant(s), attendant(s), and entry supervisor by name.
- Provides a space for the signature or initials of the entry supervisor authorizing entry.
- Lists the known or expected hazards of the permit space to be entered.
- Lists measures used to isolate the space, eliminate or control permit space hazards before entry, LOTO requirements, and procedures for purging, making inert, ventilating, and flushing the space.
- States acceptable entry conditions.
- Records the time and results of initial and periodic test monitoring, accompanied by the names or initials of the tester(s).
- Lists rescue and emergency services that can be summoned and the means of such notification,

e.g., dial 9-1-1, call a preestablished contractor, radio personnel in-plant, etc.
- States the type and method(s) of communication to be used between authorized entrant(s) and attendant during the entry.
- Lists required equipment, such as PPE, monitoring devices, communications gear, lighting, alarms, and fall restraint and worker retrieval equipment.
- Provides any other necessary information to ensure employee safety, e.g., relevant SOPs.
- References any additional permits, such as for hot work, that have been issued to authorize work in the permit space.

As with the written permit space program, plenty of free templates can be found online for entry permits. Websites for state and federal OSHA offices, OSHA On-Site Consultation services, and large institutions will have entry permit templates. Breweries using the services of safety consultants or insurance loss control professionals may also acquires templates through them. The Brewers Association has published *Best Management Practice for the Management of Confined Spaces in Breweries*, which contains three sample entry permits (Brewers Association 2015).

Evaluate the Space Prior to Entry

Permit spaces that may have an environmental hazard, like heat or an atmospheric hazard, should be evaluated prior to entry. On the best judgment of the entry supervisor or an individual with expertise in monitoring confined spaces, monitoring should continue either continuously throughout the entry activity, or periodically. Often, conditions can change once the worker is in the space. If atmospheric hazards are suspected, there is a preferred order in which gases should be monitored.

First, test for O_2. If the O_2 concentration is less than 19.5%, the space may not be entered without forced air ventilation to remove the hazard, or may not be entered without a supplied air breathing system compliant with the respiratory program requirements of 29 C.F.R. § 1910.134(d)(2). An O_2 level of less than 19.5% is deemed immediately dangerous to life or health (IDLH). If O_2 levels start out between 19.5% and 23.5% but change to be outside this range during the entry work, the entrant must immediately exit the space and the permit revised before reentry.

Consistent with other performance-oriented standards, OSHA allows a qualified employee or third party to determine, based on knowledge of the process(es) involved, what atmospheric testing should be conducted (29 C.F.R. §1910.146, Appendix B). In a brewery, the next obvious gas to test for would be CO_2. However, one cannot reliably guess or calculate the CO_2 concentration from the O_2 result, so a CO_2 monitor will be required for that measurement. Due to the vapor density of CO_2, layering can be expected and all elevations in the confined space should be evaluated. This is usually accomplished by attaching a length of vinyl tubing to the monitor and positioning the open end of the tubing in the bottom, middle, and top of the vessel by hanging it from above or attaching it to a stick. Such a monitor can also measure possible "hot spots" outside of confined spaces, such as in walk-in coolers and at specific stations in a canning or bottling line.

However, in general industry or if the entry task involves sewers, OSHA specifies that combustible gases, CO, and H_2S should be tested after O_2 (29 C.F.R. §1910.146, Appendix E). Handheld four-function gas monitors are routinely used for these sorts of confined space entries and they are usually configured for O_2, CO, H_2S, and %LEL$_{CH4}$. When gases with poor warning properties, such as O_2, CO_2, CO, H_2S, or methane (CH_4) are of concern in a confined space, it is best to monitor continuously during the entry activity.

Combustible gas meters usually report the percent lower explosivity limit (LEL) for methane, written in shorthand as %LEL$_{CH4}$. Methane is a common combustible gas wherever biological decomposition is occurring, which is why meters are calibrated to it. However, other combustible gases or vapors, for instance ethanol, acetone, or propane, may trigger a different response in the instrument, leading to errors in interpretation of the results. As a margin of safety, the alarm on an LEL meter is usually set conservatively to sound at 10% of the %LEL$_{CH4}$. Entrants are granted the right to observe the monitoring and results of atmospheric testing. If the results are thought to be inadequate or suspect for any reason, the employee has the right to more monitoring. At a minimum, during PRCS entry at least one attendant will be posted outside the permit space to communicate monitoring results and other messages. While the standard in 29 C.F.R. §1910.146 App. B has provisions for a single attendant to overwatch more than one PRCS, this is not recommended practice.

Managing Changing Conditions and Emergencies

The employer must maintain a system for the preparation, issuance, use, and cancellation of entry permits. This consistency is important to upholding strict access limitations on who can enter and under what conditions. The employer is responsible for coordinating entry operations when employees of two or more employers are conducting entry work.

Procedures for summoning rescue services, for rescuing entrants from permit spaces, for providing emergency services to rescued employees, and for preventing unauthorized personnel from attempting a rescue should be a part of the permit space programs.

If conditions have changed or a warning sign or symptom has been noted, the space will be evacuated immediately and not be reentered until either the hazards have been mitigated or the classification of the space has been redone. Situations that would necessitate review of the permit space program are: unauthorized entry to a PRCS, identifying a hazard not contemplated in the entry permit, detecting a condition prohibited under the permit, a change or reassessment of the configuration of a permit space, or receiving employee complaints about the program's effectiveness.

SPECIAL CONDITIONS REQUIRE SPECIAL CONSIDERATION

Instances where the conditions change quickly while an entrant is inside include welding, cleaning or pickling with volatile corrosives (nitric acid, acetic acid, ammonia), cleaning with solvents (isopropyl alcohol, ethoxylated alcohol), or applying paint, epoxy, or other surface coatings.

If a non-stainless steel tank is being entered (e.g., a water tank or grain silo) and it is rusty inside, be sure to monitor the O_2 levels first. In a closed tank that has rusted, atmospheric oxygen will be reduced or used up in the reaction that converts iron metal to rust.

Program Review

The employer must review entry operations whenever there it is suspected that hazard control measures may not have been adequate to protect employees. The permit space program or the entry permit form must be updated to address any deficiencies before further entries may occur. All entry permits, whether completed or cancelled, should be archived for at least one year. Any deficiencies or suggested modifications shall be marked on the permits to aid in the annual review process.

The employer must perform an annual self-review of the permit space program by reviewing completed and cancelled entry permits. If no entries occurred during the year, a review is not required.

Training

Given the technical knowhow required for confined space entry, the employer must provide training in hazard recognition, confined space peculiarities, hazard controls (engineering controls, safe work practices, PPE, and monitoring), designated roles and responsibilities, and the entry permit system. Training is provided when

- the employee is first assigned any duties covered by the PRCS standard (29 C.F.R. § 1910.146);
- before there is a change in assigned duties;
- whenever there is a change in permit space operations or hazards; or
- whenever the employer has reason to believe that workers are not adhering to entry procedures or that there are inadequacies in the employee's knowledge or use of these procedures.

The employer must document that necessary training has been accomplished. The certification shall contain each employee's name, the signatures or initials of the trainers, and the dates of training.

TYPICAL BREWERY CONFINED SPACES AND THEIR TREATMENT

From brewery to brewery, the types of confined spaces encountered remain consistent. Classes of process vessels and spaces are discussed here, with mention of the most common hazards and hazard controls used. The employer should perform a hazard assessment on each space, defining it according to figure 19.1, and develop procedures to safely enter, avoid entry, or never enter for each.

Common administrative controls include managing a complete written permit space program, marking relevant vessels in a manner appropriate for their classification, and training all staff to avoid the hazards of confined spaces.

Grain Silos, Baghouses, Grist Cases, and Spent Grain Storage

Grain silos used for the storage of bulk grain pose engulfment, sloping bottom, and asphyxiation hazards. Asphyxiation can come from being suffocated by the weight of grain during an engulfment, or from the buildup of hazardous gases like hydrogen sulfide. A hazardous atmosphere is more likely in a spent grain silo due to biological decomposition processes. Baghouses and grist cases are not always large enough to enter, but typically have converging walls or bottom. Baghouses may also pose a risk of dust explosion due to the fineness of the material.

Hazard controls for silo entries will include harness and fall protection system, air monitoring and ventilation, LOTO of mechanical hazards, safe lighting, and use of a trained entrant and attendant.

Brewhouse Vessels

Vessels used in wort production are often reclassifiable, since all hazards can be controlled from outside the space and there is not a potential atmospheric hazard. In small breweries, the mash may be stirred manually with a paddle. In systems with rakes or agitators in the mash tun, lauter tun, brew kettle, or decoction mixer, LOTO needs to be formally established and rigorously applied for each entry. Turning off the mash mixer switch does not provide adequate energy control. Other systems that will require energy control include jacket steam, process piping, CIP connections, and possibly pump motors. The section below, "Control of Hazardous Energy," details the LOTO program and procedures.

Brewhouse vessels may be too hot to enter and will need to cool off. This can be done passively by opening hatches and allowing things to cool, or by spraying cool water on heated surfaces.

Once all hazards that could exist for the entrant have been mitigated without entering the space, the worker may enter to perform their task. Upon exiting the space, the LOTO controls can be reversed and the system brought back into operation. The list of steps to secure the space, perform the work, and restore the system to operational status is written down as an SOP.

Cellar and Packaging Vessels

Fermentors, bright beer tanks, serving vessels, and large yeast brinks are confined spaces with multiple hazards, including O_2 deficiency, CO_2 above permissible limits, and sloping or inwardly converging walls or floors. Like other process vessels, incoming product, gases, rinse water, and cleaning chemicals need to be controlled.

The two main hazards with fermentors are the potential for a hazardous atmosphere and sloping or converging walls. Consequently, fermentors default to being PRCSs. Vessels with dished bottoms or horizonal lagering tanks may only have the potential hazardous atmosphere hazard, allowing for the alternate procedures approach.

Wastewater Systems

Three types of enterable spaces with chemical, thermal, and atmospheric hazards can occasionally be found in breweries. All are rarely entered and all will be classed as PRCSs.

Clean-in-place batch tanks may be large enough to enter in bigger brewery operations. Bulk tanks for CIP stock solutions often have an entrance port. Clean-in-place tank hazards can include corrosive chemicals, heating elements, and incoming process lines.

Increasingly, breweries of all sizes are adding wastewater equalization tanks upstream of the brewery wastewater outfall. These are used to detain wastewater with adverse contaminant loads so that it can be reduced or otherwise treated to come within permit limitations. Equalization tanks may host acid-base neutralization reactions to control pH or redox reactions and lower biological or chemical oxygen demand (BOD/COD). Sometimes these tanks simply moderate the flow of wastewater, such as when waste plumbing is too small or the wastewater must pass through a flow-limiting sewage lift station.

In rare situations, a brewery may have its own wastewater treatment facility or waste-to-energy fuel system. These systems can have many process vessels that are PRCSs. The confined space inventory and site plan should identify all such spaces. These spaces will need their own hazard assessments and, potentially, LOTO procedures.

Additional hazards that come with wastewater treatment vessels include flammable and hazardous gases like methane (CH_4) and hydrogen sulfide (H_2S); corrosive or oxidizing treatment chemicals; slippery surfaces; and mold and vermin. In the case of entry to perform mechanical repair or visual inspection in any chemical or wastewater treatment space, these vessels should be emptied, rinsed, tested for adequate air, and isolated from all sources of hazardous energy prior to entering. Personal protective equipment selection should be advised by consulting relevant SDSs and glove manufacturer compatibility charts.

Utility Vaults

Confined spaces can sometimes exist in the building itself. Utility vaults are often confined spaces that require a worker to climb down into a space where utility valves and meters, electrical distribution panels, or telecommunication lines are located. As with the spaces previously described, a hazard assessment will describe the hazards present and allow selection of hazard control procedures. If the services requiring attention in a vault involve electrical distribution, natural gas, or steam, work should be done by a licensed technician after the space has been cleared for entry.

Occasionally, a brewery will have a tunnel running between buildings. Tunnels may be part of older industrial campuses and were often used for running steam between buildings. Care should be taken not to damage insulation on old steam lines, which often contain asbestos, a respiratory carcinogen.

Boilers and Furnaces

Most brewery boilers are too small for entry, but boilers that serve multiple processes or businesses can be large enough to enter. This entry is done when boilers are out of service, which is usually for inspection, manual scale or rust removal, or hot work repairs. Under these work activities, boiler entry would be permit-required.

There are a few direct fired brew kettles or hot liquor tanks in use in breweries. Some of these afford a large enough space for a worker to enter, though these spaces are usually difficult to get in and out of and offer limited maneuverability inside. Hazards associated with entering these spaces include asphyxiation, hot surfaces, and the potential for a furnace to ignite. Thorough attention to LOTO and confined space permitting is essential for these spaces.

Non-Confined Spaces with Hazards

In general terms, if a space is configured to support human life during expected activities, it will not be defined as a confined space, even though the temperature, floor surface, or lack of aisle space may make working conditions more difficult. In a brewery, common non-confined spaces include attics, basements, crawlspaces, walk-in coolers and freezers, restrooms, closets, and tractor-trailer or intermodal shipping containers. However, any of these spaces could become a confined space or PRCS simply with the accumulation of a gas, vapor, or welding fume. As with any confined space, if the conditions change, all persons should immediately vacate and the space reassessed and hazards controlled before work resumes.

APPLICABLE STANDARDS

OSHA has a standard dealing specifically with confined spaces, but it is far from the only consideration for working in and around this type of hazard (table 19.2). A confined space can have many hazards, including, mechanical, ergonomic, thermal, chemical, and atmospheric hazards. Many, if not most, of these hazards will require LOTO, elimination, removal, or some other intentional effort to avoid bad outcomes. Confined spaces can also cause emotional stress, distraction, and increased injury risk for those who identify as claustrophobic.

Control of Hazardous Energy (Lockout/Tagout)

	**HAZARDOUS ENERGY – Rotatory, Linear, Gravitational Motion; Pressure, Temperature; Asphyxiation; Chemical, Thermal, Electrical, Potential and Kinetic Energy**
	Occurs: Plant-wide
	Outcomes: Electric shock, electrocution, flying objects, blunt force trauma, lacerations, connective tissue avulsions, amputation, scalping, and contact burns

Control of hazardous energy, more commonly called the lockout/tagout (LOTO) standard, is a broad set of rules and practices centered around machine safety. The general idea is that any hazardous energy that could affect a worker while they are servicing a machine shall be interrupted with a lockable device. Being protected thus, the worker carries the key to the lock with them so that while they are working on the machine it cannot be energized. The types of energy involved are varied, and include electrical, potential or kinetic mechanical energy, chemical, thermal, radiation, gravity, stored pressure, and others.

Table 19.2 **Occupational Safety and Health Standards (29 C.F.R. Part 1910) pertaining to confined spaces**

Short title of standard	Title 29 citation	Key requirements
Permit-required confined spaces	1910.146	Assess all possible confined spaces to determine space classification, hazards, controls, signage, training, permitting, and entry requirements
Control of hazardous energy (lockout/tagout)	1910.147	Control all types of hazardous energy using appropriate energy control devices and establishing a written energy control procedure
Respiratory protection	1910.134	Establishes allowable atmospheric oxygen levels; defines 19.5% O_2 as oxygen deficient and immediately dangerous to life or health (IDLH)
Hazard communication	1910.1200	Requires access to chemical safety data, PPE, and training; lists hazardous chemicals

NUANCES IN APPLYING LOCKOUT/TAGOUT

The LOTO standard applies in specific cases and not in others. It is important for compliance reasons to know when LOTO is required and when it just makes sense as a hazard control. As stated in 29 C.F.R. § 1910.147(a)(1)(i), the control of hazardous energy is required for "the servicing and maintenance of machines and equipment in which the *unexpected* energization or start-up of the machines or equipment, or release of stored energy could cause injury to employees" (emphasis in original).

The key words here are servicing and maintenance. The LOTO standard comes into play only when service and maintenance requires an employee "to remove or bypass a guard or other safety device," or "place any part of his or her body into an area on a machine or piece of equipment where work is actually performed . . . or where an associated danger zone exists during a machine operating cycle" (29 C.F.R. § 1910.147(a)(2)(ii)).

When a worker interacts with energized equipment during normal use of that equipment, the machine guarding rules take precedence (29 C.F.R. § 1910.212). One such example in a brewery is a grain mill. During regular use, the motor and pulleys are protected by machine guarding in accordance with the machine guarding standards in Subpart O of 29 C.F.R. § 1910; it is wired and powered as per the electrical standard in Subpart S. If one must service the mill, say, because someone tried to mill flaked grains through it and it got plugged, then the LOTO standard comes into play because either the guards are removed or a part of the body enters a part of the machine where the work is performed.

Lockout/Tagout Exemptions

The standard exempts specific activities that may be regulated elsewhere. These include construction activities (addressed by 29 C.F.R. § 1926.417) and electric installations (29 C.F.R. §§ 1910.302–.308). An electric installation is any building, facility, beer garden, or structure where electricity is used to power *utilization equipment*. Utilization equipment is any machine, device, heating or lighting system, appliance, or tool that requires electricity in that installation. In other words, if we are dealing with the routine use of electricity, not during servicing and maintenance, the electrical standard in Subpart S applies. When conducting service and maintenance on utilization equipment, LOTO requirements are then also required.

Another exemption from the rule is for minor tool changes and adjustments, and other minor servicing activities, taking place during normal production operations. Such routine activities are not to be covered by the LOTO standard if "routine, repetitive, and integral to the use of the equipment for production" (29 C.F.R. §§ 1910.211–.219) so long as the work is performed using appropriate protective measures in the machine guarding rules.

Two other specific cases where the LOTO standard does not apply: the use of cord and plug tools, and when hot tapping gas or fluid lines. If the worker using a cord and plug tool or other equipment has sole control over the cord, then the hazard can be controlled simply by unplugging. No energy control procedures or lockout devices are required.

Hot tapping is a special case where work needs to be performed on a pressurized line containing water, steam, or petroleum product while the line is in service. Hot tapping is used only when (a) the continuity of service is essential, (b) it is impractical to shut down the system, and (c) documented procedures, specialized equipment, and training are applied.

Conservative View of Lockout/Tagout

The following example illustrates how LOTO compliance is both the natural outcome of a thorough hazard assessment and the exercise of prudence when deciding how to apply the LOTO standard.

Consider the example of a rotating mash rake in a lauter tun. Under normal operation no one is in danger because the rake operates inside of a vessel where no one is located. However, if the knives on the rake need to be adjusted, the worker conducting that service will need to be inside the tank with the rake. The system must be deenergized and locked out prior to the work being performed. This is a clear example of LOTO applicability.

The mash rake example brings up the issue of when LOTO compliance is actually required versus when it is prudent. One can argue that cleaning the lauter tun after a brew cycle to remove adhered grain residues and flecks of spent grain is a part of normal production and is not actually servicing or maintaining the machine. Since the LOTO standard technically applies only to non-routine operations, in the case of entering the vessel as part of routine mashing operations LOTO may not be required, but its methods provide a proven means of eliminating hazards for the entrant. Conducting a hazard assessment on the lauter tun cleaning procedure would reveal the presence of confined space and mechanical hazards and would address their means of control. Those means should include LOTO-type controls. So, one arrives at LOTO one way or another.

ENERGY CONTROL PROGRAM

The employer must create an energy control program whenever workers can be exposed to hazardous energy during servicing or maintenance of equipment. The employer will provide written hazard control SOPs called energy control procedures (ECPs). Training and energy control devices shall be provided.

As is often seen with other standards, like an SDS for hazard communication and rubber gloves in a PPE program, there may be visual suggestions of a complete program but, behind the scenes, worker understanding and consistent implementation may be found wanting. A brewery will be best served by involving line-level workers in the development of ECPs and in the policing of their LOTO program.

Lockout Versus Tagout

While locking out an energy source with a physical padlock is preferred, the standard allows for the use of a tag on a nylon zip tie if the device cannot be locked out. Increasingly, tagout is considered a less effective means of energy control because the tag can be easily defeated even while work on the system is taking place. Whenever lockout is desired or system upgrades are taking place, look closely at the components and see where a lockable control can be installed. The variety of lockout devices is constantly growing and many of these new devices allow for more streamlined lockout. The sanitary valve with a built-in lock hasp, for example, means only a padlock is required to lock the valve closed.

Tags do, however, perform a valuable role in LOTO: they provide information about what, why, or who is involved in the shutdown. As an accompaniment to a lockout device, tags can indicate who is performing the service or how long the shutdown is expected to last. Adding a human element to the message is thought to improve compliance.

Energy Control Procedures

The energy control procedure (ECP) is a written procedure that applies to a specific piece of equipment. It is very much like an SOP, one that is tailored to preparing for service or maintenance to be performed on an energized system. Energy control procedures benefit from having step-by-step photographs or drawings depicting where and how control devices are installed.

Your ECPs should have a standardized template. This makes them easier to follow. It also makes producing ECPs less time consuming. There are commercial services available for a fee that offer easy creation of ECPs. Alternatively, finding a good example on the internet or from a colleague at another brewery is a good basis for creating your own.

Energy control procedures benefit from having line-level workers participate in their creation, not only by knowing how the system is energized, but also in how to develop the most efficient LOTO procedure. A common complaint with LOTO is the amount of extra time workers perceive it requires to implement.

An ECP should clearly and specifically outline the scope, purpose, authorization, rules, and techniques to be utilized for the control of hazardous energy. The ECP should clearly define and identify the equipment and what service or maintenance activities are applicable. The body of an ECP generally flows in four parts:

- **Shutdown.** Procedures for shutting down, isolating, blocking, and securing equipment to control hazardous energy.
- **Lockout.** Procedures for the placement, removal, and transfer of lockout devices and who is responsible.
- **Tryout.** Procedures to verify the effectiveness of lockout devices before work on the deenergized system begins.
- **Restart.** Procedures to confirm worker safety, device removal, and reenergization (not always the direct opposite order as shutdown).

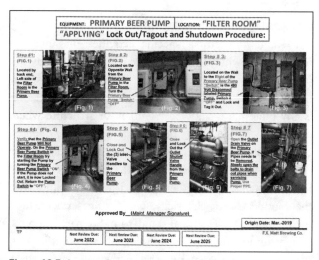

Figure 19.7. An energy control procedure for isolating a process pump. *Courtesy: F.X. Matt Brewing Co.*

As with any SOP, making the ECP available and local to workers using it will increase its effectiveness. Some employers choose to laminate the ECP and mount it right on or very close to the work area.

A documented ECP is not required only when all eight of the following elements exist:

1. The equipment has no potential for stored or residual energy or reaccumulation of stored energy after shutdown.
2. The equipment has a single energy source that is easily identified and isolated.
3. Isolating and locking out of the energy source will completely deenergize and deactivate the equipment.
4. The equipment is isolated from the energy source and locked out during servicing or maintenance.
5. A single lockout device will achieve a locked-out condition.
6. The lockout device is under the exclusive control of the authorized employee performing the servicing or maintenance.
7. The servicing or maintenance does not create hazards for other employees.
8. In utilizing this exception, the employer can have had no incidents involving the unexpected activation or reenergization of the equipment during servicing or maintenance.

Energy Control Devices

In 29 C.F.R. § 1910.147(b), an *energy isolating device* is defined as "a mechanical device that physically prevents the transmission or release of energy." An energy isolating device can be held securely in place by a *lockout device*, also defined in the same paragraph of the LOTO standard. Thus, these energy control devices are implements mounted onto any energetic system to absolutely control the energy of the system. Energy control devices act on components like switches, plugs, levers, valves, pipelines, air hoses, springs, and capacitors, allowing them to be locked into a position of zero energy flow. The energy control device can employ a padlock with a unique key as the lockout device. The operator conducting any service or maintenance then keeps the key in their possession for the duration of the work. The only way such a system can be reenergized is by first removing the lockout device or cutting the lockout device with bolt cutters, then reversing the energy isolating device.

LACK OF ENERGY CONTROL EQUALS SUREFIRE NEAR MISS

Thinking back to my own close calls with energized systems, these near misses could have been entirely avoided with better energy control procedures and training.

On a demolition job it was my task to remove all the ceiling-mounted fluorescent lights so the ceiling tiles could be replaced. I told my helper to shut off breaker number 14. He reported back that he had done so, but we did not try the circuit by flipping the light switch. When I cut the wires, the resulting short circuit blew the linesman's cutters right from my hand, melting a hole into the jaws. My helper had actually switched off breaker number 16. Moral of the story: try it before you buy it.

In another incident, I was wearing a supplied air breathing line during a hazmat cleanup. When the chemical was all cleaned up, I signaled to my coworker that I was going "off air," that is, was about to remove my respirator. I doffed the facepiece and uncoupled the 120 psi neoprene air hose and laid it on the ground. My coworker went to turn off the dual regulator diaphragm valve instead of the main cylinder valve and turned it in the wrong direction. This had the effect of letting over 1,000 psi of compressed air into the breathing line. In an instant the hose swelled to the size of a football then failed with a loud bang. The metal air chuck went sailing past my head at lightning speed. Moral of the story: better to have luck than be hit by a chuck.

Some energy control procedures do not use an energy control device per se, but instead stop the flow of power by disconnecting a mechanical or electrical linkage, unplugging a tool, disconnecting and draining a pipe, or jacking up a heavy object. Electronic multimeters, pressure gauges, and drainage valves are useful to confirm that energy has been dissipated.

In the brewery, common examples of energy control devices are padlocks on power disconnects, and valve locks on butterfly valves. Manual disconnects of air or

CO_2 lines and plug-in equipment can be placed into a lockable frame to prohibit use. Magnetic switch cabinet interlocks or light curtains on packaging system equipment falls somewhere between machine guarding and LOTO, as these devices shut down mechanical hazards when a person or body part gets too close, but they are not intrinsically lockable.

Be cautious of using logic circuitry to close off energy flows: "Push buttons, selector switches and other control circuit type devices are not energy isolating devices" (29 CFR 1910.147(b)). Logic controllers themselves rely on power, and if they lose power, the switches, pumps, and valves they control may default into an unsafe setting. For this reason, actuated steam valves should always default to a closed position during a loss of power. Nothing is as certain as an actual locked, physical disconnection or closure. Reviewing the section on power disconnection devices in chapter 15 is recommended before planning where in an electrical system to implement control measures.

OSHA intends for energy control devices to be visually consistent in terms of color, style, and font across a facility. This allows workers to become familiar with energy control devices at a glance. In addition to being standardized and identifiable, energy control devices should be made durable and substantial enough to perform as intended and to resist efforts by people to defeat the devices. Tagout tags must be attached with a self-locking, non-reusable connector that is attachable by hand and has a minimum unlocking or breakage force of at least 50 pounds (222 newtons). Generally, all-weather heavy-duty nylon zip ties are used for this purpose.

Periodic Inspection

Wherever an energy control program exists, the employer is responsible for annual review. This review will determine if ECPs are being used properly and ensure that all trained employees understand appropriate energy control methods and employee roles and responsibilities under the energy control program.

The inspection will be certified by the employer and will list each piece of equipment for which an ECP was specified, the date of inspection, the workers included in the inspection, and the identity of the inspector. If gaps are identified during the energy control program inspection, the employer needs to increase compliance, modify procedures, and train workers accordingly.

PRO TIPS FOR ENERGY CONTROL PROGRAMS

- Call it "Lockout/Tagout/Try Out"
- Always use lockable energy control devices, not just tags
- Whenever upgrading equipment, improve lockout functionality
- Build ECPs into task SOPs
- Keep ECPs laminated and mounted at point of application
- Locate lockout devices close to where they are used
- Remember to consider all potential and stored energy:
  - Gravity—use jacks and other props
  - Springs—release first, then lockout
  - Pressure—depressurize first, use multiple gauges
  - Stored electricity—discharge capacitors, uncouple batteries

---

Training and Communication

Training must be provided for all workers who will be interfacing with the energy control program, as well as adjacent workers who may encounter energy control devices in the workplace. According to the standard in § 1910.147, training shall include at least the following content:

- The purpose and function of the energy control program.
- How to safely apply, use, and remove energy control devices.
- How to recognize the type and magnitude of energy hazards and the methods to be used in the control of these hazards.
- How to dissipate stored energy in electrical, fluid, and mechanical systems.
- How to understand and implement an ECP.
- Only authorized/trained employees who are performing service or maintenance work may install energy control devices and prosecute work under an ECP.
- Adjacent workers shall learn about energy controls to the extent that they understand their purpose and do not defeat them in any way.

- Authorized employees or the employer will notify adjacent workers prior to placement and after removal of energy control devices.
- Tags are a warning device and not a control device, and have the potential to create a false sense of security.
- Tags are only to be removed by, or with authorization from, the person responsible for tag placement.
- Tags must be durable and securely attached to an energy isolating device.

Retraining shall be offered if any of the following conditions occur:

- Affected employees have experienced a job reassignment, a change in duties, a change in machines, or exposure to new hazards.
- Deviations or insufficiencies were detected during the required periodic ECP inspection.

- Employee proficiency or knowledge of a new or revised control method is found to require enhancement.

Importantly, as with any training program, the employer must certify that training has been conducted and is up to date, and must maintain records that include the names and training dates of qualified workers.

APPLICABLE STANDARDS

OSHA's control of hazardous energy (lockout/tagout) standard was adopted in 1989 and has changed little since then. The rules are fairly concise and clear. Where most energy control programs struggle is in having a written program, with the conducting of hazard assessments leading to the development of written ECPs, and in the day-to-day implementation of lockout devices by workers.

Table 19.3 **Occupational Safety and Health Standards (29 C.F.R. Part 1910) pertaining to control of hazardous energy**

Short title of standard	Title 29 citation	Key requirements
The control of hazardous energy (lockout/tagout)	1910.147	Control all types of hazardous energy using appropriate energy control devices and establishing written energy control procedures.
Machinery and Machine Guarding—General requirements for all machines	1910.212	Guards can be of many/any types as long as they protect the worker; usually fixed to the equipment.
Design safety standards for electrical systems—General	1910.303	(e) Electrical equipment is durably marked to identify manufacturer or organization, and other information such as voltage, current, wattage, or other ratings. (f) Electrical disconnecting means for motors, appliances, supply circuits; disconnects need to accept locks.
Permit-required confined spaces	1910.146	Isolation defined as the process by which a permit space is removed from service and completely protected against the release of hazardous energy with LOTO.

SECTION III
PEOPLE

Safety, when you really experience it, is comforting. It is to be cared for, even loved, by your coworkers and vice versa. We hope our brewery has a healthy level of diverse opinions, experiences, and backgrounds, because that diversity is a huge safety asset. We may have different value equations though. Some work for the money, others for the thrill of receiving a medal on stage, still others use the brewing business as a vehicle for social good. I remember hearing Kim Jordan, former CEO of New Belgium Brewing Company, say, "If it was just about the beer I wouldn't be here," or words to that effect. Yes, we all are different and come to work at a brewery with different values and experiences, but caring for your safety and the safety of your coworkers is a value we should all hold to. Our safety culture aligns us.

When I worked in hazmat response, I faced some scary tasks. These were not routine tasks, where everyone knew what to do and could do it from memory. These were one-time tasks where we did a full hazard assessment before starting, though then they were called HASPs—health and safety plans.

One time, I had to enter a 30-foot-deep manhole in the middle of a very busy intersection to put a flow monitor into a hospital sewer pipe, inserting an electromechanical sensor as far up the pipe as I could reach. This had to be done more or less while upside-down. I could list 20 hazards right off the bat, but some of the big ones included falling from height, being hit with falling tools,

asphyxiation, heat stroke from the PPE, claustrophobia, and exposure to biological and chemical disease agents. Have you ever had giant albino cockroaches crawling on you? You get the idea.

We had to pull a traffic control permit and get all the barricades and flashing lights. With the manhole lid removed, the entire scene was draped with plastic for contaminant control. It looked like a surgical field. Then we had a tripod with a winch set up in case I couldn't get myself out. Radiating out from the hole, each neat and unencumbered by anything else, were my safety line to the tripod, a breathable air supply line for my respirator, a positive flow blower blowing fresh city intersection air in to replace the sewer air, an air monitoring hose with a sensor on it, a tool bucket supply line, and a cable with the sensor ring I was installing.

I got suited up in a heavy-duty Saranex® plastic jumpsuit, two pairs of gloves, chemical boots, and a full facepiece respirator attached to bottled air. We doubly sealed any possible weak points in this ensemble with duct tape. When I came back out, there were coworkers with disinfectant sprayers that would wash everything off before I even thought about removing the PPE.

I was working with an all-star team. "J.J." was a chain-smoking industrial hygienist who was responsible for air monitoring and my air supply. He was also on top of the permit-required confined space paperwork. There was Ralph, as much a cowboy as ever you'll meet, sporting a giant handlebar moustache that he could somehow fold up into an oval around his mouth and fit under his respirator to get an airtight seal every time. Karen programmed the flow unit from a laptop. Bill rode a chrome-plated chopper and had been in the military. Jim was a former chimney sweep missing a few teeth; he was a recovering drinker and had a heart of gold.

I'm giving you these character descriptions to demonstrate that these people, my coworkers, were not like me in many ways. There were political differences, life-choice differences. We listened to different music and took our coffee different ways. But we had a functioning safety culture. Our jobs and reputation depended on how safe we remained. We were called in to do work that was deemed too sketchy for regular maintenance crews. We had a top-down safety program. We had industry knowledge and safety skills. We had hazard communication—it was written down in the HASP and the entry permit, and it was reinforced with a toolbox talk just before we started.

When I was in the interceptor sewer, every minute or so someone would yell down the manhole asking if I was OK and I'd give a visible nod or tug back on my safety line in response. Our team's diversity was part of its strength. Who better to know how to descend three stories in a narrow brick tube than a chimney sweep? Who would be better to put on traffic control than a motorcyclist with military training?

Being down in that interceptor sewer was so peaceful. It was so comforting knowing that this great support team had my back. What could have been a freakishly anxious job was more like a meditation. When did I really get scared? When I completed my task and stuck my head back out, level with the pavement, seeing the wheels of fast-moving cars shoot by in four directions. That's when the calm left me, when the randomness of unsafe people everywhere impinged on my job.

In a workplace with a strong safety culture, everyone works together. The result isn't just lower incident rates, it is mental well-being, that shot of healthy neurotransmitters that says I am rewarded, I am cared for, I am at peace, I am touched, and, along the way, I laughed.

In this last section of the book, we deal with the most vexing questions of brewery safety:

- What is a healthy safety culture and how do we get there?
- How do we know if we're improving in safety?
- How do we knit together the people in the workplace with the mechanisms for controlling hazards, all while meeting compliance requirements?

That last one really has safety professionals in a knot. Many safety experts talk about safety management systems (SMSs). These systems are a holistic approach to overall safety programs, though sometimes they deemphasize the employee having to act safely on their own. A systems approach relies a lot on engineering controls designed to protect employees regardless of where they stick their hands. Importantly, SMSs are based on the concept of continuous improvement. Then there is a minority of experts who tend to prefer behavioral safety, where we try to condition worker behavior. Hands here are just an example, of course, but behavioral safety aims at ensuring workers know exactly where to place their hands to stay safe. Chapter 22 gives you some history on this dualistic perspective, but my message is that it is not one or the other.

Neither behavioral safety nor a systems approach is achievable if not for people. Only a portion of safety is glove selection, machine guards, CO_2 monitors, and mechanical lifting devices. Safety culture is about people. It is about human resources. And yes, it is about equipment and methods and plans, too. But even these are the decisions and creations of people.

You can think of your brewery's safety journey like entering a fetid manhole. But let's take a happier and more generic example: an ocean voyage. The crew is a group of people entering a common course toward the same safety horizon, communicating thoroughly as they do it, and measuring and making course corrections as they go. It is all hands on deck, whether your brewery is a dinghy or a freighter. No freeloaders are tolerated, and you may have to ferret out safety mutineers and deal with them in accordance with current labor practices and good policies. If the crew has the collective skills and the intention to plot a course, operate the craft, and deal with changing conditions, it can navigate safely into the calm waters of a mature, successful brewery, arriving with all crewmembers uninjured and in good spirits.

20
SAFETY CULTURE

On the surface, safety culture is vexing. Within the safety world, defined by regulations and specifications, safety culture has been labeled "notoriously nebulous" (Marsden 2021).

First, we can rule out what safety culture is not. It is not separate from organizational culture. All groups of people have behaviors, beliefs, and experiences. Thus, all groups already have a culture. When we observe a business with safety problems, we can target the symptoms (behaviors) or we can go after the underlying dysfunction (culture).

Culture is not an objective thing or a binder on the shelf. It is a conceptual grouping of repeatable human thoughts and feelings. Culture is sometimes labeled the personality of an organization. It can be thought of as a basket, a filter, or a lens, but it has no physicality. It is not a program, plan, policy, procedure, or project (Barnes 2010).

Despite this, there is actually quite a bit of agreement between definitions of organizational culture. Any organizational culture is a set of beliefs, perceptions, values, assumptions, and business systems adopted by a group, which forms the foundation for behaviors exhibited by the group. Culture embodies behavioral norms; it describes what the participants *do*, not what they say or think they do. When that culture undertakes continuous improvement and is centered on the safety and well-being of the workforce, and if it is largely achieving that goal, we can say it is an effective safety culture.

What makes us behave in certain ways? You might say it is our learned experiences or because we fear punishment for certain behaviors. You would be right, but, more generally, behavior is dictated by cultural norms. In an effective safety culture, we are most inclined to wear a respirator because we *believe* it is helping us stay well. But where does this belief that a respirator works come from? It comes from the experiences and social examples provided by a workplace culture of healthy like-mindedness.

Many behaviors we see are unsafe. These behaviors are due to a number of different complicating factors related to our own thoughts and life experience, including normalization, perceived risk, social proof, prejudice, and cognitive dissonance (*see* chapter 18).

CULTURE CONNECTS EXPERIENCE TO ACTION

Culture occupies the space between the learned experiences of individuals and the behaviors they exhibit, although some experts lump experiences and culture together. The diagram in figure 20.1 illustrates this relationship.

Safe Behavior is Working Safely

Let's look at the pyramid in figure 20.1 from the top down. When we are operating and thinking safely, we are exhibiting safe behaviors. When our company has established safety systems, programs, and provided necessary equipment, we have the possibility of behaving

SAFETY CULTURE NEEDS ALL OF THE PEOPLE

When I was a kid, my parents sent me to Sunday school. We played a game with our hands by interlocking our fingers, pointing our index fingers up like a church steeple and held our thumbs like the doors. Maybe you played the same game. With our hands depicting the shape of a church, we'd say, "Here's the church, here's the steeple, open the door, and there's all the people." At this point, we turned our hands over to reveal our wriggling, interlocked fingers: the people. The message behind this game was simple. The church isn't the building, it is the people inside.

The same parable works on a secular basis when thinking about breweries. My friend and colleague Rachel Bell says, "Yeast makes beer, people make breweries." The first time I heard this, I admit I was perplexed. It went against my self-centered view that brewers make beer, and that's why we're such awesome people. But the more I thought about it, the more I realized Rachel was onto something. Technically speaking, it is the yeast that turns sugars into beer. In nature and without any human intervention, when the right yeast and carbohydrates find each other, fermentation will take place and alcohol can be produced. It is when people get involved that we can direct the process, care for the yeast, feed it, and help it make a consistent product.

A brewery is a place where fermentation happens, but people work together to optimize it and to make it agree with our tastes and preferences. Brewery people connect with customers through the beer and the social context of the brewery. If the people do not work in harmony within a system then the quality or consistency may suffer, some workers will feel dispossessed of their importance, the business may not succeed, and surely the workplace will be more unsafe. That is because it takes people acting communally toward safety for safety culture to be real and sustainable, not just the employer providing equipment or workers knowing how to be safe.

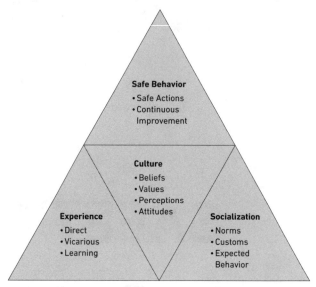

Figure 20.1. Foundations of safe behavior. Safety is the visible state of worker behaviors and management systems creating a workplace with the lowest possible risk of accident, injury, and damage to product, equipment, or the environment. Culture underpins those behaviors and is, in turn, advised by our experiences, training, and social environment.

safely because we have the necessary tools. Actions we take in the performance of our work, reducing hazards, and showing concern for the well-being of others are all behaviors. Having the right equipment and a built-in improvement process to do the work safely is part of a larger behavior—the company is employing a safety management system (SMS). This type of system is described in chapter 22.

If safety is defined as the state of being free from harm, then the way we achieve this state is to practice the avoidance of harm. We have to know how to do that, not just in terms of performing hazard assessments, but in all aspects of the work experience. The right cultural values are what direct safe behavior.

Culture is a Complicated Collection

If safety culture is the figurative vessel of ideas, what sorts of ideas make an effective safety culture? They should rest on the collective beliefs, values, principles, assumptions, perceptions, and business systems that all employees can share, even when individuals differ widely in other personal ways. We are each afforded our own politics, religion, musical preferences, diet, and racial or gender identity, among other facets of personal identity. It is generally observed that the more diverse these personal traits are in a group, the stronger the group. What is required from everyone is respect for these differences. One earns respect by being respectful of others.

The company's owners and senior managers must model this behavior. Management is also responsible for setting the example for workplace safety and expressing its vital importance in the success of the business.

Disrespect fuels division among the workforce and disempowers those who might otherwise make valuable contributions. Forms of disrespect to avoid include prejudice, misogyny, gossiping or lying, using inappropriate words or labels, demeaning others, and acting in a hostile or threatening way using words, shouting, or being physically disruptive or threatening. These are many of the same behaviors that enable psychosocial hazards in the workplace, which we covered in chapter 18.

As individuals and as a group, a blossoming organizational culture will have missteps. You may identify some form of disrespect that is commonplace in worker-to-worker communication, discover a hiring practice that results in low diversity, or identify an undercurrent of sexism. Individuals should be taught it is best to ask for clarification or simply apologize and move on with the intent to communicate better next time.

Since we're talking about safety culture, let's focus on issues centered around safety specifically. A workplace may have poor compliance with regard to wearing safety glasses along the packaging line. Workers might be regularly impaired by alcohol or drugs. While the brewery might provide the right ladder for a job, no one uses it in the manner it was designed. Brewers may routinely brew more volume than a system was designed for, thereby increasing the chances of a boilover or hop volcano. Strengthening safety culture is how we remedy these unsafe behaviors.

 WE LEARN TO ACT FROM SOCIALIZATION AND EXPERIENCE

The Fraternity Story
I attended a college with a strong "Greek" tradition and a student housing shortage. My father and grandfather were fraternity men and so I was a "legacy," a kind of preapproved candidate. Experience within my family pointed me toward pledging to the same fraternity my father and grandfather had been in. Meanwhile, a close friend of mine from high school, who was a year ahead of me at this college, sought me out the day before school began. He told me, in no uncertain terms, that fraternities were a distraction and I was far too independent of a person to conform to their traditions.

At first, I chose my family's experience over my friend's. I pledged to the fraternity, my grades went down, and I realized I didn't really like the fraternity brothers. So I quit. It took me having my own experience to reckon with what my father and friend had told me. Our personal experience factors into culture and it can be changed.

The Chainsaw Story
In high school, I was working on a trail clearing project with my cross-country ski coach and some teammates. I had brought my father's chainsaw for the job. After all, I figured, I was a country kid and had been using one for several years by that point.

We were clearing some storm damage in which a large tree had come down and pinned several other trees. The bent over trees hold a lot of potential energy and are called "spring poles." I took to them with the saw, not having learned there is a specific, counterintuitive way to deenergize these spring-loaded tree trunks. I cut in the wrong place, the tensioned side, causing the tree to break loose in the blink of an eye. Like a catapult lever, the bound tree trunk came free and swung upward to within inches of my nose. It happened so fast I couldn't have moved if I had wanted to.

I stood there, shocked, staring at the butt of this tree right in front of my eyes. My coach, who had seen the whole thing, said dryly, "I'm glad to see you still have your head." There was much I had to learn about both reading the tension in trees and safe chainsaw operation. Ever since then I have taken advantage of chainsaw safety and skills courses. We must keep learning and continuously improve our processes.

When issues that reduce the effectiveness of a workplace safety culture are found, we should seek to get better. We may have to learn to communicate better. Training, or retraining, may be necessary. We might need to develop better policies or SOPs. We will also need to measure our progress to know that we are truly improving. The goal in any culture or safety management system should be one of continuous improvement.

Right here we have identified five key cultural drivers: management leadership, healthy communication, accountability/measurability, respectful involvement of everyone, and technical knowhow. These are discussed in more detail later in this chapter under "Drivers of Safety Culture."

Experience and Socialization

The foundation of culture is the collective experience, knowledge, and social environment of those within it. Experience can be direct, that is, a lived experience, or indirect, the experience of others. Experience is a diverse category that includes life events, personal recollection of those events, training received, media content consumed, and much more. Experience may be vicarious. It can be obtained by hearing the story of a coworker or through an industry webcast. It can also include professional development, safety training, and the use of resources like safety data sheets and equipment ratings or more expansive resources like the internet and mass media. The experience of others may influence the values of the one, and therefore the culture.

I may tell you that wearing a respirator made me feel better around malt dust. To the degree that you and I are peers and have mutual trust, my experience can affect your behaviors. But if I'm a seller of respirators and not a personal acquaintance, you may not find the same value in my recommendation. Indeed, you may completely distrust me. An SOP written by you and your team will be more personal and believable than one dropped in your lap by a consultant.

Loosely speaking, experience is strongest when it is those firsthand experiences like a peer or mentor's recommendation or personal history. Let's say you are a newer brewer who has never entered a national beer competition. You seek advice from a brewer you know has won a great many national awards. Chances are, you will listen carefully to that brewer's experiences. You sought out the information, you have a lot at stake, and you trust the source. These types of interactions are where the social component of experience comes into play.

DRIVERS OF SAFETY CULTURE

As mentioned at the outset, an effective safety culture within an organization is a set of beliefs, perceptions, values, assumptions, and business systems adopted by the company, which forms the foundation for behaviors exhibited by the group. One of the great ironies of an effective safety culture is that it does not have that much to do with specific safe actions, safety knowledge, or institutionalized hazard controls. Those competencies are a part of safety culture, but a common mistake for brewery owners and workers is to focus on just the safety part and miss the human connection. We know that a sound safety culture results in safer worker actions, employee retention, lower workers' compensation premiums, and a host of other good things. So the question becomes, what beliefs, values, and business systems truly improve safety?

In 2019, researcher Elizabeth Embry and colleagues from the Leeds School of Business in Boulder, CO conducted a survey of brewery owners and workers regarding injury rates, the state of safety programs, and individual perceptions to safety in their operations (Embry, Rollings-Taylor, and York, unpublished). The design of the study was based on an investigation into safety culture in the construction industry and was found to translate readily to a manufacturing setting (Embry and Stinchfield 2020).

In the Leeds School of Business study, participants were asked a series of questions pertaining to their perceptions of various cultural factors. For every statement given, the respondent reported their degree of agreement or disagreement with the statement using a 1-to-5 Likert scale. These responses were collated and compared to discover relationships between various perceptions. For example, the researchers found that 90% of management respondents said employees were able to discuss safety issues with them, while only 63% of employees agreed.

One interesting finding was that technical or compliance knowledge of safety was not significantly correlated with safe performance. In other words, even when employees knew a lot about safety, that did not mean they behaved more safely than those who lacked that knowledge. This is particularly important, because many breweries try to build a safe workplace by ensuring employees have a strong knowledge of safety procedures to achieve safe behaviors. This is shown to be ineffective.

What was also revealing is that there is a strong positive correlation between the presence of an established safety culture and the practice of safe behaviors. Furthermore, the researchers found very strong correlations between a safety culture that is effecting safe outcomes and each of the following drivers: management values, communication, accountability, inclusivity, and competence (fig. 20.2). The correlation coefficients reported by Embry's group are quite high considering they represent relationships between human behaviors and a conceptual safety culture.

Going back to the notion of safety culture as a vessel of ideas containing our collective beliefs, values, principles, assumptions, perceptions, and business systems, we can now see a direct pathway from an effective safety culture to safe outcomes in the workplace. The remainder of this chapter looks at these five drivers—expressed management values, communication, accountability, inclusivity, and competence—and how they can be improved in the brewery workplace.

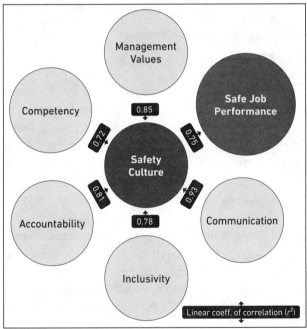

Figure 20.2. Drivers of safety culture. Safe job performance is significantly correlated with a healthy safety culture in craft breweries. Safety culture is driven by well-implemented business principles applied consistently. *Adapted from Embry and Stinchfield (2020, slide 35).*

Expressed Management Values

In this discussion about management, let us understand managers to be the person or persons who run the business. They may be participating owners, designated executives, or operations managers. They are the people who we identify as "in charge."

If we define management as the coordination and administration of operations to achieve goals, we position management where it should be in the culture discussion: top and center. Regardless of how big a shift a company

needs to make to improve its safety culture, management is still management. Professional and thoughtful managers will set the desired tone for the overall cultural setting.

When we say "management is responsible for" it may mean that the managers have to do something or that they collect and organize the doings of others. Whereas most employees are responsible for doing their own jobs, a manager's job is to ensure that employees are doing their jobs in an effective, productive, and safe manner that does not harmfully impinge on the performance of other employees. Managers often focus on desired results as a collective achievement. That same approach applies when it comes to safety culture.

Setting Objectives

Managers set objectives for the rest of the workforce. Objectives are wide-ranging and might include production goals, human resource allocation, and sales and marketing targets, among others. Specific to safety, stated management objectives can include many or all of the following:

- Safety is to be valued as a top priority.
- Operating safely is to be understood to increase productivity and enhance overall business operations.
- The company will work actively to prevent workplace injuries and illnesses.
- The company will leverage compliance resources, such as using OSHA on-site consultation services, participating in OSHA alliances, and attending technical safety presentations.
- A safety committee comprising employees from all departments will be established and maintained.
- The company will endeavor to comply with relevant health and safety rules and regulations.
- Employees will be required to engage in process improvement and safety programming.
- The company will encourage near miss reporting and will not punish whistleblowers.
- The company will provide and document necessary training and protective systems.
- The company will stay current with best practices and emerging safety technologies with the understanding that safety culture is evolutionary.
- Management and ownership will abide by the same safety and health requirements as the workforce at large.
- The company will establish social responsibility goals and measure progress.

Ensuring Success of the Other Drivers

Besides the honest implementation of management's stated objectives regarding safety, owners and managers also shepherd the other four drivers. They need to communicate effectively and be good listeners. Safety communication should be frequent and consistent. Motivating the workforce can be a challenge when the outcomes of an effective safety culture may only become apparent slowly at first. If the previous style of management did not emphasize safety as a priority value, the first thing management must do is declare there is a new direction or a new appreciation for safety.

People often resist change, and an entrepreneurial brewery owner may have their head in the "I'm living my dream" space, not the "I'm responsible for all of these people and processes" space. In an effective safety culture, management has to exemplify the behavior they expect from the workforce. If management expects changes in worker behavior, it needs to set the tone. It could be as minor as something said at a staff meeting, like, "Going forward, I want us all to be more focused on safety, and that begins today by putting a safety discussion on every staff meeting agenda." It could be a dramatic, fall-on-your-sword sort of revelation, like, "I realize I have set a reckless example of unsafe behavior, which I confused with having fun. From now on, we're going to run this place as safely and conscientiously as possible, starting with me."

After modeling good communication, managers have three responsibilities to oversee. They must involve the entire workforce in safety, not simply make company dictates. Management needs to develop, improve, and allow contributions from its full complement of human resources. This inclusive engagement allows ideas to circulate and increases employees' feelings of self-worth. It helps create fertile ground for workplace equity and diversity. This stimulates worker satisfaction and retention.

Inclusion also boosts the development of processes and procedures. Those included in conducting hazard assessments and outlining SOPs will be more likely to follow procedures and rules and are more likely to report near misses. Significantly, when management listens to all employees and is sincere in involving them, it sows the seeds for trust to emerge. Trusting that the company will act with employee safety in mind helps to alleviate a common anxiety that many workers experience.

PROVERBS FOR A SOUND SAFETY CULTURE

"The fish rots from the head," said Rumi in his epic thirteenth-century poem, the *Masnavi*. If we think of management as the head of the fish, we realize corruption of the organization's safety culture begins here.

"Not the cry, but the flight of a wild duck leads the flock to fly and follow" is a Chinese proverb. We realize from this that management must set the example for safety.

"If you want to go fast, go alone. If you want to go far, go together" is a proverb purportedly originating on the African continent. Safety culture requires everyone—in that way, it can be lasting and self-sustaining.

Once the workforce is engaged and participating, management can help build safety expertise in every employee. Programming and adhering to a training schedule and encouraging workers to become part of safety teams are just two ways to build competence; more are discussed below.

Finally, once management has expressed safety as a value and involved others, goals for safety program improvement can be set. These goals express companywide aspirations, systems of improvement, and how those improvements will be measured over time. Resources should be made available for the measurability and accountability of the safety program.

Communication

Of the five drivers, communication most obviously threads through the others. The company's safety values need to be communicated. Goals and achievements have to be communicated once measured. Competency in safety is rooted in communication, whether verbally to apprentices and trainees, through formal training and the reading of SDSs, or from coworkers sharing their experiences.

It seems fair to say that most people could use some improvement in their communication skills. Building a safety culture needs everyone to be present and communicating. A workforce may be made up of people who are different in many ways, but they are similar in that they all wish for a safe workplace. The following are some suggestions for successful communication:

- Talk about safety frequently and consistently, in person, in writing, visually, and by exhibited behaviors.
- Keep open communication within the organization.
- Make efforts to bridge any communication gaps between management and the workforce.
- Express what is known, expected, or required when working together.
- Seek to understand what is unknown by asking questions and respect those who ask questions of you.
- Establish and communicate schedules.
- Make sure that goals, timelines, expectations, and rewards are communicated.
- Communicate policies and procedures in writing; this should also include verbally reviewing and acknowledging the following:
  - Employment policies
  - Required signs and posters
  - SOPs, SDSs, and other safety documentation
  - Emergency, fire prevention, and first aid plans
  - Near miss, suggestion, and injury reporting procedures
- Publicize expected behaviors and rules and the consequences for their violation.
- Conduct check-ins for mental health.
- Avoid disparaging or derogatory attacks, and avoid shaming, embarrassing, or otherwise demeaning any person.

Know that miscommunication is bound to happen. People have differing styles of communication: business-like, casual, locker room, trash-talking, and so on. In a work context, communicating professionally is a good way to start until you become more familiar with coworkers.

Some people create rules about communicating, such as "I never apologize" or "My way or the highway." These are counterproductive. If you make a mistake in the words you use, check immediately with the other person if it was indeed a problem and apologize and move on. Try to do better next time. Some situations where we might stumble include using the wrong pronoun when speaking to or referring to a person, or using a derogatory trope that we have unconsciously normalized.

Accountability

Accountability means taking responsibility for personal or corporate actions. The first step is to describe the desired action or outcome and who will be responsible for it. This can be done on an executive level, say, charting the company priority for safety. Or it can be done on a granular level, for instance, assigning an individual to ensure fire extinguishers are all accounted for and regularly inspected. Participating parties develop a mutual understanding of what the outcome looks like. The parties involved may then collectively chart a path toward accomplishing the goal, or they may leave the planning up to an individual or work group.

No matter how well the plan is laid out, one should always couple it with accountability measures. Having measured progress on a certain thing, the team can then renew or increase the goal and in this way continuously improve. This cyclical process of goal setting, measuring progress, and pushing for further improvement is the basis for safety management systems discussed in chapter 22.

Worker and Manager Solidarity

One of the most importance principles of accountability is avoiding actual or perceived dual standards. If workers have to wear certain PPE in the warehousing area, then so do managers and visitors. Failing to abide by safety protocols that (one hopes) were set following a hazard assessment creates mistrust and calls into question the validity of the underlying assessment process.

If managers are permitted a drink after their shift, employees should be given this privilege also. If management determines that employees are not to have after-shift drinks, then it should adhere to that same standard.

Personal Accountability:
Reward, Improvement, and Discipline

When thinking about worker safety accountability, we quite often go right to the topic of reprimand, discipline, and termination. It is true that employers are responsible for a safe workplace, and that workers are legally responsible for adhering to training, using provided safety equipment, and complying with safety laws and regulations. But to focus on these responsibilities without providing needed supplies, training, and good workplace examples sets up the employee for failure.

It is crucial that expected behaviors are modeled by management and expressed in writing, otherwise management's expectations may be applied haphazardly.

Table 20.1 **Accountability examples**

	INWARD FACING	OUTWARD FACING
INDIVIDUAL	• Wears assigned PPE, uses safety equipment • Responds favorably to instances of discipline • Follows through on safety assignments in a timely manner	• Reports near misses • Reports accidents, injuries, and illnesses • Watches out for the safety of coworkers, customers, and general public • Participates in safety as a safe coworker and, possibly, as a safety team member
GROUP / COMPANY	• Establishes meaningful forward-looking key performance indicators • Safety improvements are prioritized, made measurable, and are time-specified • Clearly states expectations of safety compliance and safe behaviors and the consequences for failing to adhere to them • Applies a system of praise and discipline fairly and consistently, while maintaining privacy where appropriate • Safety spending is budgeted and utilized	• Workforce and management are held to the same level of accountability • Management lives up to its stated safety values and commitment to safety • Company acts true to its safety identity as portrayed to customers, suppliers, and community stakeholders • Applies a system of praise and discipline fairly and consistently, while maintaining privacy where appropriate

Two good modeling techniques here are (1) reward or recognition for doing a task safely, as established by training or SOP, and (2) demonstrating beliefs, values, and administrative processes by modeling desired behaviors and creating experiences and social learning opportunities that model positive cultural values.

Reward or recognition is straightforward. Follow the old management aphorism, "Praise in public, punish in private." Verbal praise is very effective when given before a group of the individual's peers. This can happen at a shift meeting, a toolbox talk, or at a larger company meeting. Other ways include mentioning the employee in a company bulletin or on the website, giving out quarterly safe worker awards, or maybe with a gift. Such a gift might be a business gift card, a special article of clothing that denotes the award, or a cash award. If an entire group of employees, a department say, has earned recognition, it is common to provide a meal to the group, like a pizza party or a barbecue.

Positive recognition often doesn't occur because management has reasoned the meritorious behavior was expected of the employee as part of their job. On the contrary, this is exactly when praise should be given.

When a task is not performed safely or the employee's behavior toward others is not in line with cultural norms and expressed expectations, managers normally try to change the behavior. Behavior is action. It is where the rubber meets the road, as the saying goes. But behavior rests on culture: collective beliefs, values, principles, assumptions, perceptions, and business systems. So, to get meaningful behavioral improvement, one should target changes on these culture underpinnings.

Another approach to building safe behavior is to ask the employee why they aren't complying with a requirement or conforming to an established procedure. The brewery culture creates an opportunity for a peer-to-peer interaction as you work through the thought process behind the worker's existing behavior. Is their reason for not doing the job as specified because they have normalized a contrary behavior, do they carry a cultural myth learned socially, or do they believe the specified way is a waste of time, causes a quality problem, or is unsafe? Working through the issue in this way can uncover the root cause, allowing the parties to see where the truths and misconceptions are. This provides a platform on which improved behavioral outcomes can be built.

One cannot expect to foresee every possible cause for reprimand or discipline. However, general guidelines should be developed and printed in the employee manual. General rules and examples can provide context for dealing with unforeseen behavioral problems.

In the case of a particular unforeseen undesirable behavior, managers should step in quickly and provide ad hoc guidelines to amend the behavior. If the business has a human resources (HR) manager, involve them from the earliest possible point. An HR manager can also help establish the company's response regarding the severity of the safety issue. For example, entering a confined space without proper training and controls could qualify for immediate dismissal.

Progress Accountability: Key Performance Indicators

In addition to personal accountability, there is the wider monitoring of stated safety goals to see whether they are being met. This where key performance indicators (KPIs) should be assigned to each goal. The KPIs outline how progress will be measured, how often it will be measured, and how to interpret the results. Here is a simple example around the idea of starting a safety committee:

Goals

1. Create a company safety committee that includes someone from each department, and with both supervisors and front line employees.
2. Meet six times per year to identify unaddressed safety concerns throughout the plant.
3. For every identified safety concern, implement a workplace safety improvement.

Measurement

1. Did the committee have representation from every department? Yes or no. Desired answer: yes.
2. How many times did the safety committee meet in the year? Desired answer: six times or more.
3. What percentage of identified safety concerns resulted in a workplace safety improvement? Desired answer: 75 percent or more.

Once a period of measurement has taken place, the results advise the company if they are on track or if corrective measures are needed. If the KPIs show the goals have been achieved, the plan might be to do the same for the next period or, possibly, to elevate the goals to push for even greater improvement.

In the case of falling short, the goals might be reemphasized or reframed to better clarify the objectives. Ideally, the KPI metrics do not change, because having the same measurements over a long period allows one to know if there is a legitimate trend emerging. Chapter 21 is entirely dedicated to measuring safety culture improvements.

Inclusivity

Inclusivity, as it is used in the context of safety culture, is the state of providing equal access, opportunity, and involvement in workplace safety. It means equal access to training and learning opportunities related to safety. This might involve a worker undergoing cross-training or task rotation, or participating in a safety committee. It should absolutely mean that all workers have parity in task risk and access to PPE when performing the same task.

Inclusivity is a powerful tool to help eliminate workplace disempowerment fueled by socioeconomic status, race, gender expression, or any other differentiating factor. One way to boost inclusivity is by using a buddy system. Pair up a new employee, or an existing employee who has formerly been left out, with an established employee in a visible, prominent role. Have them work together on solving one or more safety problems from start to finish. The newly included employee will learn essential business processes, will engage with employees outside their sphere, and will hear and be heard in a wider, more diverse business setting. The buddy employee or manager will learn more about the employee, their communication processes, and perspectives from their work assignments.

Once everyone has been afforded visibility and a voice, they are encouraged to participate in safety development processes. This might include participating in hazard assessments, advising SOP writers of the current procedures used for a task, or joining the safety committee. Involving everyone in near miss and injury reporting is the best way to help prevent a future accident or injury.

Inclusivity goals can and should be part of your safety culture's measurement and accountability process. It has been said that you can't call it culture if everyone isn't included. There are easy KPIs to establish, such as making sure everyone who needs training for a certain task has received it, and asking if the demographics of the safety committee is similar to the workforce as a whole.

Competence

Sound management, good communication pathways, accountability, and inclusivity are all core tenets of any endeavor, whether a brewery, a service organization, or a kids' soccer league. Competence is where the specific required safety skillset becomes a part of your company's safety culture.

Most businesses are going to have a range of safety competencies in their workforce. There may be some individuals working unsafely because they are uninformed or reckless. You can also have hidden expertise, like a warehouse employee who worked for a major logistics company or a bartender who was an emergency medical technician. First, take inventory of your existing workforce. Invite people with a history or interest in safety to participate in hazard assessments, develop company policies, or form a safety committee.

Second, include safety in your hiring practices. Add application or interview questions that allow you to assess a candidate's perspective on safety. How do they perceive safety? Have they been trained or certified in specific safety subjects? Ask the candidate about a close call they might have had and what they learned from it. When checking references, ask the respondent about how safe the candidate was to work with. It is a big job to reorient a chronically unsafe worker, no matter how good of a brewer they may be. Hire wisely.

Where do we learn how to work safely and keep others from harm? In a small to medium-sized brewery, it usually comes down to learning from those we work with. If workers are not learning from safety-focused individuals, it can be easy for unsafe practices to become normalized over time.

If there is bravado toward hazardous tasks, that attitude will often be adopted due to the strong influence of socialization on safety culture. Some common examples of machismo in the brewery include comparing battle scars, calling others "wimps" when they are trying to work safely, and refusing to ask for help with heavy lifting tasks or labeling those who ask for help as weaklings.

Embry's study revealed that most of what brewers learn about safety is through oral and experience-based learning (Embry, Rollings-Taylor, and York, unpublished). It seems most learning is not from adhering to written programs or receiving formalized safety training, even though such training is often required by OSHA. However, safety and health training has been shown to be effective in reconciling workplace risk mythologies arising from culture dysfunction (Rodríguez-Garzón et al. 2016; Halkitis and Parsons 2000; Colligan and Cohen 2004).

Some individuals will have gone to school for brewing. Those with brewing diplomas or certificates will have received various amounts of safety education depending on their program. It is best not to assume that someone with a brewing degree has a vast knowledge of safety topics or that they will work safer than everyone else. Some brewery employees enter the industry as an intern or a candidate of a mentoring program. Just as with brewing schools, instilling safety will depend on who is doing the teaching and how much weight the teaching program puts on safety.

Many people in brewing come from a background other than manufacturing and may have little appreciation for the amount of energy carried by industrial hazards. Finding training resources through OSHA, trade associations, and safety institutes is part of the employer's responsibility owed to employees.

21
MEASURING PROGRESS

The story goes that an OSHA inspector asks an employer if they can see company training records. The employer explains that even though they conduct lots of training, they don't have written records of it . . . but training definitely took place! To this, the OSHA person says, "If it ain't written down, it didn't happen." Of course, the employer could have written down lots of training that never occurred, but you get the point.

Documenting conditions, maintenance, inspections, and training is a business fundamental. Many OSHA standards require documentation in the form of written programs, injury recording, air monitoring, safety data sheets and labels, medical evaluations, training, and more. See Table 2.1 for more on these requirements.

The thing is, just because we have some sort of document on a subject it doesn't mean we are in good shape or improving. It doesn't mean we are reading or following what is written down. These records are just a snapshot. What would be more helpful is to see how we have progressed over time, and whether we could set goals for the company and know if we are meeting them.

KEY PERFORMANCE INDICATORS

The way we track the trends in safety over time is with *key performance indicators* (KPIs). The best KPIs fit the SMART formula: specific, measurable, accountable, reasonable, and time bound.

Safety-related KPIs must support specific actions needed to reduce risk and improve outcomes. They need to be measurable in some way, whether that is with a rate, frequency, percentage, a finite number, a yes or no, an up or a down, or similar. KPIs are really measurements attached to a decision rule, so the measurement needs to be reliable and reproducible so the decision made will be valid.

Accountable indicators track a meaningful quantity, behavior, or circumstance that is known to track with the desired improvement. Indicators can track positively or inversely with the occurrence, and either in linear or non-linear relationships. Sometimes it is unknown whether a certain metric is a valid indicator for an observed phenomenon. In that case, validating the relationship after some data is collected is an opportunity to continue or to revise the measurement.

Reasonableness cuts two ways. Can you reasonably achieve the goal that you set for your leading indicator and is the metric reasonably obtainable and reliable? If you arrive at a particular KPI because of intimate knowledge of the outputs, you will likely already know a reliable metric.

Finally, KPIs need to be expressed within a time context. Are you taking measurements often enough to identify trends? Does your decision rule have a specified time in which remedies are to be applied?

Table 21.1 **How to use key performance indicators and some examples**

Measurement purpose	Safety program example(s)
Input Tracking	
Track resource usage	Establish a budget for PPE and track expenditures against the budget.
Measure human performance	Track months/quarters/years without a recordable injury.
	Observe and reward efforts to reduce trip hazards in the cellar.
Output Tracking	
Assess progress toward a goal	Measure percent hazard assessments completed, with a long-term goal of having a written hazard assessment for every brewery task.
Process Tracking	
Track completions	Track near miss reporting and resolution processes.
	Schedule and track completions in preventive maintenance.
Improve business decision-making	Track overtime hours in relation to production in order to determine future hiring needs and avoid employee burnout.
Directional Indicators	
Track unsafe normalized behaviors	Follow barrel or batch records to identify unsafe practices that, e.g., could lead to boilovers or dry hopping incidents.
Culture	
Worker mental health and personal safety, i.e., culture growth	Evaluate the results of worker wellness surveys to track gains in safety culture or in consideration of increasing workforce.
	Create a safe retail environment by developing front of house safety program, training staff, welcoming diversity, and investing in security methods.
Practical change (often unique to company's existing culture)	Track success in initiatives that change the course of an entrenched behavior, like eliminating wearing of shorts or personal listening devices.
Planning	
Build strategy for the future	Define and measure investment in human resources by way of training, provisioning, and performance recognition.
	Measure involvement in outside safety initiatives, like membership in state brewers guild safety committee, attending or presenting safety presentations at conferences, or mentoring emerging safety and health students.
Financial Indicators	
Expense/savings of safety program	Track workers' comp experience modification rate (EMR, essentially an experienced-based insurance premium multiplier).

Sources: Gacioch and Stinchfield (2022), Perez (2021).

KPI VARIABLES

At their core, KPIs are measurement rules. They establish what is to be measured, how it is to be measured, and often how the results are to be construed. Key performance indicators often include an expressed goal or minimum desired achievement. Some KPIs are considered more valuable than others, and it is these that we should prioritize. For example, does the KPI in question lead us to a useful understanding? Does it allow us to detect a trend in our safety system? Is the data relatively easy to acquire and will those data inputs continue to be available across a span of time? Are the KPIs unambiguous and do they motivate the people that management is trying to reach?

Leading versus Lagging

Leading indicators allow projections into the future. They are sometimes called forward-looking indicators and that explains well what they do. They often involve setting of milestones, checking the progress on those milestones, then renewing them for further future cycles. Leading indicators are "proactive, preventive, and predictive measures that provide information about the effective performance of your safety and health activities" (US Department of Labor 2019b).

In a brewery safety program, a leading indicator might be the rate of safety meeting attendance. The KPI states that the company wants 95% of the workforce to attend their department's monthly toolbox talks. Leading indicators like this help brewery safety evolve by framing a goal that is believed will have a positive outcome for safety culture and will also yield specific benefits in areas like injury rates, employee retention, and worker morale.

Leading indicators must come from somewhere. If we have a past trend that we know indicates a certain outcome, then we can logically assign a measurement to that trend going forward. We can estimate that if X goes down, Y goes up, for example. We don't know for sure because we're just setting up how and where to collect data. These heuristic or intuitively determined metrics may or may not be actually related to the desired outcome. It is wise to review KPI decision rules as time goes by. This allows faulty assumptions to be replaced with something more accurate.

Lagging indicators, or backward-looking indicators, look at past data to confirm trends. The major complaint about lagging indicators is that they do not establish a direct pathway to improvement. Sure, we can say, "Well, that was a bad year, let's do better next year," but that statement is not giving us tangible, relevant measurements that we think will monitor improvement.

Lagging indicators "can alert you to a failure in an area of your safety and health program or to the existence of a hazard" (US Department of Labor 2019b, 2) Some examples are the number of injuries sustained, the number of OSHA citations received, or the rate of absenteeism.

Both lagging and leading indicators have a potential pitfall. This occurs if we create an incentive that alters the measurements for some reason. Let's say a brewery is interested in reducing lost time due to injuries. The brewery decides to track the number of consecutive days without time lost to injury and it incentivizes the employees with a reward—let's say free pizza—if they make it through a quarter without any. Given that the crew all like pizza, they collude in not reporting injuries that occur. Not only is the measurement incorrect, but we are rewarding staff for making the safety program less reliable.

Safety programs benefit from having both leading and lagging indicators. Leading indicators are harbingers of change. They set goals and measure progress toward those goals. Lagging indicators measure the effectiveness of the overall program. As a safety program evolves over time, trends can be observed over the lifetime of lagging metrics. A correlation between a leading KPI and its related lagging KPI demonstrates the effectiveness of the leading indicators that have been chosen.

There are also some intermediate indicators called coincident KPIs. These fall somewhere between lagging and leading indicators. They can be thought of as real-time measurements. Employee wellness is a good area for coincident indicators, as long as the metrics can be made specific and measurable like every other kind of indicator. As an example, the sign-in sheet for a safety meeting has the worker's name, signature, and then a space for one word to describe how they are feeling that day. The words used are counted into positive, negative, and neutral categories. Over time, these form a trend that can be the basis for further worker well-being initiatives, but on the day the results can be used as a discussion point with the aim of improving the workforce's attitudes.

Quantitative versus Qualitative

Key performance indicators can be expressed as values, percentages, true/false, yes/no, it happened/it didn't happen, and shades of gray.

Quantitative measurements involve a sole number. The number may represent something that is continuous, periodic, or discrete. A continuous example is the record of CO_2 concentrations in the brewery cooler from a datalogging gas monitor. The results are used in real time to determine if a gas leak inspection is warranted. Periodic measurements are just that, they occur on a regular basis according to a schedule. One could be the quarterly inspection of fire extinguishers. The measurement is the percentage of extinguishers that are currently up to date on their certification, with the goal of the measurement to always be 100% compliance. A discrete measurement might be the predicted high temperature for the day. Above a predetermined threshold, the workforce undertakes specific precautions against heat stress.

Qualitative measurements most often are of a yes/no format. Did we accomplish that objective? Did something bad occur? But there are also ranges of agreement with a premise. The Likert scale puts data into discrete bins along a scale (Likert 1932). For example, in assessing safety culture progress and processes, we might ask employees to express their level of agreement with a statement. An example is: "I can comfortably discuss my safety concerns with ownership. __ Strongly Agree, __ Agree, __ Neither Agree nor Disagree, __ Disagree, __ Strongly Disagree."

Inward versus Outward

We can design KPIs to measure internal or external processes or trends. Examples of inward metrics are the number of workers wearing appropriate safety footwear since a footwear allowance was instituted or if the pest control contractor has come every six months as planned by their contract.

The most common outward-facing KPIs deal with meeting of delivery schedules, but can be established for compliance agencies, corporate responsibility, or for other initiatives. In safety programs, one outward metric might be tracking of refusal of service episodes in the taproom. Cutting off customers due to age, behavior, or level of intoxication benefits both public safety and company liability.

Business as Usual versus Attaining New Objectives

We can frame KPIs as either control measurements or as objectives and key results (OKR). As control measurements, we might be checking that we've maintained a status quo. For instance, with the lagging indicator of recordable injury and illness (RII) rate, we could establish a KPI measure to compare each prior year's RII rate for the company with the RII rate for the whole brewing industry. All we really get out of this is a sense of how well we did at staying under OSHA's radar and keeping workers' comp premium multipliers stable. Our real effort here wasn't to innovate, but to make sure we've kept to the status quo.

Objectives and key results are driven by the idea of improving things. This is accomplished by setting milestones that are indicative of the desired change. Using the above example of RII rate, let's say we set an objective of having an annual RII rate that is less than 25% of the national industry average. We frame this by creating KPIs that measure progress of the initiatives necessary to achieve this result. We might prescribe some leading indicator measurements in near miss reporting, SOP development, safety training, and a recognition and reward program. We choose these KPIs because we believe they will collectively measure the conditions that will give rise to our RII rate objective.

KPI APPLICATIONS

With KPIs, we can plan and track aspects of the company's strategic plan by tracking product offerings, sales and marketing, market position, financial metrics, skills development and organizational capacity, and diversity, equity, and inclusion and social good. In some organizational hierarchies, strategic measures would be the province of directors, executives, and departmental managers.

A good place to start in your brewery's safety and health management program is to develop KPIs that move the company forward in all aspects of an effective safety culture. We can look for KPIs that measure trends in the five key drivers of safety culture (*see* chapter 20). These are: expressed management values, communication, accountability, inclusivity, and competence. We can also add a catch-all category, hazard assessment and control, which represents essential safety processes that transcend all five of the culture drivers. Any of these indicators may be inward or outward facing, numerical or heuristic, and leading or lagging. The KPIs shown in the next six subsections are derived primarily from OSHA's bulletin *Using Leading Indicators to Improve Safety and Health Outcomes* (US Department of Labor 2019b).

Management Engagement

One important thing about cultural evolution and change management, is that the company's owners and managers are still in charge. No one is stripping them of their authority. The way a company does things may have to change, depending on the company and the people involved. This change in direction, attitude, and awareness will guide and support institutional change. If the leadership can't improve, it shouldn't be asking for change from the workforce.

In discussing KPIs relevant to management, it comes down simply to how well we are managing and if we are creating and overseeing an environment fruitful for safety improvement. This is done by dedicating the company to the ideas that safety is of great value to everyone's well-being, it makes the company better and more successful, and it requires everyone to participate. To meet those objectives, management pledges to communicate, include, measure, and support safety. Supporting safety involves providing necessary training, monitoring, workplace hazard controls, PPE, and other things that may have a fiscal cost and benefit.

Table 21.2 **Key performance indicators for improving safety management**

Expected leadership practices	
• Make and communicate a commitment to a safe and healthful workplace • Define objectives for the safety and health (S&H) program • Motivate the team • Develop human resources • Monitor, oversee, and set examples for all other areas (communication, accountability, inclusion, competency, and hazard assessment and control)	• Allocate resources toward safety • Set an example for job performance and avoid double standards • Monitor performance and progress on specific initiatives • Devise systems to verify that the S&H program is implemented and functioning • Detect and correct initiative shortcomings, and identify opportunities to improve
KPI examples	
• Average score on survey questions related to workers' perception of management's S&H commitment • Percentage of managers who attend mandatory S&H training for workers • Frequency with which management initiates discussion of a safety and health topic • Number of times management praises safe work performance • Average time between worker report of a hazard or concern and management acknowledgement of the report	• Number of safety-related line items in budget and percentage of these fully funded each year • Number of inspections completed to identify hazards or S&H program weaknesses • Number of lagging indicators that have improved as a result of using leading indicators to take action (e.g., total RIIs, number of OSHA inspections) • Number of goals achieved, number of goals in need of revision, or new goals that you should set for the following year

 KEY ADVICE

These maxims and acronyms track closely with the concept of KPIs. Use them to reinforce KPI principles in your company's safety culture.
  • Plan the work, work the plan.
  • What gets measured gets done.
  • Plan, do, check, act.
  • SMART: specific, measurable, accountable, reasonable, time-bounded
  • Measure twice, cut once.

Communication

Good workplace communication means expressing what is known, expected, or required to perform our work. Communication allows us to state what is unknown and express the risks that come with it. And communication is how we engage in an interpersonal context. This socialization is a key factor that underpins safety culture values.

Good communication is dynamic, meaning anyone can engage in it with anyone else using whatever form is most appropriate. Workers need to be able to speak openly with managers and owners when giving safety input and advice. While there are laws against reprisals for whistleblowing, oftentimes the fear of retribution keeps individuals from sharing safety insights. Workers must be also encouraged and trained by example to communicate peer-to-peer to identify, disclose, and resolve stresses such as worry, pressure, workload, task inequity, and social engagement, to the best of their abilities. It should run vertically and laterally throughout an organization. Good communication means expected performance requirements can be outlined, with publicized rewards and consequences known in advance. Communication can be face-to-face and verbal, or written, visual, or symbolic. What really matters is not the form of communication as much as its effectiveness.

There are often built-in biases that can work against good communication. It is commonly held that two different groups communicate differently, have different levels of intelligence, and possess all sorts of other differences. Just look at the preconceived notions of management versus workforce, brewers versus cellar people, production versus front of house. When healthy and effective communication exists between any such divided population—and there's no reason why it shouldn't—then cohesiveness, productivity, and job satisfaction improve. Skillful communication helps remove doubt and worry, diffuses prejudice and harassment, and is the currency of all forms of training.

Another liability with communication is presuming it happened when it didn't, or not listening carefully and projecting that you heard what you wanted to hear. Be present in communication, listen carefully, express yourself clearly, and know that you will sometimes get it wrong. Miscommunications are a fact of life. It is best to deal with misunderstandings or misstatements in the moment. Clarify or revise the message, express appreciation or regret (as warranted), and then simply move on.

Accountability

Accountability KPIs can be specific to a certain behavior or can be used to summarize the effectiveness or accomplishments of an entire initiative. Most often in safety KPIs, we are interested in (1) did we do what we set out to do on a program level, (2) did individuals do what was desired or required of them, and (3) did we take appropriate action to praise, reprimand, or renew the results?

An importance principle of accountability is avoiding actual or perceived dual standards. If workers have to wear certain PPE in the warehousing area, so do managers when they are in the warehouse. If cellar workers are not permitted to wear shorts, then neither are brewers or warehouse workers.

Inclusion

Worker participation, occupational stress, and suitability for certain roles are good candidates for KPIs. Such measurements will often come from the results of standardized surveys or questionnaires. Human resource professionals and diversity, equity, and inclusion (DEI) officers are good resources for developing tools that can be used in the right setting.

Workers may be most suited for, and happiest with, job assignments that align with their own personality or work style preferences. Inclusion begins with recruiting a diversified workforce in terms of job skills, backgrounds, cultural origins, and gender expressions. Staff should

Table 21.3 **Key performance indicators for improving safety communication**

Expected communication practices	
• Establish, maintain, and exemplify effective communication • Practice active listening • Communicate positive and supporting messages through all levels of the organization, including for policies, procedures, schedules, and performance expectations	• Establish, maintain, and exemplify coordination of tasks to achieve measurable results • Make acknowledgments and corrections when poor communication has occurred
KPI examples	
• Percentage of employees who received orientation training • Number of workers trained on hazard identification and control prior to starting an assignment • Number of concerns, suggestions, or near misses communicated to management by staff • Frequency of public praise for a worker's safe behavior	• Frequency of checking in with others for well-being assessment • Frequency of private communication addressing a worker's unsafe behavior • Frequency of discussions between the brewery employer and contractor addressing the work environment at the brewery

Table 21.4 **Key performance indicators for measuring safety systems**

Expected accountability practices	
• Hold the workforce and management to the same level of accountability • Set goals, support pursuit of those goals, and ensure achievement of goals • Using forward-looking KPIs	• System of praise and discipline exists and is applied consistently • Management lives up to its stated safety values and commitment to safety
KPI examples	
• Number/percentage of identified hazards resolved through hazard assessment, deployment of controls, repairs or replacement, or process substitutions • Number or frequency of near miss reports evaluated and resolved • Frequency of complaints over differential or inequitable treatment of workers • Frequency of utilizing OSHA On-Site Consultation or other third-party program reviewer	• Growth in spending for safety controls like PPE • Growth in per capita hours of safety training logged • Decline of post-incident spending on worker care, replacement, or retraining • Frequency of first aid events and recordable injuries. • Successful completion and/or submittal of essential recordkeeping like OSHA 300 forms, training records, and medical surveillance

have an expressed willingness to communicate with others and operate with safety in mind. Management will set the tone by developing and modeling inquisitive communication and procedurally availing all employees of opportunities to advance, participate in hazard assessment, and be assisted with both hard skills training and being successful in a complex psychosocial setting.

Personalities and, presumably, good job fit can be assessed during recruiting and hiring, with check-ins regularly scheduled. Accommodations should be made for life circumstances, if possible. Occasionally, a business may use a standardized questionnaire to identify desired work style preferences or stressful work conditions. These only measure the individual. They do not generally assess relationships between people, nor do they set company goals for inclusion or provide actionable steps for maintaining an inclusive environment.

Larger brewery businesses may wish to consider discovering an employee's preferences with the Myers-Briggs Type Indicator® (MBTI®) test, or "personality inventory." Although opinions range on the effectiveness or scientific legitimacy of the MBTI test, results do typically align with an individual's self-identity. It may be more helpful as part of a larger program of employee care (Kammonen 2017). There are numerous MBTI-type tests available online for free.

The Job Content Questionnaire (JCQ) has a broader scope than the MBTI test. It also looks at things like a job's physical and psychological demands, decision independence, social support, and job insecurity (Karasek et al. 1998). Key performance indicators specific to worker stress can be derived from scholarly works and with assistance from qualified HR professionals.

In addition to the JCQ, there is the occupational stress indicator (OSI) in the *Occupational Stress Indicator Management Guide* (Cooper, Sloan, and Williams 1988). While it is designed to show how an individual would react to job stress, the OSI overall is "less of an indicator of stressful working conditions and more of an indicator of personality characteristics and personal well-being" (Evers, Frese, and Cooper 2000, 235).

Competence

Competence is a broad term referring to how well we know how to do our job, and in the context of safety, how we do the job safely. To work safely, we require a sufficiency of knowledge, skills, abilities, and behaviors. Training is a big part of competency, but not just the kind of training that is sitting in front of a computer screen for an hour.

An individual's safety skills and knowledge can come from traditional training, apprenticeships, a job history in brewing, a job history in occupations different from brewing, certification programs, committee involvement, and their own personal history of near misses and tragedy. It is important to have hiring policies that

Table 21.5 **Key performance indicators for employee inclusion and wellness**

Expected inclusion and wellness practices	
• Encourage workers to participate in safety and health (S&H) program • Give workers access to S&H information • Engage workers in all aspects of the program	• Provide tools, examples, and processes for equity and inclusion and for the resolution of conflicts related to same • Remove barriers to participation
KPI examples	

• Number of workers asked for feedback on good safety goals ahead of safety meetings
 	
• Number/percentage of workers involved in developing SOPs
 	
• Number/percentage of workers participating in toolbox talks
 	
• Safety perception surveys
 	
  - Rate of participation in such surveys
 	
  - Perception of management's sincerity regarding S&H
 	
  - Scores for stress, anxiety, fatigue, fear of workplace, etc.
	• Number of workers involved in developing task-specific hazard assessments and determination of hazard controls
	
	• Number of workers participating in accident investigation teams and helping to identify/implement corrective actions to eliminate hazards

Table 21.6 **Key performance indicators for employee competence**

Expected Competency Practices	
• Provide safety and health (S&H) program orientation or awareness training • Train workers on their specific roles in the S&H program • Train workers on hazard assessment and hazard controls	• Develop a workforce with relevant industrial safety experience • Provide necessary hazard controls for safety to be achievable
KPI Examples	
• Number of trainings provided on hazard recognition and control compared to attendance rates • Percentage of workers receiving mandatory training on schedule • Percentage of workers receiving refresher training • Percentage of workers receiving HazCom training • Percentage of workers cross-trained to at least one other brewery job • Percentage of workers certified for specific activities, for example, forklifts or confined spaces • Number of safety presentations at conferences attended	• Percentage completed online S&H professional development trainings • Percentage of incident investigations concluded • Number of workers trained on how to recognize and report a hazard or near miss as compared to the number of workers that report understanding the training they have received. • Percentage improvement on post-training assessment versus pre-training assessment scores • Training investment per capita, i.e., total training hrs/yr multiplied by pay rate • Dual-use KPIs between competency and communication, accountability, and inclusion can be developed

identify safety skills required for a job and actively recruit and hire with safety experience as a core requirement.

Competence is most often closely tied with communication, since we learn from being communicated to and teach by communicating with others. Normalized behavior (i.e., bad habits) are learned environmentally from coworkers and our private life and upbringing. Normalization runs contrary to safety competence, and the remediation of normalized behaviors comes from individuals pointing out unsafe behavior to others.

Hazard Assessment and Control

Hazard assessment and hazard control are the core principles behind knowing how to perform a job safely. Chapters 3, 4, and 5 detail this process. Such a critical stage in workplace safety should certainly be able to be measured and good and bad outcomes described.

For hazard assessments to be valuable, we have to conduct them, document them, and develop the findings into useful procedures. These are all countable pursuits. Hazard controls are set based on the findings of hazard assessments, and the rate of completion of enacting controls is also countable.

KPI TIPS FOR SUCCESS

Data Reliability

When planning a measurement program, start with the questions you are trying to answer. Is our incident rate going down? Are we engaging employees in line-level safety decisions? Have we ensured there is no way we will serve caustic line cleaner to a customer?

From the chosen question, ask yourself what or where a meaningful measurement might be. For the above examples these could be tracking the annual OSHA 300 reporting, adding a question to employee's annual performance evaluation, and keeping records of rinse water pH after each cleaning, respectively.

When determining what data to collect, consider the ease and reliability of the data collection process. Except when data collection is automated in some way, KPIs

Table 21.7 **Key performance indicators for hazard assessment and control**

Expected hazard assessment and control practices	
• Collect existing information about workplace hazards	• Develop and update a hazard control plan
• Inspect the workplace for safety hazards	• Identify control options, including prioritized and interim controls
• Identify health hazards that could result in occupational illnesses	• Select controls to protect workers during nonroutine operations and emergencies
• Conduct incident investigations	• Select, implement, and follow up with effective controls
• Identify hazards associated with emergency and non-routine situations	
KPI examples	
• Frequency with which preventive maintenance tasks are initiated and completed on schedule	• Length of time interim controls have been in place
• Percentage of hazard assessments completed compared to a master list	• Percentage of recommendations implemented that pertain to PPE hazard controls, administrative controls, engineering controls, substitution, and elimination
• Percentage of SOPs completed compared to a master list	
• Percentage of incident investigations that include a root cause investigation	• Number of special work permits filled out
• Percentage of daily/weekly/monthly inspections completed	• Number of hazards identified where you used leading indicators to control the hazard
• Percentage of inspections that include a follow-up inspection to ensure that the hazard has been controlled	• Percentage of hazards abated on the same day, week, or month in which the hazard was identified
	• Number of workers required to wear respiratory protection

require a person or persons to record measurements or review records of events. It can be easy for individuals to be too busy, too forgetful, or too distracted to collect meaningful data. They may also simply be reporting the wrong data because they do not understand what is required. Work with those collecting data so that they understand the importance of it. Be sure to share findings and appreciation with them.

Next, determine what metric is relevant to the indicator and whether it is easy to obtain and record. Will measurements be able to be collected for the foreseeable future? If the metric meets these requirements, create a process for collecting, recording, and analyzing the data. This can be done in the data spreadsheet itself or in an SOP or similar document. Schedule data collection in a calendar so you do not miss timely opportunities to collect measurements.

Most smaller businesses take the approach of using spreadsheet applications. Data stored this way rarely encounters storage or compatibility problems. Spreadsheet applications also offer a variety of built-in graphing and data analysis tools. For those who move beyond spreadsheets, there are both generic and specific relational databases on the market that can provide more in terms of data reduction and reporting. Relational databases also make it easier to evaluate relationships between different sets of data. This can reveal underlying correlations that might not have been previously suspected.

How one collects, stores, and interprets this data is as important as the KPIs themselves. Measurement programs span years or decades, and the data needs to be managed as any other important business documents.

Data Sufficiency

A common mistake in establishing KPIs is to collect data that does not readily support useful decisions or identify important trends. Beginning by defining what decision you are trying to make or identifying a trend and its desired direction can help you lay out the data requirements. Collecting data for the sake of it does not benefit anyone, wastes resources, and distracts from efforts to track progress.

In some cases, people will just collect tons of data without a clear sense of what to do with it. Collect the volume of data that is necessary for your purposes and avoid collecting too much or too little. Trends are observed conditions over time. If data is collected without a time, date, or sequence reference, it cannot be used in trend analysis.

Evaluate how difficult the data will be to collect and whether it truly serves a useful purpose. Collect sufficient measurements or data points, not a surplus. Collecting wastewater parameter results daily for a year is likely excessive and expensive if all you are doing is trying to establish your average wastewater loads. Monitoring for three typical days will likely provide equivalent results and will be faster and cheaper. Similarly, a paucity of data will make conclusions unreliable.

Data Collection Design

In designing a measurement program, be careful about being too granular. A fine level of detail may not be as effective as a broad stroke. For example, monitoring the percentage of workers wearing safety glasses in packaging may cause you to miss other areas where workers are not protecting themselves.

Some programs will be designed by an isolated person or group and end up not being representative of the workforce. This is sometimes called the bubble effect. Similarly, focusing on only one area, for instance, fiscal performance, can lead to greater imbalance and uncertainty in safety administration.

Avoid subjective or biased decision rules. Collecting data that is easy to get but does not support the decision being made can only provide misleading results. Do not avoid a particular decision because it reveals there is still a lot of room for improvement. Key performance indicators that inform you when things are out of whack are at least as important as those that tell you are on course.

SDS
STATION

Safety
Data
Sheets

22

SAFETY MANAGEMENT SYSTEMS

The contemporary view of a great company safety program is one in which a system of safety policies, recordkeeping, procedures, legal compliance, hazard control equipment, and persons with designated roles coalesce into an environment with a low risk of injury and property damage. The organization-wide application and continuous review and improvement of a safety program to foster this environment is essentially what defines a *safety management system* (SMS). Safety management systems vary in complexity depending on the size of the company, the company's access to resources, its own level of management bureaucracy, and its dedication to safety as a key business principle (Paoletta 2020; Dağdeviren and Yüksel 2008). An SMS is usually not directly linked to a particular government's requirements and, therefore, works well regardless of which state or province the business is located. However, the SMS must not conflict with the company's regulatory jurisdiction and needs to reference the local requirements directly or as an addendum.

SAFETY MANAGEMENT SYSTEMS BACKGROUND

The history of management systems, and specifically those applied to safety, is colored by the period in which theories were developed and tested, and whether the focus was more on the individual worker or the company administration. The issue of safety has often been split into a binary question: is it up to the employer to leverage resources to manage its legal responsibility; or is it up to the employee, who has to know how to be safe and to behave safely?

Early Safety Management and Behavioral Safety

We were introduced to the notion of incident versus accident in chapter 3. In safety terms, accidents are not random or without fault. Every accident has a root cause, and with a little deduction it is almost always possible to determine the cause or causes. The relationship between observable near misses and the rate of serious and minor injuries in a workplace was explored by H.W. Heinrich, which was elaborated upon further in the 1960s through the 1980s (Heinrich 1931; Bird and Germain 1986). The central tenet of this work is that the number of accidents involving some kind of loss or injury are only a fraction of the number of observable incidents. By studying data sets on chemical and manufacturing industry accident and incident rates, the "safety pyramid" was born, which correlates the number of serious and minor injuries that occur to the number of observable near misses. Figure 22.1 depicts the pyramid as further described in this century.

A major limitation of the safety pyramid is that it does not stipulate whether the root causes of these accidents are related to worker behavior or to the failure or absence of engineering controls. Heinrich and those who furthered his work proposed that worker behavior was responsible for the vast majority of accidents. In direct opposition to Heinrich, W. Edward Deming proposed that almost all

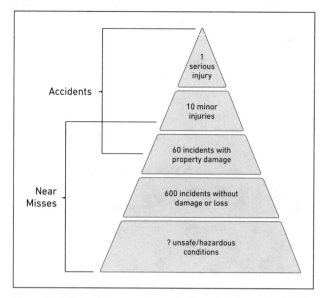

Figure 22.1. The safety pyramid, adapted from Phimister et al. (2003).

workplace accidents were the fault of management not putting into place systems to protect the worker (Smith 2014). This systemic view of safety also has merit and the theory and practice has many proponents (Oedewald 2014). But this reduction to a binary between the worker and management or ownership misses the point. It perpetuates an enduring workplace dividing line, creating a fertile climate for blame and blind adherence to bureaucratic compliance as opposed to the growth and evolution of safety programs that place workers' safety and health at the center.

Over a century ago, Frederick Taylor wrote of *scientific management*, which was based on workers doing what they did best and rewarding them on the basis of their production, rather than basing everyone's performance on the best performer (Taylor 1911). Taylor thought that all workers were motivated by money, so his management style included rewards based on productivity and did not highly value teamwork, worker independence, or business adaptability (Mind Tools Content Team n.d.). Once a task was defined, managers largely stayed out of the mix. Taylor also recommended an early systems approach, which is still referred to as Taylorism. Here are the four principles of his thesis:

- Scientifically study the work and develop the most efficient way to perform a task.
- Assign workers to a task based on their motivation and performance to create a more efficient workforce.
- Monitor individual performance and provide training and guidance to maintain peak performance.

- Specialize tasks so workers work and managers manage.

In these earlier system approaches, there was a lower value put on the individual but also more blame, with less worker safety education and a lower sense of worker safety self-determination (Howe 2001).

So where is the middle ground? We know that workers exist within the sort of complex psychosocial milieu that characterizes any individual place of business. The contemporary model is that worker behavior derives from beliefs based on lived experience, the experiences of peers, and the social context they live and work in. In chapter 18 we looked at behavioral safety, including breaking down cognitive dissonance, learning self-determination to tackle safety problems on the spot, and reinforcing effective peer-to-peer communication. But this can cover up larger problems, much like Heinrich's pyramid failed to address rare and severe events that happened even in safety-driven organizations (Paoletta 2020). Behavioral safety is nowadays a widely applied principle, but if an SMS is entirely focused on the individual, behavior programs rarely succeed. Limitations include mismatched levels of trust and responsibility between management and workers, a higher-than-expected time and resource commitment, and even that the behavior program may run counter to an otherwise well-functioning safety culture (Anderson 2007).

Those who disagree with behavioral safety argue it is clearly easier to formalize safety with a machine guard and good management than expect consistently safe worker behavior. Again, the middle ground eludes us in this hotly debated subject (Frederick and Lessin 2000; McSween 2003; Mitchell n.d.). How, then, should an SMS be run?

Modern Systems Management

In 1918, a physicist named Dr. Walter Shewhart joined the staff of the Western Electric Co. in Illinois. He was working on quality problems with Bell telephone components, and ultimately became famous for what we today call quality control charts (Smith 2009).

Control charts are widely used in brewing for everything from wort extract efficiency to brand flavor analysis to dissolved oxygen content. There is a popular quality management system called Six Sigma that gets its name from the idea of keeping manufacturing tolerances within ±3 standard deviation units (sigma) from the expected measurement (American Society for Quality 2022a).

Importantly, during the 1920s, an intern by the name of W. Edward Deming worked with Shewhart at Western

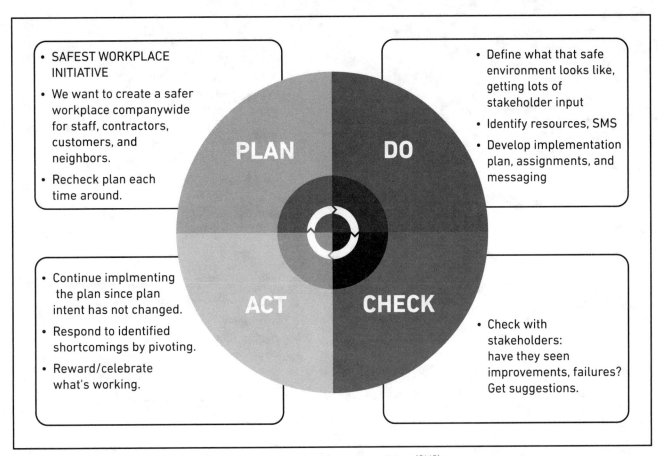

Figure 22.2. The Plan–Do–Check–Act (PDCA) cycle applied to a safety management system (SMS).

Electric. Deming later helped form the American Society for Quality Control, but he was a fairly obscure scientist and lecturer until he went to Japan in 1947. During his time there, he learned about Japanese culture and manufacturing methods. He taught Japanese engineers his theories of scientific management and statistical quality control, and these caught on with Japanese business leaders. Eventually, around 1980, Deming became esteemed in his home country too, long after he had been instrumental in the formation of the modern Japanese quality manufacturing ethic (Columbia School of Business 2022).

Deming emphasized taking the blame off of individuals and examining and repairing failures in the system in a less punitive way. Or, as Smith puts it, "Apparent performance actually is attributable mostly to the interaction between workers and the system" (Smith 2014). In essence, Deming took the best of behavioral science and combined it with Taylor's management theory, Shewhart's quality control, and Japan's continuous improvement ideals to develop the foundation of today's total quality management (TQM) programs.

Continuous Improvement

Safety programs don't materialize out of thin air. They require adaptation to a company's regulatory setting, cultural values, and prioritization of safety concerns, all of which are ever-evolving. The idea of continuous improvement—we often use the Japanese term *kaizen*—requires the detection of deficient processes and a means to effect improvements and test them for effectiveness. Oftentimes, kaizen is described as the cyclical process of Plan–Do–Check–Act (PDCA), sometimes called the Deming Cycle. The PDCA process subsequently spawned Define–Measure–Analyze–Improve–Control (DMAIC). Being quantitative by nature, DMAIC fits well into the Six Sigma approach, while PDCA can fit both quantitative and qualitative contexts.

A great feature of PDCA is that it can be used at any level of detail in an organization. At an executive level it can express the overall way in which company safety goals are described and the general processes needed to get there, as depicted in figure 22.2, for example.

The PDCA process is also great for approaching systemic issues around the plant (fig. 22.3). One outcome

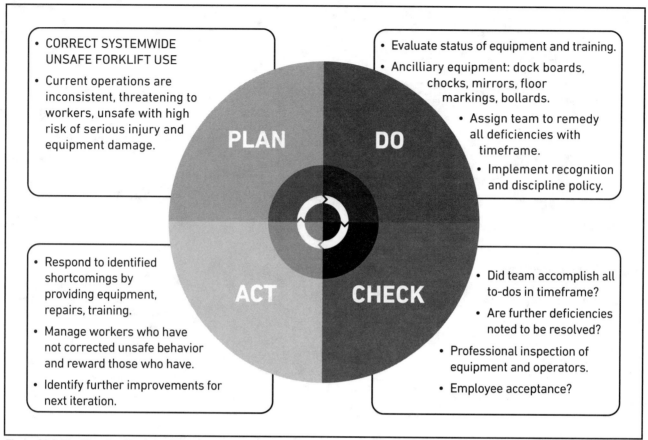

PLAN
- CORRECT SYSTEMWIDE UNSAFE FORKLIFT USE
- Current operations are inconsistent, threatening to workers, unsafe with high risk of serious injury and equipment damage.

DO
- Evaluate status of equipment and training.
- Ancilliary equipment: dock boards, chocks, mirrors, floor markings, bollards.
- Assign team to remedy all deficiencies with timeframe.
- Implement recognition and discipline policy.

ACT
- Respond to identified shortcomings by providing equipment, repairs, training.
- Manage workers who have not corrected unsafe behavior and reward those who have.
- Identify further improvements for next iteration.

CHECK
- Did team accomplish all to-dos in timeframe?
- Are further deficiencies noted to be resolved?
- Professional inspection of equipment and operators.
- Employee acceptance?

Figure 22.3. The Plan–Do–Check–Act (PDCA) cycle applied to a systemic workplace issue, this example being the issue of unsafe forklift use.

of changing a systemic problem is the need to modify the behavior of many people. Be prepared to go through the cycle several times. You may need to be creative, for example, by trying different procedural approaches, changing incentives, escalating rewards and discipline, or trying different hazard controls.

PDCA can also apply on a very task-specific, granular level. If we want to improve the safety of a particular task, we want to know what is keeping it from being safe in the first place, what we can try to make it safer, and assessing if we achieved that. Figure 22.4 shows PDCA applied to a specific brewery task: dry hopping. This example considers a small brewery that is currently dry hopping from a stepladder and the sole cellar operator is very concerned for her safety. She brings this concern to the head brewer, who assures her that together they are going to solve the problem and make the job as safe as possible.

SAFETY MANAGEMENT SYSTEMS

We will now take a systems approach to safety, but bear in mind the previous discussion of worker behavior.

This section will also draw associations with the five key cultural drivers laid out in chapter 20.

It may appear that an SMS is a way of institutionalizing culture. Since safe behavior by the worker is based on values, beliefs, and perceptions, which in turn come from experience and social context, any SMS will still be involving some parts of behavioral safety. But as behavioral safety by itself emphasizes modifying the symptomatic behavior rather than the underlying experiences and socialization, a systems approach is needed to build the foundation of a workplace safety culture. An SMS requires sound business principles, institutional hazard controls, iterative improvement, and a workplace lifestyle of desired behaviors derived from a healthy safety culture.

Safety Management Systems Consensus Standards

There are several common consensus standards that employers can use as resources (table 22.1). These standards can appear bulky and bureaucratic, but there are options to guide you through it, including software

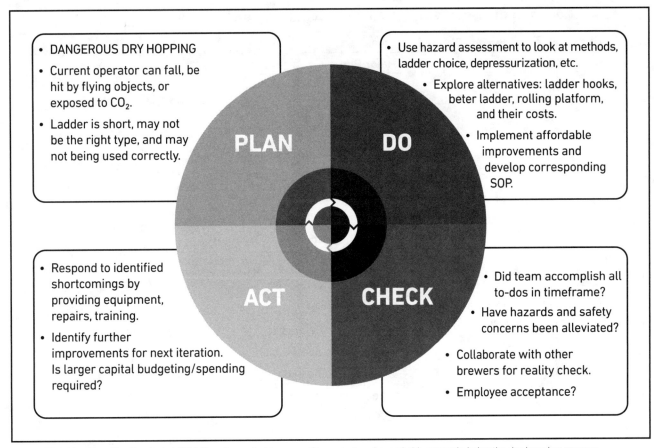

Figure 22.4. The Plan–Do–Check–Act (PDCA) cycle applied to a specific workplace hazard, this example being the dry hopping process.

applications and templates. The benefit of using an established standard is that it simplifies safety auditing and provides more of a readymade roadmap. But, like any company program or policy, employee buy-in and ongoing management and maintenance are a must.

SAFETY MANAGEMENT SYSTEMS FOR A SMALL BREWERY

A rudimentary SMS typically has a minimum of four components to it: a safety mission statement, ongoing risk identification and management, communication and education, and review and improvement. We will look at each of these in turn. How these are recorded, put into action, and adapted over time will depend on an organization's existing cultural expressions of management, communication, and competency. A small brewery that has a written safety mission statement, conducts hazard assessments, develops SOPs, trains its workforce, and checks on the progress of these things has, in effect, the backbone of an SMS, even if it is just known as the company safety program.

An SMS cannot succeed as a document on a shelf; it will only work if the principal drivers of an effective

⚠ CORNERSTONES OF A SAFETY MANAGEMENT SYSTEM

- A determination by ownership to make a safe workplace
- An intent to comply with applicable rules, regulations, and industry best practices
- Identifying and managing hazards in the workplace
- Leveraging resources to implement effective controls
- Building and reinforcing a culture of safety within the workforce
- Enabling improvements and success by monitoring and improving safety processes
- Involving and recognizing individual contributions
- Transmitting awareness of safety programs to stakeholders

Table 22.1 **Common management systems consensus standards**

Directly Applicable to Company Safety Programs		
Reference	**Title**	**Major Components**
ANSI Z10[a]	*Occupational Health and Safety Management Systems*	1. Scope, Purpose, and Application 2. References 3. Definitions 4. Context of the Organization – Strategic Considerations 5. Management Leadership and Worker Participation 6. Planning 7. Support 8. Implementation and Operation 9. Evaluation and Corrective Action 10. Management Review
ISO 45001[b]	*Occupational Health and Safety Management Systems — Requirements with Guidance for Use*	1. Scope 2. Normative References 3. Definitions and Terms 4. Context of the Organization 5. Leadership and Worker Participation 6. Planning 7. Support 8. Operation 9. Performance Evaluation 10. Improvement
OSHA[c]	*Recommended Practices for Safety and Health Programs* Essentially, this is the same as an injury and illness prevention program (IIPP)[d]	1. Management Leadership 2. Worker Participation 3. Hazard Identification and Assessment 4. Hazard Prevention and Control 5. Education and Training 6. Program Evaluation and Improvement 7. Communication and Coordination for Host Employers, Contractors, and Staffing Agencies

safety culture are operational. As a reminder, these are: expressed management intent, communication, accountability, inclusivity, and safety competence. These culture building blocks are detailed in chapter 20.

When looking over the consensus standards and related programs outlined in table 22.1, you might wonder, "Where are all the OSHA rules? Shouldn't this be a list of things we have to do to comply?" These rules and standards are in the SMS or, more precisely, are referenced in the SMS. The SMS is the overall guiding program that specifies your safety goals and the systems you will use to achieve, maintain, and monitor safety success. The safety rules and standards are applied through the framework of your SMS.

Safety Mission Statement

Ownership should use a cross-section of management and the workforce to develop and transmit a safety mission statement. Once developed, this will be first

Table 22.1 **Common management systems consensus standards** (cont.)

Related Systems Management Programs		
Reference	**Title**	**Major Components**
FSMA	21 C.F.R. § 117 "Current Good Manufacturing Practice, Hazard Analysis, and Risk-Based Preventive Controls for Human Food" Rules pertaining to the FDA Food Safety Modernization Act of 2011 (Pub. L. No. 111–353, 124 Stat. 3885).	1. General Provisions 2. Current Good Manufacturing Practice 3. Hazard Analysis and Risk-Based Preventive Controls 4. Modified Requirements 5. Withdrawal of a Qualified Facility Exemption 6. Records to be Established and Maintained 7. Supply-Chain Program
GMPs[e]	Good Manufacturing Processes	1. Plants and Grounds 2. Equipment and Utensils 3. Sanitary Facilities and Controls 4. Processes and Controls 5. Personnel
HACCP[f]	Hazard Analysis and Critical Control Point Principles and Application Guidelines	1. Hazard Analysis 2. Determine Critical Control Points (CCPs) 3. Establish Critical Limits 4. Monitoring Procedures 5. Corrective Actions 6. Verification Procedures 7. Record-keeping and Documentation
Five S[g]	Five S	1. *Seiri* – organize, sort 2. *Seiton* – orderliness, set in order 3. *Seiso* – cleanliness, shine 4. *Seiketsu* – standardize 5. *Shitsuke* – discipline, sustain
Six Sigma[h]	Six Sigma	1. Define 2. Measure 3. Analyze 4. Improve 5. Control

[a] American Society of Safety Professionals (2020); [b] International Organization for Standardization (2018); [c] US Department of Labor (2016b); [d] Safety.BLR.com (2014); [e] Brewers Association (2016); [f] NACMCF (1997); [g] American Society for Quality (2022b); [h] American Society for Quality (2022a).

and foremost in the employee handbook and often conveyed elsewhere throughout the business on posters or the company website. It can be entirely inward facing, directed only at employees, but it can also be outward facing to varying degrees. For example, it might be outward facing to vendors, suppliers, and distributors, or it could be reprinted in the company's annual report or broadcast to everyone on a webpage.

It is best if the statements made are not wistful, but that they state that the company will do certain things.

In this way, the company and its management will be held to account for accomplishing them.

This statement is vital to setting the tone for workplace activities and it is the way that ownership and senior management say that safety is vitally important for the business and the workforce. Stating the company safety intentions is key to ensuring management buy-in and accountability. Furthermore, this can be an important message to transmit to customers, distributors, contractors, and any other stakeholders.

The safety mission statement for any company should be personalized and appropriate with the overall culture of the business. Several examples are provided here to show the wide range of wording. Some are more accountable than others.

Meadowlark Brewing
– Sidney, Montana

A Culture of Safety

Working Environment

Breweries are rife with industrial hazards. We must hold the physical safety of our personnel and patrons as paramount to our bottom line.

Mental Environment

Bigotry and sexual objectification are alienating and divisive. We must provide an atmosphere where everyone can feel welcome to a seat in our public house or a position in our company.

Consumption Practices

Alcohol has a notorious history as a social vice. We must encourage responsible alcohol consumption with respect to local laws and cultural mores.
(Meadowlark Brewing n.d.)

Cape May Brewing
– Cape May, New Jersey

Work Safe, Work Smart:

See something, say something | No shortcuts | No one is above safety
(Cape May Brewing Co. n.d.)

CANarchy Craft Brewery Collective
– Colorado, Utah, Texas, Michigan,
Florida, and North Carolina

It is our policy to provide safe working conditions preventing accidents and injuries. All jobs and tasks must be performed in a safe manner as safety is crucial to the quality of our products and services. Our business operates with a goal of zero damage to people, property and product. At CANarchy we care about the safety, health and well-being of our employees. We value the contributions our employees make toward our success.
(CANarchy 2022)

Heineken N.V.
– Amsterdam, Netherlands

Promoting health & safety

Our goal is simple: zero fatalities, and 'Safety First' is our number one company behaviour.

Why does it matter?

Nothing matters more than the safety of our people. We believe every single person who works for HEINEKEN should benefit from a safe working environment and return home safely at the end of each day. That's why 'Put Safety First' is our number one company behaviour.

Safety standards and regulations vary greatly in different regions around the world and can be a particular challenge in emerging markets. Our job is to apply the same high standards across all our operating companies. The 12 HEINEKEN Life Saving Rules help us to achieve it. They set out clear and simple 'do's and 'don'ts' for our highest-risk activities. These must be followed across all our operations and our companies are required to assess their safety performance against these goals and invest in projects to close any gaps.
(Heineken n.d.)

Hazard Assessment and Hazard Control

The hazard assessment and hazard control part of an SMS should include the following three sections: a stipulated process for conducting hazard assessment (*see* chapter 5), a dynamic list of individual responsibilities, and a compendium of how hazards are managed. An important piece of documentation, this part of the SMS should show the following:

- All parts of the process have been looked at and hazards and compliance issues identified and documented.
- The hazards identified are prioritized and managed using a hierarchy of controls.
- Roles and responsibilities have been assigned.
- Legal requirements in the process are being complied with; evidence of these compliance efforts is enclosed.

Hazard Assessment

The fundamental way the employer creates a safe workplace is to conduct hazard assessments to discover and document hazards. You can't manage what you don't know is there. In your company's SMS there can be a statement that affirms the company aims to reduce workplace hazards and is doing so through this hazard assessment process. The process itself should also be outlined in this section; maybe there's a diagram that depicts the PDCA cycle, for example.

You don't have to go it alone on conducting company-wide hazard assessments. Utilize the resources of OSHA's consultation branch, the local university, an insurance provider, colleagues from other breweries, your state brewers guild, and private safety consultants. Like any big endeavor, seek to break it into parts and then work through the parts while tracking their progress toward completion.

Delegation of Responsibilities

A key part of any safety program is accountability. Who or what team was seeing to a certain safety improvement? What timeline or completion criteria were established? Did the team(s) assigned achieve those goals? If not, how do we get the process back on track? Decide how you want to handle these questions, then document the plans.

Hazard Control Strategies and Compliance

If a company's SMS was in a printed book, this would be the thickest chapter. It would contain the hazard assessment procedures and resources used. It would include documentation of each hazard assessment of every task, and it would have SOPs that describe the best and safest ways to perform those tasks. It would also have all the company's required written programs, like hazard communication, permit-required confined spaces program, hearing conservation program, and so on.

Refer to the following parts in this book to help you understand what you need to do for any particular task or compliance question:

- Table 2.1 in chapter 2, pp. 14–15
- The appendix "Job Activity Hazards Index" on p. 304
- All of section 2, where compliance requirements and specific hazards and effective controls are described (pp. 55–259)

Look for additional help and resources, such as OSHA consultation services, safety presentations at conferences, listening to podcasts, and talking with other brewers.

An exceptional resource is the NIOSH/OSHA publication *Small Business Safety and Health Handbook* (National Institute for Occupational Safety and Health 2022). It is loaded with easy-to-navigate checklists and much less emphasis on the chapter and verse of regulatory requirements. Every brewery, regardless of size, would be well advised to go through the checklists. This exercise will help you discover where you really need work, where you are ahead of the game, and get you closer to full compliance without having to read the *Code of Federal Regulations*.

Communication and Training

Communication

Of the five key cultural drivers, good all-around communication is often the most evasive. Communication comes in many forms. It must include top-down safety messaging from ownership, lateral communication between coworkers about how to do a job safely, and recordkeeping to document operational procedures, training, injury reporting, and process improvements.

First and foremost, good communication begins by example from ownership and senior management. If the brewery owner or a manager yells angrily at an employee for a mistake, it is likely that employee will not want to communicate back and will also tell other employees about their negative experience. A manager who can give constructive advice and asks the employee to explain it back to them is building trust. Safety culture demands effective communication.

Presuming you've begun to set good examples and have communicated a company safety mission statement, the next step is to deploy information about *how* to stay safe according to company policies. Ways to do this include SOPs, familiarity with SDSs, developing and using checklists, and casual check-ins. A check-in might mean asking a coworker if they have looked for mold in the ice maker recently or one worker telling another, "Hey, I had a free minute, so I vacuumed the mill room." Do not leave things to telepathy.

Equally important to transmitting information is receiving feedback. Good communication requires good listening. When a near miss is identified, even if your brewery does not have a formal procedure, the manager should make a note and give the employee feedback that it will be first on the agenda at the next staff meeting. Trust is built when sound actions follow the feedback that good communication engenders.

Some companies have found it useful to bring in a communications expert, and this doesn't have to be a stuffy or expensive process. Find a successful business-person in your area that the workforce can relate to and engage them to come in for a casual chat at a time when all the staff can be there. Ideas include a local media personality, a successful entrepreneur, or a well-known sports coach. There are, of course, consultants that specialize in communication. Find someone who can discuss what makes communication work, explain why it is vital to business (and safety) success, and provide some examples of communication pitfalls. Training like this gives people permission to accept that they need improvement and opens a dialogue in the workforce specifically about communication.

Training

Another tenet of a successful safety culture is competence. Do workers know how to identify hazards, eliminate them, keep others safe, and comply with requirements? This knowledge comes from many sources, as outlined in chapter 3: on-the-job training, safety courses, mentorship, etc. Communication is vital to competence, because we are, or should be, constantly discussing our work with colleagues.

What is the content of the training we all need? In production terms, it is about how to do the job at hand in a way that is productive, cost-effective, doesn't cause equipment breakdowns, and meets quality and sustainability needs. But let's not leave safety off this list. For many companies, safety is carved out separately, where safety training is all about safety and not so much about all of the other production aspects. This creates a dynamic where safety has its own communication, or lack thereof. When training is focused on a task and safety is built into the conversation, communicating about safety becomes a more natural part of workplace dialogue.

Review and Improvement

Any SMS will employ a feedback cycle that (1) promotes continuous improvement by detecting unsafe task aspects and (2) improves safety by utilizing a hierarchy of controls. In its simplest terms, this process is PDCA. It will necessarily involve programmed use of the hazard assessment process and developing consistent procedures in the form of SOPs.

Culture and technical sophistication change over time, and so will the SMS. As with any system needed to operate a successful brewery, the SMS will need review and modification. Where management is supportive of creating a safe and healthy workplace, these changes are generally in the direction of a safer workplace, more control over unexpected events, and better cost efficiencies. Investing in existing human resources and bringing in new people with diversified experiences fuels this evolution. Change, in this regard, need not be difficult if some structure for an SMS is already in place.

APPENDIX
JOB ACTIVITY
HAZARDS INDEX

You may want to create a reading list or training regimen for workers in a certain department. What are all the typical hazards encountered by a canning line operator or a warehouse person or a cook? Search the following chart for the job activity to find the most important sections on relevant hazards and their controls. This chart can also be helpful when creating orientation or training agendas according to a worker's expected duties.

Table A.1 **Job Activity Hazards Index**

Activity Class	Job Assignment	Company Safety Culture Orientation, Statement, Or Employee Handbook	Emergency Action: Egress, Fire, Injury, Disaster, Bad Actor	Housekeeping, Walking And Working Surfaces, Sanitation	Hazard Awareness For Non-Production Staff & Tours	Hazard Assessment And Standard Operating Procedures	Hazards are an Exchange of Energy	Repair And Maintenance, Portable Tools	Manual Materials Handling	Powered Industrial Trucks, Forklifts, Powered Lifts (if present)
See Chapter:		8	8	8	8	3, 4, 5	7	8	11	11
Production	Prod./Ops. Mgr.	X	X	X		X	X	X	X	X
	Principal Brewer	X	X	X		X	X	X	X	X
	Lead Brewer/Shift Brewer	X	X	X				X	X	X
	Cellar Mgr./Person	X	X	X				X	X	X
	Filtration/Carbonation	X	X	X				X		
	Lab/Quality Mgr. or Tech.	X	X	X		X	X	X		
	EHS/Sustainability Mgr.	X	X	X		X	X	X	X	X
	Safety Team Member	X	X	X		X	X	X	X	X
Packaging	Packaging Mgr./Lead	X	X	X		X	X	X	X	X
	Packaging Tech.	X	X	X					X	X
	Manual Packaging Tech.	X	X	X					X	
Warehousing	Warehouse/Logistics Mgr.	X	X	X		X	X	X	X	X
	Warehouse Tech.	X	X	X					X	X
Maintenance	Maintenance/Trades Mgr.	X	X	X		X	X	X	X	X
	Trades Tech./Welder	X	X	X				X	X	
	Janitor/Grounds	X	X	X				X	X	
Front Of House & Other Retail	General Mgr.	X	X	X	X	X	X	X	X	
	Bartender	X	X	X	X				X	
	Server, Busser	X	X	X					X	
	Host, Hostess, Clerk	X	X	X						
	Door Security	X	X	X						
	Draught System Tech.	X	X	X						
	Head Chef, Kitchen Mgr.	X	X	X	X	X	X		X	
	Line Cook	X	X	X	X					
	Dishwasher	X	X	X	X				X	
	Festival Server	X	X						X	
Ownership & Administration	Managing Owner	X	X		X	X	X			
	Senior Mgr.	X	X		X	X	X			
	HR/DEI/Legal	X	X							
	Office/Finance Admin.	X	X							
	Sales/Marketing/PR/Tours	X	X		X					

Table A.1 **Job Activity Hazards Index** (cont.)

Activity Class	Job Assignment	Personal Protective Equipment, Respiratory Protection (if needed)	Documentation: Injury Reporting, Medical Records, Training	First Aid, CPR, and AED Training	Lacerations, Abrasions, Punctures, Avulsions, Amputations	Musculoskeletal Disorders: Strains, Sprains, Back Injury, Repetitive Motion, Posture, Arthritis, Hernias	Slips, Trips, And Falls From Ground Level	Falls from Height and Falling Objects	Engulfment (if silos present), Grain Handling, Dust Explosion Hazard	Noise, Hearing Protection (if needed)
See Chapter:		9	2, 22	8	10	10	10	10	11	10
Production	Prod./Ops. Mgr.	X	X		X	X	X	X	X	X
	Principal Brewer	X	X		X	X	X	X	X	X
	Lead Brewer/Shift Brewer	X			X	X	X	X	X	X
	Cellar Mgr./Person	X			X	X	X	X		X
	Filtration/Carbonation	X			X	X	X	X		X
	Lab/Quality Mgr./Tech.	X	X		X	X	X	X		X
	EHS/Sustainability Mgr.	X	X	X	X	X	X	X	X	X
	Safety Team Member	X	X	X	X	X	X	X	X	X
Packaging	Packaging Mgr./Lead	X	X		X	X	X	X		X
	Packaging Tech.	X			X	X	X			X
	Manual Packaging Tech.	X			X	X	X			X
Warehousing	Warehouse/Logistics Mgr.	X	X		X	X	X	X		
	Warehouse Tech.	X			X	X	X	X		
Maintenance	Maintenance/Trades Mgr.	X	X		X	X	X	X		
	Trades Tech./Welder	X			X	X	X	X		
	Janitor/Grounds	X			X	X	X	X		
Front Of House & Other Retail	General Mgr.	X	X	X	X	X	X			
	Bartender	X			X	X	X			
	Server, Busser	X			X	X	X			
	Host, Hostess, Clerk				X	X	X			
	Door Security				X	X	X			
	Draught System Tech.				X	X	X			
	Head Chef, Kitchen Mgr.	X			X	X	X			
	Line Cook	X			X	X	X			
	Dishwasher	X			X	X	X			
	Festival Server				X	X	X			
Ownership & Administration	Managing Owner		X							
	Senior Mgr.		X							
	HR/DEI/Legal		X							
	Office/Finance Admin.									
	Sales/Marketing/PR/Tours									

Table A.1 **Job Activity Hazards Index** (cont.)

Activity Class	Job Assignment	Thermal: Burns, Boilovers, Heat Disorders	Motion Hazards: Machinery, Vehicles, Flying Objects	Pressurized Systems: Compressed Gases, Pneumatic, Hydraulic, Dry Hopping	Electrical: Shock, Electrocution, Arc Flash	Flammability and Explosivity	Radiation: Non-Ionizing, Hazardous Light Energy, and Ionizing (if present)	Hazard Communication (HazCom), Chemical Terms, Safety Data Sheets, Compatability, Spills	Corrosives: Acids, Bases, Wastewater	Oxidizers, Disinfectants, Sanitizers, and Cleaners
	See Chapter:	14	11	12	15	13	15	16	16	16
Production	Prod./Ops. Mgr.	X	X	X	X	X	X	X	X	X
	Principal Brewer	X	X	X	X	X	X	X	X	X
	Lead Brewer/Shift Brewer	X	X	X					X	X	X
	Cellar Mgr./Person			X				X	X	X
	Filtration/Carbonation		X	X				X	X	X
	Lab/Quality Mgr. or Tech.			X		X		X	X	X
	EHS/Sustainability Mgr.	X	X	X	X	X	X	X	X	X
	Safety Team Member	X	X	X	X	X	X	X	X	X
Packaging	Packaging Mgr./Lead		X	X	X		X	X	X	X
	Packaging Tech.		X	X			X	X	X	X
	Manual Packaging Tech.		X							
Warehousing	Warehouse/Logistics Mgr.		X		X					
	Warehouse Tech.		X							
Maintenance	Maintenance/Trades Mgr.	X	X	X	X	X		X		
	Trades Tech./Welder	X	X	X	X	X		X		
	Janitor/Grounds	X	X					X		
Front Of House & Other Retail	General Mgr.				X			X		
	Bartender									
	Server, Busser									
	Host, Hostess, Clerk									
	Door Security									
	Draught System Tech.									
	Head Chef, Kitchen Mgr.	X								
	Line Cook	X								
	Dishwasher									
	Festival Server									
Ownership & Administration	Managing Owner							X		
	Senior Mgr.							X		
	HR/DEI/Legal									
	Office/Finance Admin.									
	Sales/Marketing/PR/Tours									

Table A.1 **Job Activity Hazards Index** (cont.)

Activity Class	Job Assignment	Other Chemicals: Solvents, Process Aids, Laboratory	Facility Chemicals: Boiler Additives, Refrigerants, Kitchen/Bar, Maintenance	Human and Food Transmitted Illnesses	Animal and Plant Illnesses and Allergies	Individual and Group Dysfunctions in Behaviors and Beliefs	Individual Psychological Factors, Neurodiversity, Designations, and Substance Use Disorders	Workplace Interactions, Toxic Workplace, Disempowerment	Societal Dysfunction, Prejudice, Violence	Permit-Required Confined Spaces
	See Chapter:	16	16	17	17	18	18	18	18	19
Production	Prod./Ops. Mgr.	X	X	X	X	X	X	X	X	X
	Principal Brewer	X	X	X	X	X	X	X	X	X
	Lead Brewer/Shift Brewer	X								X
	Cellar Mgr./Person	X								X
	Filtration/Carbonation	X								X
	Lab/Quality Mgr. or Tech.	X		X	X					
	EHS/Sustainability Mgr.	X		X	X	X	X	X	X	X
	Safety Team Member	X		X	X	X	X	X	X	X
Packaging	Packaging Mgr./Lead					X	X	X	X	X
	Packaging Tech.									
	Manual Packaging Tech.									
Warehousing	Warehouse/Logistics Mgr.					X	X	X	X	
	Warehouse Tech.									
Maintenance	Maintenance/Trades Mgr.		X	X	X	X	X	X	X	X
	Trades Tech./Welder		X		X					X
	Janitor/Grounds		X	X	X					
Front Of House & Other Retail	General Mgr.		X	X	X	X	X	X	X	X
	Bartender		X							
	Server, Busser									
	Host, Hostess, Clerk									
	Door Security									
	Draught System Tech.									
	Head Chef, Kitchen Mgr.			X	X					
	Line Cook			X	X					
	Dishwasher									
	Festival Server									
Ownership & Administration	Managing Owner					X	X	X	X	X
	Senior Mgr.					X	X	X	X	X
	HR/DEI/Legal					X	X	X	X	
	Office/Finance Admin.									
	Sales/Marketing/PR/Tours									

Table A.1 **Job Activity Hazards Index** (cont.)

Activity Class	Job Assignment	Control of Hazardous Energy (Lockout/Tagout)	Safety Culture	Goal Setting and Progress Measurements	Safety Management Systems	Suggested Training or Certification Not Covered in This Book
	See Chapter:	19	1, 20	21	2, 22	
Production	Prod./Ops. Mgr.	X	X	X	X	ServSafe Allergens recommended
	Principal Brewer	X	X	X	X	ServSafe Allergens recommended
	Lead Brewer/Shift Brewer	X				ServSafe Allergens recommended
	Cellar Mgr./Person	X				
	Filtration/Carbonation	X				
	Lab/Quality Mgr. or Tech.		X	X	X	
	EHS/Sustainability Mgr.	X	X	X	X	
	Safety Team Member	X	X	X	X	
Packaging	Packaging Mgr./Lead	X	X	X	X	
	Packaging Tech.	X				
	Manual Packaging Tech.					
Warehousing	Warehouse/Logistics Mgr.		X	X	X	
	Warehouse Tech.					
Maintenance	Maintenance/Trades Mgr.	X	X	X	X	
	Trades Tech./Welder	X				
	Janitor/Grounds					
Front Of House & Other Retail	General Mgr.					ServSafe Alcohol or TIPS recommended
	Bartender					ServSafe Alcohol or TIPS recommended
	Server, Busser					ServSafe Alcohol or TIPS recommended
	Host, Hostess, Clerk					ServSafe Alcohol or TIPS recommended
	Door Security					ServSafe Alcohol or TIPS recommended
	Draught System Tech.					Brewers Association Draught Safety
	Head Chef, Kitchen Mgr.					ServSafe Manager recommended
	Line Cook					ServeSafe Food Handler recommended
	Dishwasher					
	Festival Server					ServSafe Alcohol or TIPS recommended
Ownership & Administration	Managing Owner	X				
	Senior Mgr.	X				
	HR/DEI/Legal		X	X	X	
	Office/Finance Admin.					
	Sales/Marketing/PR/Tours					

BIBLIOGRAPHY

ACGIH. 2021. *TLVs and BEIs Based on the Documentation of Threshold Limit Values for Chemical Substances and Physical Agents and Biological Exposure Indices.* Cincinnati: American Conference of Governmental Industrial Hygienists.

Ahsan, H., A. Ali, and R. Ali. 2003. "Oxygen free radicals and system autoimmunity." *Clinical and Experimental Immunology* 131(3): 398–404.

Aittomäki, A., E. Lahelma, and E. Roos. 2003. "Work conditions and socioeconomic inequalities in work ability." *Scandinavian Journal of Work, Environment & Health* 29(2): 159–165.

Akiomi, I., H. Eguchi, Y. Kachi, S. McLinton, M. Dollard, and A. Tsutsumi. 2021. "Reliability and Validity of the Japanese Version of the 12-Item Psychosocial Safety Climate Scale (PSC-12J)." *International Journal of Environmental Research and Public Health* 18(24): 12954. https://doi.org/10.3390/ijerph182412954.

American Cleaning Institute. 2022. *Quaternary Ammonium Compounds: FAQ on Common Disinfectant Ingredients.* Accessed January 12, 2022. https://www.cleaninginstitute.org/understanding-products/disinfectants /quaternary-ammonium-compounds-faq-common-disinfectant.

American Optometric Association. n.d. "Protecting your eyes at work." Accessed September 19, 2022. https://www.aoa.org/healthy-eyes/caring-for-your-eyes/protecting-your-vision.

American Psychological Association. 2022. *Work, Stress, and Health & Socioeconomic Status.* Last updated April 2022. https://www.apa.org/pi/ses/resources/publications/work-stress-health.

American Society for Quality. 2022a. "What is Six Sigma." Last accessed July 9, 2022. https://asq.org/quality-resources /lean/five-s-tutorial.

American Society for Quality. 2022b. "What Are the Five S's (5S) of Lean." Last accessed July 11, 2022. https://asq.org/quality-resources/lean/five-s-tutorial.

American Society of Safety Professionals. 2020. *Occupational Health and Safety Management Systems.* ANSI/ASSP Z10.0-2019. Park Ridge, IL: ASSP, approved August 22, 2019.

American Society of Safety Professionals. 2021. *Prevention through Design Guidelines for Addressing Occupational Hazards and Risks in Design and Redesign Processes.* ANSI/ASSP Z590.3-2021. Park Ridge, IL: ASSP.

Anderson, Martin. 2007. "Human factors: Behavioural safety approaches - an introduction (also known as behaviour modification)." Health and Safety Executive. Last accessed July 9, 2022. https://www.hse.gov.uk/humanfactors /topics/behaviouralintor.htm.

Arturson, G. 1980. "Pathophysiology of the Burn Wound." *Annales Chirurgiae et Gynaecologiae* 69(5): 178–90.

Austenitex. n.d. "Sanitary Tri-Clamp® Gaskets." Last accessed November 11, 2022. https://www.austenitex.com /shop_by_category/SBC-Tri-Clamp-Gaskets.

Babitsch, B., L. Bretz, H. Mansholt, and N.A. Götz. 2020. "The relevance of cultural diversity on safety culture: a CIRS data analysis to identify problem areas and competency requirements of professionals in healthcare institutions." *GMS Journal for Medical Education* 37(2): Doc14. https://doi.org/10.3205/zma001307.

Barnes, Valerie. 2010. "What is Safety Culture? Theory, Research, Challenges." PowerPoint slides presented at US Nuclear Regulatory Commission Meeting to Reach Alignment on Safety Culture Definition, Concepts and Description/Traits, Rockville, Maryland, February 2, 2010. https://www.nrc.gov/about-nrc/regulatory/enforcement /barnes.pdf.

Bates, M. n.d. "Dyslexia Statistics and Myth Busting Facts." The Reading Well. Last accessed August 20, 2022. https://www.dyslexia-reading-well.com/dyslexia-statistics.html.

Baumer, N., and J. Frueh. 2021. "What is Neurodiversity?" Harvard Health Publishing. November 23, 2021. https://www.health.harvard.edu/blog/what-is-neurodiversity-202111232645.

Bauto, Loida. 2022. "A Guide to the Hierarchy of Controls." SafetyCulture. December 2, 2022. https://safetyculture.com/topics/hierarchy-of-controls/.

BioExplorer. 2021. "Top 14 Most Infectious and Deadliest Diseases Caused by Bacteria." BioExplorer.Net. Last updated June 12, 2021. https://www.bioexplorer.net/bacterial-diseases.html/.

Bird, Frank E., and George L. Germain. 1986. *Practical Loss Control Leadership.* Loganville, GA: International Loss Control Institute.

Blank, R., A.L. Barnett, J. Cairney, D. Green, A. Kirby, H. Polatajko, et al. 2019. "International Clinical Practice Recommendations on the Definition, Diagnosis, Assessment, Intervention, and Psychosocial Aspects of Developmental Coordination Disorder." *Developmental Medicine and Child Neurology* 61(3): 242–85. https://doi.org/10.1111/dmcn.14132.

Brewers Association. 2015. *Best Management Practice for the Management of Confined Spaces in Breweries.* Educational Publications. Brewers Association Safety Subcommittee, April 23, 2015. https://www.brewersassociation.org /educational-publications/confined-spaces/.

Brewers Association. 2016. *Good Manufacturing Practices for Craft Brewers.* March 24, 2016. https://www.brewersassociation.org/educational-publications/good-manufacturing-practices-for-craft-brewers/.

Brewers Association. 2021a. "Brewery Pressure-Rated Vessels FAQ." Industry Updates. Brewers Association Engineering Subcommittee, June 16, 2021. https://www.brewersassociation.org/brewing-industry-updates /brewery-pressure-rated-vessels-faq/.

Brewers Association. 2021b. *Preventing Kettle Boilovers*. Best Practice Guidance. Brewers Association Safety Subcommittee, August 11, 2021. https://www.brewersassociation.org/educational-publications /best-practice-guidance-for-preventing-kettle-boilovers/.

Brewers Association. 2022. "Don't Let the Heat Stress You Out: Tips for Keeping Employees Safe." Industry Updates. Brewers Association Safety Subcommittee, June 22, 2022. https://www.brewersassociation.org /brewing-industry-updates/dont-let-the-heat-stress-you-out-tips-for-keeping-your-cool/.

Broussard, Ian M., and Chadi I. Kahwaji. 2022. "Universal Precautions." StatPearls. Updated September 1, 2022. Treasure Island, FL: StatPearls Publishing; January 2022. https://www.ncbi.nlm.nih.gov/books/NBK470223/.

Bureau of Labor Statistics. 2005. "Sprains and strains most common workplace injury." *The Economics Daily*, April 1, 2005. US Department of Labor. Accessed August 7, 2022. https://www.bls.gov/opub/ted/2005/mar/wk4/art05.htm.

Bureau of Labor Statistics. 2018. "Back injuries prominent in work-related musculoskeletal disorder cases in 2016." *The Economics Daily*, August 28, 2018. US Department of Labor. Accessed February 22, 2023. https://www.bls.gov /opub/ted/2018/back-injuries-prominent-in-work-related-musculoskeletal-disorder-cases-in-2016.htm.

Bureau of Labor Statistics. 2020a. "Fatal work injuries from unintentional overdose increased on average 24 percent per year, 2011–18." *The Economics Daily*, August 31, 2020. US Department of Labor. Accessed January 14, 2022. https://www.bls.gov/opub/ted/2020/fatal-work-injuries-from-unintentional-overdose-increased-on-average-24 -percent-per-year-2011-18.htm.

Bureau of Labor Statistics. 2020b. "Fact Sheet | Occupational injuries and illnesses resulting in musculoskeletal disorders (MSDs) | May 2020." *Injuries, Illnesses, and Fatalities*. Last accessed May 22, 2022. https://www.bls.gov/iif /oshwc/case/msds.htm.

Bureau of Labor Statistics. n.d. "IIF Databases." Accessed May 22, 2022. https://www.bls.gov/iif/data.htm.

Canadian Centre for Occupational Health and Safety. 2022. "Aging Workers." Last accessed May 12, 2022. https://www.ccohs.ca/oshanswers/psychosocial/aging_workers.html.

CANarchy. 2022.

Cape May Brewing Co. n.d. "Our Core Values." Accessed July 11, 2022. https://capemaybrewery.com/about-us /our-values.

Centers for Disease Control and Prevention (CDC). 2006. *Principles of Epidemiology in Public Health Practice: An Introduction to Applied Epidemiology and Biostatistics*. SS1978. Updated May 2012. Atlanta: U.S. Department of Health and Human Services. https://www.cdc.gov/csels/dsepd/ss1978/ss1978.pdf.

Centers for Disease Control and Prevention. 2019. "Types of Fungal Diseases." Last reviewed May 6, 2019. https://www.cdc.gov/fungal/diseases/index.html.

Centers for Disease Control and Prevention. 2020a. "Work-Related Musculoskeletal Disorders & Ergonomics." Last reviewed February 12, 2020. https://www.cdc.gov/workplacehealthpromotion/health-strategies/musculoskeletal-disorders/index.html.

Centers for Disease Control and Prevention. 2020b. "Osteoarthritis (OA)." Last reviewed July 27, 2020. https://www.cdc.gov/arthritis/basics/osteoarthritis.htm.

Centers for Disease Control and Prevention. 2020c. "Four Steps to Food Safety: Clean, Separate, Cook, Chill." Last reviewed March 3, 2020. https://www.cdc.gov/foodsafety/keep-food-safe.html.

Centers for Disease Control and Prevention. 2021. "Health Effects of Cigarette Smoking." Last reviewed October 29, 2021. https://www.cdc.gov/tobacco/data_statistics/fact_sheets/health_effects/effects_cig_smoking/index.htm.

Centers for Disease Control and Prevention. 2022. "Gout." Last reviewed April 28, 2022. https://www.cdc.gov/arthritis/types/gout.html.

Centers for Disease Control and Prevention. n.d. *Picture of America Heat-Related Illness Fact Sheet.* Accessed May 22, 2022. https://www.cdc.gov/pictureofamerica/pdfs/picture_of_america_heat-related_illness.pdf.

Cleveland Clinic. 2022. "Neurodivergent." Last reviewed June 2, 2022. https://my.clevelandclinic.org/health/symptoms/23154-neurodivergent.

Cleveland Clinic. 2023. "Hernia." Last reviewed February 2, 2023. https://my.clevelandclinic.org/health/diseases/15757-hernia.

Colligan, M.J., and A. Cohen. 2004. "The role of training in promoting workplace safety and health." In *The Psychology of Workplace Safety*, edited by J. Barling & M. R. Frone, 223–248. Washington DC: American Psychological Association.

Columbia School of Business. 2022. "Deming Philosophy and Principles: About Dr. W. Edwards Deming." The W. Edwards Deming Center for Quality, Productivity, and Competitiveness. Last accessed July 9, 2022. https://business.columbia.edu/demingcenter/about/deming-philosophy-and-principles.

Cooper, C.L., S.J. Sloan, and S. Williams. 1988. *Occupational Stress Indicator Management Guide.* Windsor: NFER-Nelson.

Cooper, J.S., P. Phuyal, and N. Shah. 2022. "Oxygen Toxicity." In: StatPearls [Internet]. Treasure Island (FL): StatPearls Publishing, January 2022. Updated August 10, 2022. https://www.ncbi.nlm.nih.gov/books/NBK430743.

Cornell Law School. n.d. "29 CFR Part 1910 - Occupational Safety and Health Standards." Legal Information Institute. Accessed July 6, 2020. https://www.law.cornell.edu/cfr/text/29/part-1910.

D&D Engineered Products. n.d. *Sanitary Tri-Clamp® Gaskets.* Last accessed November 11, 2022. https://ddenginc.com/wp-content/uploads/sanitary-tri-clamp-gaskets.pdf.

Dağdeviren, Metin, and İhsan Yüksel. 2008. "Developing a fuzzy analytic hierarchy process (AHP) model for behavior-based safety management." *Information Sciences* 178(6): 1717–33.

Danielson, M.L., R.H. Bitsko, R.M. Ghandour, J.R. Holbrook, M.D. Kogan, and S.J. Blumberg. 2018. "Prevalence of Parent-Reported ADHD Diagnosis and Associated Treatment Among U.S. Children and Adolescents." *Journal of Clinical Child and Adolescent Psychology* 47(2): 199–212.

Darley, J. M., and B. Latané. 1968. "Bystander intervention in emergencies: Diffusion of responsibility." *Journal of Personality and Social Psychology* 8 (4, Pt. 1): 377–83.

Doyle, N. 2020. "Neurodiversity at Work: A Biopsychosocial Model and the Impact on Working Adults." *British Medical Bulletin* 135(1): 108–25. https://doi.org/10.1093/bmb/ldaa021.

Druley, Kevin. 2018. "The ROI of Safety: What to consider when analyzing the economic benefits of safety." *Safety+Health*. December 20, 2018. https://www.safetyandhealthmagazine.com/articles/17819-the-roi-of-safety.

Druley, Kevin. 2020. "Diversity, equity and inclusion in the workplace: 'A safety issue.'" *Safety+Health*, September 27, 2020. https://www.safetyandhealthmagazine.com/articles/20307-diversity-equity-and-inclusion-in-the -workplace-a-safety-issue.

Elsabbagh, M., G. Divan, Y. Koh, Y.S. Kim, S. Kauchali, C. Marcín, et al. 2012. "Global Prevalence of Autism and Other Pervasive Developmental Disorders." *Autism Research* 5(3): 160–79.

Embry, Elizabeth and Matt Stinchfield. 2020. "Praise and Paradox: What We Learned from the Brewers Association Safety and Injury Survey." Online seminar presented for the Craft Brewers Conference® and BrewExpo America® Online 2020. https://www.brewersassociation.org/seminars/praise-and-paradox-what-we-learned-from-the -brewers-association-safety-and-injury-survey/.

Embry, Elizabeth, Jessica Rollings-Taylor, and Jeffrey York. Unpublished. "Health and Safety Practices in the Craft Beer Industry: A Survey of Brewery Management from Members of the Brewers Association." Unpublished manuscript, last saved March 3, 2021. PDF from Microsoft Word file.

Engineers Edge. n.d. "Particle Size and Distribution Air / Fluid Filter - Filtration." Last accessed July 13, 2022. https://www.engineersedge.com/filtration/filtration_particle_size.htm.

Environmental Protection Agency. 2022. "Integrated Pest Management (IPM) Principles." Last updated August 2, 2022. https://www.epa.gov/safepestcontrol/integrated-pest-management-ipm-principles.

European Dyslexia Association. n.d. "What is Dyslexia." Last accessed August 20, 2022. https://eda-info.eu /what-is-dyslexia/.

Evers, A., M. Frese, and C.L. Cooper. 2000. "Revisions and further developments of the Occupational Stress Indicator: LISREL results from four Dutch studies." *Journal of Occupational and Organizational Psychology* 73(2): 221–40.

Ferguson, Alan. 2019. "OSHA's General Duty Clause." *Safety+Health*, December 20, 2019. https://www.safetyandhealthmagazine.com/articles/19258-oshas-general-duty-clause.

Ferrari, Luca, Michele Carugno, and Valentina Bollati. 2019. "Particulate matter exposure shapes DNA methylation through the lifespan." *Clinical Epigenetics* 11:129. https://doi.org/10.1186/s13148-019-0726-x.

Food Safety and Inspection Service. 2008. *Kitchen Companion: Your Safe Food Handbook.* Revised March 2015. US Department of Agriculture. https://www.fsis.usda.gov/sites/default/files/media_file/2020-12/Kitchen -Companion.pdf.

Fox, M.A., R.T. Niemeier, N. Hudson, M.R. Siegel, and G.S. Dotson. 2021. "Cumulative Risks from Stressor Exposures and Personal Risk Factors in the Workplace: Examples from a Scoping Review." *International Journal of Environmental Research and Public Health* 29(18): 5850. https://doi.org/10.3390/ijerph18115850.

Frederick, James, and Nancy Lessin. 2000. "Blame the worker: The rise of behavioral-based safety programs." *Multinational Monitor* 21(11): 10.

Gacioch, Matt, and Matt Stinchfield. 2022. "Track & Yield: A Winner's Guide to Safety and Sustainability Key Performance Indicators." Audiovisual recording of PowerPoint presentation given at the Craft Brewers Conference and BrewExpo America, Minneapolis, MN, May 3, 2022. https://www.brewersassociation.org/seminars /track-yield-a-winners-guide-to-safety-and-sustainability-key-performance-indicators/.

Gastil, Raymond D. 1961. "The Determinants of Human Behavior." *American Anthropologist* 63(6): 1281–91.

Gerba, Charles P. 2013. "Occurrence and Transmission of Food- and Waterborne Viruses by Fomites." In *Viruses in Food and Water: Risks, Surveillance and Control,* edited by Nigel Cook, 205–216. Cambridge: Woodhead Publishing.

Gerba, Charles P. 2015. "Environmentally Transmitted Pathogens." In *Environmental Microbiology,* edited, by Ian L. Pepper, Charles P. Gerba, and Terry J. Gentry, 509–550. San Diego: Academic Press, an Imprint of Elsevier.

Ghahramani, A., and H. R. Khalkhali. 2015. "Development and Validation of a Safety Climate Scale for Manufacturing Industry." *Safety and Health at Work* 6(2): 97–103.

Great Plains Center for Agricultural Health. 2015. "Grain Engulfment Entrapment." January 23, 2015. https://gpcah.public-health.uiowa.edu/grain-engulfment-entrapment/.

Guha, Suvajyoti, Prasanna Hariharan, and Matthew R. Myers. 2014. "Enhancement of ICRP's lung deposition model for pathogenic bioaerosols." *Aerosol Science and Technology* 48(12): 1226–1235. https://doi.org /10.1080/02786826.2014.975334.

Halkitis, Perry N., and Jeffery T. Parsons. 2000. "Oral Sex and HIV Risk Reduction." *Journal of Psychology & Human Sexuality* 11(4): 1–24.

Harmer, Jake. n.d. "How Many People Die Rock Climbing?" Last accessed January 23, 2021. https://therockulus.com /rock-climbing-deaths/.

Heineken. n.d. "Promoting Health and Safety." Accessed July 11, 2022. https://www.theheinekencompany.com /our-sustainability-story/our-strategy-and-achievements/promoting-health-and-safety.

Heinrich, H.W. 1931. *Industrial accident prevention: a scientific approach.* New York: McGraw-Hill.

Henkel Corporation. n.d. "BONDERITE® Surface Cleaning." Accessed October 8, 2021. https://www.henkel-adhesives .com/us/en/industries/metals/metal-pretreatment/cleaning-degreasing.html.

Hernia Clinic. n.d. "Types of Hernias." Last accessed November 21, 2022. https://www.herniaclinic.co.nz/information /types-of-hernias/.

Hillenbrand, B. 1983. "Instruction and warning labels for ladders." OSHA Archive Document pertaining to Standard Number 1910.27. Standard Interpretations (Archived), Occupational Safety and Health Administration. July 25, 1983. https://www.osha.gov/laws-regs/standardinterpretations/1983-07-25.

Hopkins, A. 2006. "What are we to make of safe behaviour programs?" *Safety Science* 44(7): 583–87. https://doi.org /10.1016/j.ssci.2006.01.001.

Howard, Gillian S. 2019. "'Legal aspects of fitness for work." In *Fitness for Work: The Medical Aspects*. 6th ed. Edited by John Hobson and Julia Smedley, 25–49. Oxford University Press. https://doi.org/10.1093 /med/9780198808657.003.0002.

Howe, Jim. 2001. "Warning! Behavior-Based Safety Can Be Hazardous To Your Health and Safety Program! A Union Critique of Behavior-Based Safety." UAW, International Union. November 2001. http://www.uawlocal974.org /BSSafety/Warning!_Behavior-Based_Safety_Can_Be_Hazardous_To_Your_Health_and_Safety_Program!.pdf.

Huang Y.H., T.B. Leamon, T.K. Courtney, P.Y. Chen, and S. DeArmond. 2007. "Corporate financial decision-makers' perceptions of workplace safety." *Accident Analysis & Prevention* 39(4): 767–75. https://doi.org /10.1016/j.aap.2006.11.007.

Ikeda, H., and K. Kobayashi. 1998. "Pathophysiologic changes in patients with severe burns: role of hormones and chemical mediators." [In Japanese.] *Nihon Geka Gakkai Zasshi* (*Journal of the Japanese Surgical Association*) 99(1): 2–7.

Interactive Learning Paradigms. 2020. "Mist." Interactive Learning Paradigms, Incorporated (ILPI). Last updated February 23, 2020. http://www.ilpi.com/msds/ref/mist.html.

Interactive Learning Paradigms. 2020. "Smoke." Interactive Learning Paradigms, Incorporated (ILPI). Last updated October 8, 2020. http://www.ilpi.com/msds/ref/smoke.html.

Interactive Learning Paradigms. 2022. "Vapor." Interactive Learning Paradigms, Incorporated (ILPI). Last updated February 26, 2022. http://www.ilpi.com/msds/ref/vapor.html.

International Organization for Standardization. 2011. *Industrial trucks — Safety requirements and verification — Part 1: Self-propelled industrial trucks, other than driverless trucks, variable-reach trucks and burden-carrier trucks.* ISO 3691-1:2011. Paris: ISO.

International Organization for Standardization. 2018. *Occupational health and safety management systems — Requirements with guidance for use.* ISO 45001:2018. Paris: ISO.

IOSH. 2018. "Psychosocial Hazards (including stress)." Institution of Occupational Safety and Health. Accessed August 18, 2022. https://iosh.com/health-and-safety-professionals/improve-your-knowledge/occupational -health-toolkit/psychosocial-hazards-including-stress/.

Johnson, Dana. 2009. "Don't Lose Your Head! A Look at Using Antifoams in the Brewery." *The New Brewer*, July/August 2009.

Kammonen, Eveliina. 2017. "The Myers Briggs Type Indicator – What Is It and Is It Any Good?" Better Than Sliced Bread. September 25, 2017. https://btsbzine.com/blog/2017/9/25/the-myers-briggs-type-indicator -what-is-it-and-is-it-any-good.

Karasek, R., C. Brisson, and N. Kawakami, I. Houtman, P. Bongers, and B. Amick. 1998. "The Job Content Questionnaire (JQA): an Instrument for Internationally Comparative Assessment of Psychosocial Job Characteristics." *Journal of Occupational Health Psychology* 3(4): 322–55. https://doi.org/10.1037//1076-8998.3.4.322.

Korey Stringer Institute. 2019. "Heat Stroke Recognition." University of Connecticut. Last updated August 5, 2019. https://ksi.uconn.edu/emergency-conditions/heat-illnesses/exertional-heat-stroke/heat-stroke-recognition/.

Lamontagne, A.D., T. Keegel, A.M. Louie, A. Ostry, and P.A. Landsbergis. 2007. "A Systemic Review of the Job-Stress Intervention Evaluation Literature, 1990–2005." *International Journal of Occupational & Environmental Health* 13(3): 268–80.

Leonard, Jayne. 2019. "Cognitive dissonance: What to know." *Medical News Today*. October 21, 2019. Last accessed May 13, 2022. https://www.medicalnewstoday.com/articles/326738.

Likert, R. 1932. "A Technique for the Measurement of Attitudes." *Archives of Psychology* 22(140): 5–55.

Lim, L.T., E.Y. Ah-Kee, and C.E. Collins. 2014. "Common eye drops and their implications for pH measurements in the management of chemical eye injuries." *International Journal of Ophthalmology* 7(6): 1067–1068.

Madl, P., and S. Egot-Lemaire. 2015. "The field and the photon from a physical point of view." In *Fields of the Cell*, edited by D. Fels, M. Cifra, and F. Scholkmann, 29–54. Kerala, India: Research Signpost.

Marsden, Eric. 2021. "Safety culture: A contentious and confused notion." Risk Engineering. Last updated January 18, 2021. https://risk-engineering.org/concept/safety-culture.

Mayo Clinic. 2021. "Food Allergy." December 31, 2021. https://www.mayoclinic.org/diseases-conditions/food-allergy /symptoms-causes/syc-20355095.

Mayo Clinic. 2021. "Mold Allergy." June 21, 2021. https://www.mayoclinic.org/diseases-conditions/mold-allergy /symptoms-causes/syc-20351519.

Mayo Clinic. 2022a. "Muscle strains." October 11, 2022. https://www.mayoclinic.org/diseases-conditions /muscle-strains/diagnosis-treatment/drc-20450520.

Mayo Clinic. 2022b. "Hypothermia: First aid." April 16, 2022. https://www.mayoclinic.org/first-aid/first-aid -hypothermia/basics/art-20056624.

McSween, Terry. 2003. *Values-based Safety Process: Improving Your Safety Culture With Behavior-Based Safety.* 2nd Edition. Hoboken, NJ: John Wiley & Sons.

Meadowlark Brewing. n.d. Vision and Values. Accessed July 10, 2022. https://meadowlarkbrewing.com/vision-values/.

Mehmet, A.T. 2013. "Factors affecting mortality in burn patients admitted to intensive care unit." *Eastern Journal of Medicine* 18:72–5.

Michie, S. 2002. "Causes and Management of Stress at Work." *Occupational and Environmental Medicine* 59:67–72.

Mind Tools Content Team. n.d. "Frederick Taylor and Scientific Management: Understanding Taylorism and Early Management Theory. Last accessed July 11, 2022. https://www.mindtools.com/pages/article/newTMM_Taylor.htm.

Minnesota Department of Health. n.d. Histoplasmosis (*Histoplasma capsulatum*). Accessed January 12, 2022. https://www.health.state.mn.us/diseases/histoplasmosis/index.html.

Mitchell, Steve. n.d. "The Origin and Fallacies of Behavior-Based Safety: The UAW Perspective." UAWlocal974. Last accessed July 3, 2022. http://www.uawlocal974.org/safetyArticle/Origin%20and%20Fallacies%20of%20BS%20 Safety%20(11.04).htm.

Moore, R.A., A. Waheed, and B. Burns. 2022. "Rule of Nines." In: StatPearls [Internet]. Treasure Island (FL): StatPearls Publishing, January 2022. Updated May 30, 2022. https://pubmed.ncbi.nlm.nih.gov/30020659/.

NACMCF. 1997. "HACCP Principles & Application Guidelines." US Department of Agriculture, National Advisory Committee on Microbiological Criteria for Foods. Last updated February 25, 2022. https://www.fda.gov/food /hazard-analysis-critical-control-point-haccp/haccp-principles-application-guidelines.

National Fire Protection Association. 2020. "Second Revision No. 1-NFPA 53-2020." National Fire Protection Association Report, June 16, 2020, pp. 4/23–9/23. https://www.nfpa.org/assets/files/AboutTheCodes/53 /53_F2020_OXY_AAA_SD_SRStatements.pdf.

National Fire Protection Association. n.d. "Codes & Standards." NFPA. Last accessed February 16, 2023. https://www.nfpa.org/Codes-and-Standards.

National Institute for Occupational Safety and Health. 1998. *Occupational Noise Exposure: Revised Criteria 1998*. Criteria document, DHHS (NIOSH) Publication No. 98–126. Available from "Occupational Noise Exposure," Centers for Disease Control and Prevention, last updated June 6, 2014, https://www.cdc.gov/niosh/docs/98-126/default.html.

National Institute for Occupational Safety and Health. 2010. "Aerosols." Centers for Disease Control and Prevention, Department of Health and Human Services. Last reviewed June 29, 2010. https://www.cdc.gov/niosh/topics/ aerosols/default.html.

National Institute for Occupational Safety and Health. 2013. "Eye Safety." Centers for Disease Control and Prevention. Last reviewed July 29, 2013 (archived). https://www.cdc.gov/niosh/topics/eye/default.html.

National Institute for Occupational Safety and Health. 2015. "Productive Aging and Work." Centers for Disease Control and Prevention, Department of Health and Human Services. Last reviewed September 11, 2015. https://www.cdc.gov/niosh/topics/productiveaging/safetyandhealth.html.

National Institute for Occupational Safety and Health. 2016. *NIOSH criteria for a recommended standard: occupational exposure to heat and hot environments*. DHHS (NIOSH) Publication No. 2016-106. Centers for Disease Control and Prevention, US Department of Health and Human Services. https://www.cdc.gov/niosh/docs/2016-106/pdfs/2016-106.pdf.

National Institute for Occupational Safety and Health. 2017. *Heat Stress: Acclimatization*. DHHS (NIOSH) Publication No. 2017-124. Centers for Disease Control and Prevention, Department of Health and Human Services. https://www.cdc.gov/niosh/mining/userfiles/works/pdfs/2017-124.pdf.

National Institute for Occupational Safety and Health. 2020. *NIOSH Manual of Analytical Methods (NMAM).* 5th ed. Ronnie Andrews and Paula Fey O'Connor (eds.). Centers for Disease Control and Prevention, February 2020.

National Institute for Occupational Safety and Health. 2022. *Small Business Safety and Health Handbook.* DHHS (NIOSH) Publication No. 2021-120 (revised 07/2022). OSHA Publication No. 2209-07R 2022. Centers for Disease Control and Prevention, Department of Health and Human Services; Occupational Health and Safety Administration, US Department of Labor. https://www.osha.gov/sites/default/files/publications/small-business.pdf.

National Institute for Occupational Safety and Health. 2023. "MRSA and the Workplace." Centers for Disease Control and Prevention, Department of Health and Human Services. Last reviewed February 13, 2023. https://www.cdc.gov /niosh/topics/mrsa/default.html.

National Institute of Neurological Disorders and Stroke. n.d. "Repetitive Motion Disorders." NINDS, National Institutes of Health (NIH). Last accessed November 22, 2022. https://www.ninds.nih.gov/health-information /disorders/repetitive-motion-disorders.

National Library of Medicine. n.d. "Spores." MedlinePlus. Accessed January 14, 2022. https://medlineplus.gov/ency /article/002307.htm.

National Research Council. 1994. *Under the Influence?: Drugs and the American Work Force.* Edited by J. Normand, R.O. Lempert and C.P. O'Brien. National Research Council and Institute of Medicine US Committee on Drug Use in the Workplace. Washington, DC: National Academies Press.

National Safety Council. n.d.[a] "Systemic Risks." NSC [website]. Accessed July 2, 2021. https://www.nsc.org/workplace /safety-topics/work-to-zero/hazardous-situations/systemic-risks.

National Safety Council. n.d.[b] "Work Injuries and Illnesses by Age." Industry Incidence and Rates, NSC Injury Facts [website]. Accessed May 22, 2022. https://injuryfacts.nsc.org/work/industry-incidence-rates /work-injuries-and-illnesses-by-age/.

Nattrass, C., C. J. Horwell, D. E. Damby, A. Kermanizadeh, D.M. Brown, and V. Stone. 2015. "The global variety of diatomaceous earth toxicity: a physiochemical and in vitro investigation." *Journal of Occupational Medicine and Toxicology* 10:23. https://doi.org/10.1186/s12995-015-0064-7.

Ochs, Matthias, Jens R. Nyengaard, Anja Jung, Lars Knudsen, Marion Voigt, Thorsten Wahlers, Joachim Richter, and Hans Jørgen G. Gundersen. 2003. "The Number of Alveoli in the Human Lung." *American Journal of Respiratory and Critical Care Medicine.* 169(1): 120–4. https://doi.org/10.1164/rccm.200308-1107OC.

Oedewald, Pia. 2014. "History of Systemic Approach to Safety." PowerPoint presentation at Technical meeting on the Interaction between Individuals, Technology and Organization — A Systemic Approach to Safety in Practice, Vienna, June 10–13, 2014.

OSHA (Occupational Health and Safety Administration). 2016. *Hazard Classification Guidance for Manufacturers, Importers, and Employers.* Hazard Communication OSHA 3844-02 2016. US Department of Labor.

OSHA (Occupational Health and Safety Administration). n.d. "Estimated Costs of Occupational Injuries and Illnesses and Estimated Impact on a Company's Profitability Worksheet." Last accessed February 6, 2021. https://www.osha.gov/safetypays/estimator.

Paoletta, Dave. 2020. "The Ultimate Guide to Safety Management Systems." Safesight. April 26, 2020. https://safesitehq.com/safety-management-systems/.

Parker, Douglas L. 2022. "Heat Injury and Illness Prevention in Outdoor and Indoor Work Settings Rulemaking." United States Department of Labor, Occupational Safety and Health Administration. Directive CPL 03-00-024, April 8, 2022. https://www.osha.gov/sites/default/files/enforcement/directives/CPL_03-00-024.pdf.

Partnership for Food Safety Education. 2022. "The Core Four Practices." Partnership for Food Safety Education [website]. Accessed January 6, 2022. https://www.fightbac.org/food-safety-basics/the-core-four-practices/.

Perez, Susan. 2021. "Why You Need to Use Qualitative and Quantitative KPIs to Grow Your Business (+ Examples)." BrightGauge blog. July 7, 2021. https://www.brightgauge.com/blog/quick-guide-to-11-types-of-kpis.

Petrovic, Dragan, Časlav Mitrović, Natasa Trisovic, and Zorana Golubovic. 2011. "On the particles size distribution of diatomaceous earth and perlite granulations." *Strojniski Vestnik (Journal of Mechanical Engineering)* 57(11): 843–50.

Phimister, James R., Ulku Oktem, Paul R. Kleindorfer, and Howard Kunreuther. 2003. "Near-Miss Management Systems in the Chemical Process Industry." *Risk Analysis* 23(3): 445–59.

Psychology Notes HQ. 2018. "What is the Social Proof Theory?" The Psychology Notes Headquarters, March 10, 2018. https://www.psychologynoteshq.com/social-proof/.

Ranjit, P.S., and A. Thakur. 2020. "Essential Aspects of Day to Day Life and Its Influence on Industry 4.0." In *LoRA and IoT Networks for Applications in Industry 4.0*, edited by A. Gehlot, K.K. Sharma, R. Singh, and R. Sharma. Hauppauge, NY: Nova Science.

Reid, Lesley Williams, and Miriam Konrad. 2004. "The Gender Gap in Fear: Assessing the Interactive Effects of Gender and Perceived Risk on Fear of Crime." *Sociological Spectrum* 24(4): 399–425.

Rhodes, Nancy, and Kelly Pivik. 2011. "Age and gender differences in risky driving: The roles of positive affect and risk perception." *Accident Analysis & Prevention* 43(3): 923–931.

Robinson, James C. 1988. "Labor Union Involvement in Occupational Safety and Health, 1957–1987." *Journal of Health Politics, Policy and Law* 13(3): 453–68. https://doi.org/10.1215/03616878-13-3-453.

Rodríguez-Garzón, I., M. Martínez-Fiestas, A. Delgado-Padial, and V. Lucas-Ruiz. 2016. "An Exploratory Analysis of Perceived Risk among Construction Workers in Three Spanish-Speaking Countries." *Journal of Construction Engineering and Management* 142(11): 04016066. https://doi.org/10.1061/(ASCE)CO.1943-7862.0001187.

Safety.BLR.com. 2014. "6 key elements of an effective I2P2." Last accessed July 13, 2022. https://safety.blr.com/workplace-safety-news/safety-administration/safety-plans/6-key-elements-of-an-effective-I2P2/.

Schuster, John G. 2011. "Training and its Value in Reducing Maintenance Costs." *EHS Today*, July 1, 2011. https://www.ehstoday.com/training-and-engagement/article/21913641/training-and-its-value-in-reducing-maintenance-costs.

Singh, P., M. Tyagi, K.K. Gupta, and P.D. Sharma. 2013. "Ocular chemical injuries and their management." *Oman Journal of Ophthalmology* 6(2): 83–6.

Sliney, D.H., and B.E. Stuck. 2021. "A Need to Revise Human Exposure Limits for Ultraviolet UV-C Radiation." *Photochemistry and Photobiology* 97(3): 485–92.

Smith, J.L. 2009. "Remembering Walter A. Shewhart's Contribution to the Quality World." *Quality Magazine*, March 2, 2009. https://www.qualitymag.com/articles/85973-remembering-walter-a-shewharts-contribution-to-the-quality-world.

Smith, P.M, and J. Berecki-Gisolf. 2014. "Age, occupational demands and the risk of serious work injury." *Occupational Medicine* 64(8): 571–576.

Smith, Thomas A. 2014. "What Dr. Deming Can Teach Us About Safety Management." EHS Today. Jun 9, 2014. https://www.ehstoday.com/safety-leadership/article/21916397/what-dr-deming-can-teach-us-about-safety-management.

Solow, M., S. Chheng, and K. Parker. 2015. "Culture and engagement: The naked organization." *Deloitte Insights*. February 27, 2015. https://www2.deloitte.com/us/en/insights/focus/human-capital-trends/2015/employee-engagement-culture-human-capital-trends-2015.html.

StateFoodSafety. n.d. "Fridge Storage for Food Safety Chart." AboveTraining, Inc. Accessed March 15, 2022. https://www.statefoodsafety.com/Resources/Resources/fridge-storage-for-food-safety.

Stillman, Peter. 1975. "The Limits of Behaviorism: A Review Essay on B. F. Skinner's Social and Political Thought." *American Political Science Review* 61(1): 202–13.

Takahashi, M. 2017. "Tackling psychosocial hazards at work." *Industrial Health* 55(1): 1–2.

Talbot, L.A., R.J. Musiol, E.K. Witham, and E.J. Metter. 2005. "Falls in young, middle-aged and older community dwelling adults: perceived cause, environmental factors and injury." *BMC Public Health* 5:86.

Taylor, Frederick W. 1911. *The Principles of Scientific Management.* Norwood, Massachusetts: The Plimpton Press.

Tidy, Colin. 2021. "Fungal Groin Infection: Tinea Cruris." Last reviewed March 16, 2021. https://patient.info/infections/fungal-infections/fungal-groin-infection-tinea-cruris.

Trinity J.D., M.D. Pahnke, J.F. Lee, and E.F. Coyle. 2010. "Interaction of hyperthermia and heart rate on stroke volume during prolonged exercise." *Journal of Applied Physiology* 109(3): 745–51. https://doi.org/10.1152/japplphysiol.00377.2010.

Tuchin, V. 2015. "Tissue Optics and Photonics: Biological Tissue Structures." *Journal of Biomedical Photonics & Engineering* 1(1): 3–21.

Ulloa, R. Zamorano, M.G. Hernandez Santiago, and V.L. Villegas Rueda. 2019. "The Interaction of Microwaves with Materials of Different Properties." In *Electromagnetic Fields and Waves*, edited by K.H. Yeap and K. Hirasawa. London: IntechOpen.

US Department of Labor. 2000. "Compliance Assistance for the Powered Industrial Truck Operator Training Standards." United States Department of Labor, Occupational Safety and Health Administration. Updated directive CPL 02-01-028, November 30, 2000. https://www.osha.gov/enforcement/directives/cpl-02-01-028.

US Department of Labor. 2002. *Hearing Conservation*. OSHA 3074, 2002 (Revised). United States Department of Labor, Occupational Safety and Health Administration. https://www.osha.gov/sites/default/files/publications /osha3074.pdf.

US Department of Labor. 2003. "Protecting Young Workers: Prohibition Against Young Workers Operating Forklifts." SHIB 03-09-30. Occupational Safety and Health Administration, September 30, 2003. https://www.osha.gov/sites /default/files/publications/shib093003.pdf.

US Department of Labor. 2016a. "Occupational Safety and Health Administration (OSHA) Inspections." OSHA Fact Sheet 3783, United States Department of Labor, Occupational Safety and Health Administration, August 2016. https://www.osha.gov/OshDoc/data_General_Facts/factsheet-inspections.pdf.

US Department of Labor. 2016b. *Recommended Practices for Safety and Health Programs*. OSHA 3885. United States Department of Labor, Occupational Safety and Health Administration, October 2016. https://www.osha.gov/sites /default/files/publications/OSHA3885.pdf.

US Department of Labor. 2019a. "U.S. Department of Labor Issues Final Rule to Protect Privacy of Workers." OSHA Trade Release, January 24, 2019. https://www.osha.gov/news/newsreleases/trade/01242019.

US Department of Labor. 2019b. *Using Leading Indicators to Improve Safety and Health Outcomes*. OSHA 3970. United States Department of Labor, Occupational Safety and Health Administration, June 2019.

US Department of Labor. 2021. "Grain Dust (Oat, Wheat, Barley)." OSHA Occupational Chemical Database. United States Department of Labor, Occupational Safety and Health Administration, last updated January 29, 2021. https://www.osha.gov/chemicaldata/789.

US Department of Labor. n.d. "Business Case for Safety and Health." Occupational Safety and Health Administration: Safety and Health Topics. Last accessed February 6, 2021. https://www.osha.gov/businesscase/benefits.

US Department of Labor. n.d.[a] "Selected OSHA Recordkeeping Q & A." OSHA Injury and Illness Recordkeeping: Q & A Search. Accessed 8 3, 2020. https://www.osha.gov/recordkeeping/faq_search/index.html.

US Department of Labor. n.d.[b] "Develop your Safety + Health Program." Safe + Sound. Occupational Safety and Health Administration. Accessed August 23, 2022. https://www.osha.gov/safeandsound/safety -and-health-programs.

US Department of Labor. n.d.[c] "Powered Industrial Trucks (Forklift) eTool." United States Department of Labor, Occupational Safety and Health Administration, last accessed August 21, 2022. https://www.osha.gov/etools /powered-industrial-trucks.

US Department of Labor. n.d.[d] "Heat Illness Prevention." Occupational Safety and Health Administration. Last accessed March 1, 2023. https://www.osha.gov/heat/general-education.

US Department of Labor. n.d.[e] "Heat Injury and Illness Prevention in Outdoor and Indoor Work Settings Rulemaking." Occupational Safety and Health Administration. Last accessed December 21, 2022. https://www.osha.gov/heat-exposure/rulemaking.

US Energy Information Administration. 2020. "What is Energy? Forms of Energy." Updated June 17, 2020. https://www.eia.gov/energyexplained/what-is-energy/forms-of-energy.php.

US Food and Drug Administration. 2022. "Allergens in Cosmetics." Last updated February 25, 2022. https://www.fda.gov/cosmetics/cosmetic-ingredients/allergens-cosmetics.

US Food and Drug Administration. 2023. "Food Allergies." Last updated January 10, 2023. https://www.fda.gov/food/food-labeling-nutrition/food-allergies.

US Office of Technology Assessment. 1985. *Preventing Illness and Injury in the Workplace*. OTA-H-256. Washington, DC: US Government Printing Office.

von Hayek, H. 1960. *The Human Lung*. New York, NY: Hafner Publishing.

Washington State Department of Labor and Industries. 2009. "Burn Injury Facts: Scald Burns in Restaurant Workers." Hazard prevention report # 86-7-2009. Safety and Health Assessment and Research for Prevention Program at the Washington State Department of Labor and Industries, April 2009. https://lni.wa.gov/safety-health/safety-research /files/2009/RestaurantScaldBurns.pdf.

Wasserman D.D., J.A. Creech, and M. Healy. 2022. "Cooling Techniques for Hyperthermia." In: StatPearls [Internet]. Treasure Island (FL): StatPearls Publishing, January 2022. Updated October 17, 2022. https://www.ncbi.nlm.nih.gov /books/NBK459311/.

Way, Kirsten. 2020. "Psychosocial Hazards." In *The Core Body of Knowledge for Generalist OHS Professionals*. 2nd ed. Tullamarine, VIC: Australian Institute of Health and Safety.

Weibel, E. 1980. "Design and structure of the human lung." In *Pulmonary Diseases and Disorders*, edited by Alfred P. Fishman, 224–71. New York: McGraw-Hill.

Weibel, E.R. 2009. "What makes a good lung?" *Swiss Medical Weekly* 139(2728): 375–86. https://doi.org/10.4414 /smw.2009.12270.

World Health Organization. 2002. *IARC Monographs on the Evaluation of Carcinogenic Risks to Human*. Vol. 81, *Man-Made Vitreous Fibres*. Lyon, France: IARC Press.

World Health Organization. 2003. *Health Aspects of Air Pollution with Particulate Matter, Ozone and Nitrogen Dioxide*. Report on a WHO Working Group, Bonn, Germany, January 13–15, 2003. EUR/03/5042688. Copenhagen: WHO Regional Office for Europe.

World Health Organization. 2020. "Occupational health: Stress at the workplace." Questions and answers, October 19, 2020. https://www.who.int/news-room/questions-and-answers/item/ccupational-health-stress-at-the-workplace.

Wright, M. Imelda, Barbara Polivka, Jan Odom-Forren, and Becky Christian. 2021. "Normalization of Deviance." *Advances in Nursing Science* 44(2): 171–180.

Zabkowicz, John. 2020. "Tri-Clamp Gasket Material Indicators." Sanitary Fittings. October 20, 2020. https://sanitaryfittings.us/tri-clamp-gasket-material-indicators.

Zimmerman, Kim A. 2018. "Skin: The Human Body's Largest Organ." Live Science, October 22, 2018. https://www.livescience.com/27115-skin-facts-diseases-conditions.html.

INDEX